COMBINED POWER PLANTS

Including Combined Cycle Gas Turbine (CCGT) Plants

Thermodynamics and Fluid Mechanics Series

General Editor: W. A. WOODS

Pergamon Titles of Related Interest

DIXON
Fluid Mechanics: Thermodynamics of Turbomachinery, 3rd Edition

DUNN & REAY
Heat Pipes, 3rd Edition

HAYWOOD
Analysis of Engineering Cycles, 4th Edition
Analysis of Engineering Cycles—Worked Problems

JAPAN SOCIETY OF MECHANICAL ENGINEERS
Visualized Flow

LAI *et al.*
Introduction to Continuum Mechanics

REAY & MacMICHAEL
Heat Pumps: Design and Applications, 2nd Edition

XIN MINGDAO
Advances in Phase Change Heat Transfer

Pergamon Related Journals

(free specimen copy gladly sent on request)

Chinese Journal of Mechanical Engineering

Computers and Fluids

Computers and Industrial Engineering

International Communications in Heat and Mass Transfer

International Journal of Engineering Sciences

International Journal of Heat and Mass Transfer

International Journal of Multiphase Flow

Heat Recovery Systems and CHP

Previews of Heat and Mass Transfer

COMBINED POWER PLANTS

Including Combined Cycle Gas Turbine (CCGT) Plants

J. H. HORLOCK, F.Eng., F.R.S.
Whittle Laboratory, Cambridge, U.K.

PERGAMON PRESS
OXFORD · NEW YORK · SEOUL · TOKYO

U.K. Pergamon Press Ltd, Headington Hill Hall, Oxford OX3 0BW, England

U.S.A. Pergamon Press, Inc, 660 White Plains Road, Tarrytown, New York 10591-5153, U.S.A.

KOREA Pergamon Press Korea, KPO Box 315, Seoul 110-603, Korea

JAPAN Pergamon Press Japan, Tsunashima Building Annex, 3-20-12 Yushima, Bunkyo-ku, Tokyo 113, Japan

---

First edition 1992

British Library Cataloguing in Publication Data
A catalogue record for this book is available from the British Library

Library of Congress Cataloging-in-Publication Data
Horlock, J. H.
Combined power plants : including combined cycle gas turbine plants / J. H. Horlock.—1st ed.
p. cm.—(Thermodynamics and fluid mechanics series)
1. Gas-turbine power-plants. 2. Steam power-plants—Binary vapor systems. 3. Gas-turbines. I. Title. II. Series.
TK1076.H67 1992 621.31'21—dc20 92-7177

ISBN 0 08 040502 9

Printed in Great Britain by B.P.C.C. Wheatons Ltd, Exeter

In memory of

DR CLAUDE SEIPPEL

TECHNICAL DIRECTOR OF BROWN BOVERI

Contents

Editorial Introduction

THE books in the Pergamon Thermodynamics and Fluid Mechanics Series were originally planned as a set of texts for three-year undergraduate courses in Mechanical Engineering. Subsequently, the aims of the series were broadened and several volumes were introduced which catered not only for undergraduates but also for postgraduate students and engineers in practice. The present volume on combined power plants, written by Professor Horlock, is a companion text to his *Cogeneration: Combined Heat and Power*, published in 1987.

This book appears at a time of an unprecedented increase in installations of combined power plants, on a worldwide basis. This is being stimulated by the thermal advantages offered, the extensive reserves of natural gas in various parts of the world and the environmentally friendly nature of the plants. The author explains how these plants are cost effective and why they are environmentally attractive.

In future, customers for combined plant will demand even greater technical and commercial performance and the present volume will help engineers to achieve this.

Professor Horlock's work is rigorous and uses modern thermodynamics, together with contemporary methods used in economics, in the design and optimisation of combined thermal plants. He has set out some of his views about notation in thermodynamics in the Preface.

August, 1992 W. A. WOODS

Preface

LIKE ITS companion volume, *Cogeneration: Combined Heat and Power* [1], this is a book for engineers. It uses the first and second laws of thermodynamics to estimate the efficiency of combined *power* plants—in which two types of power-producing plants are linked together but no useful heat is delivered—and to discuss the merits of different forms of combination. As in the earlier book, there is also a brief discussion of the economics of the resulting plants, but I do not discuss CHP, referring the reader to the earlier book.

The history of combined power plants goes back to the early part of this century. As my predecessor at Liverpool University, Professor W. J. Kearton, once put it, the search for an alternative fluid to water (and, one could add, its possible use in a combined cycle plant) seems almost as old as the steam engine itself. In 1913 Emmet proposed a mercury/steam combined plant, with the heat rejected by the "higher" mercury cycle used to supply heat to the "lower" steam cycle. The General Electric Company in the U.S.A. built several of these plants, culminating in the 40 MW Schiller power station in New Hampshire in the late forties. These mercury/steam power stations, while magnificent engineering achievements, proved to be uneconomic in comparison with the developing conventional steam plants, which were gaining in efficiency as steam pressures and temperature were increased.

Meanwhile, by the 1960s the gas turbine (open circuit) plant was beginning to benefit from developments in the aircraft jet engine. European and American manufacturers had designed and built so-called CCGT plants, combined cycle gas turbines, mainly but not entirely on a relatively small scale (in the 10–25 MW range), although the use of the word "cycle" for an open circuit gas turbine plant was and remains unfortunate. Gains in efficiency from the use of the lower steam cycle plant were relatively small and further major developments were to be delayed for several years.

Two factors have been critical in the recent explosion of interest in the combined power plant—the increase in the gas turbine top temperature (and consequently the specific work) and the increasing freedom to burn natural gas. The latter not only enables higher efficiency to be obtained (mainly through a reduced temperature of the exhaust stack gases), but also leads to low production of CO_2 with consequent environmental benefit (related to possible reduction in global warming).

The competitive advantages of CCGT plants are now widely recognised; the slow-down in nuclear power development and the increasing environmental difficulties associated with coal-fired plants (even for plants with high overall efficiency) have enabled the CCGT to become a front-runner in electric power generation. The National Grid Company plc, of which I am a director. It is responsible for connecting new generating stations to the U.K. national grid. It is at present involved in connection of some 15 GW of plant (some 30–35% of the maximum demand) and 90% of that newly connected plant will be gas-fired CCGTs. This is a massive shift in electrical generation.

We discuss these historical matters in the last chapter of this book, after we have studied the thermodynamics and economics of combined power plants. The reasons for the "stop-start" development of these plants over the past 50 years then become clearer.

But firstly we review some basic thermodynamics of power plants, emphasising the search for higher temperatures of heat supply, coupled with low temperature of heat rejection, in attempts to approach Carnot cycle efficiency. We also emphasise in Chapter 1 the importance of reversibility, both internal and external, introducing the concept of exergy, more widely used in recent years.

In the second chapter we describe, with limited thermodynamic analysis, various proposals for combined plant. Some have been developed, others have remained on the thermodynamic drawing board.

In Chapter 3 we introduce more detailed thermodynamic analyses of combined power plants. Firstly, we calculate the effect of putting two plants of known individual efficiency together, one thermodynamically on top of the other, but we then assess the effects of various factors such as temperature drop and heat loss between the two plants, supplementary intermediate heating, etc.

Subsequently, in Chapter 4 we emphasise how variations in the thermodynamic parameters in one cycle can affect the efficiency of the other, and hence of the whole plant.

In Chapter 5 we introduce exergy analysis of combined plants, giving detailed numerical examples—of a mercury/steam plant and of a CCGT plant, with and without feed heating in the steam cycle.

Chapter 6 comprises economic analyses of combined plant. The approach followed is similar to that adopted in *Cogeneration*, with an emphasis here on electricity pricing. My experience with NGC these past three years has taught me how important tax regimes are in such economic assessments, and I include reference to this point in this chapter.

Some practical schemes are described in Chapter 7. This has involved me in considerable literature searches—particularly for the mercury/steam plant developed so long ago—and once again I am grateful to Mr C. Hunter-Brown of the Open University Library for his considerable help and skill in tracking down relevant papers. The Brown Boveri Company (now ASEA Brown Boveri) has also been a major source of helpful information.

The book concludes with both the brief historical summary to which I have referred above and an assessment of the current "state of play".

There are a few academic niceties that I should mention in this Preface. In referring to the heat supplied and the work output in a process I write

$$[Q]_X^Y = \int_X^Y dQ \text{ and } [W]_X^Y = \int_X^Y dW,$$

where X and Y are the states at the beginning and end of the process; this is a convenient notation. On sign convention for heat and work transfer there has always been much discussion and there still is (see Mayhew [2]). In general, here I have followed Haywood's practice [3], writing both heat supplied (Q_B) and heat rejected (Q_A) as positive quantities, and work output (W) also as positive. This is a pragmatic approach to cycle and plant analysis, so that, simply,

$$(\text{Heat supplied, } Q_B) - (\text{Heat rejected, } Q_A) = (\text{Work output, } W)$$

where each of the quantities in parentheses is positive. However, in the discussions on exergy in Chapter 5 I have also followed Haywood [4] in occasionally using a separate symbol ($\mathscr{W}$) for a (positive) work input.

My editor, Professor Woods, has drawn my attention to my somewhat unusual use of the word "flux", which the *Oxford English Dictionary* defines as "continued motion" or "rate of flow of any fluid across given area". I find it a useful word to describe the flow (or rate of flow) of various quantities, not only of mass (e.g. the "flux" of exergy into a control surface).

After long discussions with R. W. Haywood I persuaded him to accept my use of "criteria of performance" (e.g. for efficiency) and "parameters" (quantities held constant in a performance analysis)—see the explanation in the footnote on page 5 in Chapter 1.

Finally, I have again followed Haywood [3] in referring to "the lost work due to irreversibility" rather than shorter terms such as "irreversibility" or "exergy loss"; it is a clumsier expression than the others but it has the virtue of clarity.

As with the other four books I have written on engineering thermodynamics and fluid mechanics, this volume is one aimed not only at engineers who are learning about the subject for the first time (possibly in the final year of first degree courses, more likely at graduate or post-experience level) but also at practising engineers who may use it in their day-to-day work. (It has been a great source of satisfaction to see my earlier books on the shelves in design offices round the world.) The book should be read in conjunction with *Cogeneration*, for there is considerable overlap between the two and in the approach I have followed. Many combined heat and power plants are also combined power plants in that two power "cycles" are employed. Indeed I now wish that I had put the two volumes into one, because it would have led to some economy in presentation.

I am grateful to many people in the preparation of this book, but I owe a particular debt to four distinguished engineers.

The late Dr Claude Seippel, to whom the book is dedicated, was in many ways the father of the gas turbine/steam turbine combined plant. I met him first at Baden in 1956, when Brown Boveri was considering whether to build the Field cycle plant (see Chapter 2). Seippel impressed me greatly as an industrial engineer of the highest quality. He showed great courtesy to me as a young engineer and encouraged me to write my books on axial compressors and turbines. His paper with Bereuter in 1960 was the authoritative statement of the position on CCGT plants at that time. Seippel was a gentleman engineer of an earlier age.

My teacher, colleague and friend, the late R. W. Haywood, was for me a source of encouragement, constructive criticism and wisdom for many years. As the first editor of this series of books, I persuaded him to write *Analysis of Engineering Cycles*, which recently reached its fourth edition. We discussed together in 1990 whether we should combine forces to write an enhanced version of that book, concentrating on combined power plants, but eventually decided to keep our two volumes separate and help each other in the writing of them. This book owes much to his influence as a result.

Professor W. A. Woods, a colleague and friend from Liverpool days, succeeded me as editor of this series. He has read this script with great diligence and common sense, and helped considerably to improve its clarity. He remains as industrious as ever.

Finally, my requests for assistance to the Brown Boveri Company led to useful discussions with Mr R. Kehlhofer, who has very substantial experience with that company in the field of combined power plants. His practical handbook, mainly on CCGT plants, preceded the publication of this book. My approach is, perhaps naturally, more academic than Kehlhofer's but that is not to detract from the soundness of his work. I am most grateful to him for his help and enlightenment (particularly on the use of regenerative feed-heating in the steam cycle, the arguments for and against which mystified me for some time). I hope I have been accurate in attributions to Kehlhofer's work and I am most grateful to him for his assistance.

There are many others who have given help. They include Mr E. M. Curtis, of the Whittle Laboratory where I now work once again, who read part of the manuscript; Dr F. Hala, of the Korneuburg power station in Austria, who advised me on the details of the Korneuburg A plant; Mr G. Plumley, formerly of the General Electric Company, U.S.A., who was helpful with information on the Cool Water IGCC plant; Dr T. Kotas, who gave me some direct assistance with reading part of the manuscript and whose excellent book on energy analysis I have called upon frequently; Messrs K. R. Keeley and B. J. Davidson, of National Power plc, who commented usefully on Chapter 8, my summary of the historical and current perspective; and Dr M. El masri, a prodigious publisher in the field of combined power plants, of Thermoflow Inc. In my descriptions of economic analyses of combined plant I have been assisted considerably by colleagues at the National Grid Company plc, namely Mr John Uttley and Mr Duncan Innes. Mr Simon Cowan of Worcester College, Oxford was also generous in his advice on the implications of tax regimes on rate of return.

I must also thank the Council of the Open University for granting me occasional leave of absence to complete this book. The life of a Vice-Chancellor is a rum one, to quote the late Sir Alec Merrison, since it involves a grasshopper-like existence, in which one hops frequently from one task to another. For my part I always felt it wise to maintain some academic work going, in order to keep in touch with academic colleagues; it also helped this Vice-Chancellor to retain a sense of balance at a time of enormous and continuing government pressure over the eighties. The OU was generous in allowing me time to remain an academic, although most of this volume was nevertheless written in evenings and at weekends.

Once again I record my thanks to two ladies without whose help and forbearance this book would not have been written. My former secretary and personal assistant, Mrs Sheila Watts, again suffered from my incomplete and untidy manuscripts, converting draft after draft into successively more respectable form, with great patience and skill. Finally, my wife has again put up with me writing yet another book, and the demands on home life that it entails, even in my so-called retirement. Each book leads to a promise from me to her that I shall not write another, but the promise gets broken after three or four years.

Cambridge, 1991 J. H. HORLOCK

References

[1] Horlock, J. H. *Cogeneration: Combined Heat and Power*. Pergamon Press, Oxford, 1987.
[2] Mayhew, Y. R. Does the Methodology of Teaching Thermodynamics to Engineers Need Changing for the 1990's? *Proc. Inst. Mech. Engrs.*, **205**, 283–286, 1991.

[3] Haywood, R. W. *Analysis of Engineering Cycles*, 4th edition. Pergamon Press, Oxford, 1991.
[4] Haywood, R. W. *Equilibrium Thermodynamics for Engineers and Scientists*. Wiley, Chichester, 1980.

Acknowledgements

Where figures given by other writers have been reproduced, appropriate reference is made in the titles. The author is grateful to the following bodies for permission to include these figures in this book: the American Society of Mechanical Engineers, the Alsthom Company, the *Brown Boveri Review*, Her Majesty's Stationery Office [for the Department of Energy], the Institution of Mechanical Engineers, *Modern Power Systems*, *Proceedings of the American Power Conference*, Springer Verlag and the Von Karman Institute.

Notation

Note: Lower case symbols for properties represent specific quantities (i.e. per unit mass)

Symbol	Meaning	Typical Units
A	annual cash flow	£, \$ p.a.
A, B, C	areas on T,s diagram	kJ/kg
b, B	steady flow availability	kJ/kg, kJ
c	cost per unit of exergy	£/kJ
C	capital cost	£, \$
c_p	specific heat capacity, at constant pressure	kJ/kg K
$(CV)_0$	calorific value at temperature T_0	kJ/kg
e, E	exergy	kJ/kg, kJ
E^Q	work potential of heat transferred, thermal exergy	kJ
$\mathscr{E}=E+E^Q+W$		kJ
EUF	energy utilisation factor	(—)
f	fuel/air ratio; also exergoeconomic factor	(—); (—)
f'	fuel/gas ratio; also modified exergoeconomic factor	(—); (—)
${}_Nf_{AP}$	present worth factor	(—)
F	fuel energy supplied	kJ
g, G	Gibbs function	kJ/kg, kJ
h, H	enthalpy	kJ/kg, kJ
H	plant utilisation	h/year
i	interest or discount rate	(—)
I	lost work due to irreversibility (total)	kJ
I^{CR}	lost work due to internal irreversibility	kJ
I^Q	lost work due to heat transfer to the atmosphere	kJ
l	latent heat	kJ/kg

m	mass of bled steam (per unit boiler flow)	(—)
M	mass of circulating fluid; also fuel cost per annum; also molecular weight; also Mach number	kg/s; £, \$ p.a.; (—); (—)
N	plant life	years
OM	annual operational and maintenance costs	£, \$ p.a.
p	pressure	N/m^2
P	electricity cost per year	£, \$ p.a.
q, Q	heat supplied or rejected	kJ/kg, kJ
r	pressure ratio; also rate of return	(—)
R	heat recovery (defined in text); also fraction of maximum feed heating; also gas constant	kJ; (—); kJ/kg K
$\bar{R}$	universal gas constant	kJ/kmol/K
S	fuel cost per unit mass	£, \$/kg
s, S	entropy	kJ/kg K, kJ/K
t	time	s
T	temperature	°C, K
U	constant operating costs	£, \$ p.a.
V	variable operating costs	£, \$ kW h
w, W	specific work output, work output	kJ/kg, kJ
$\mathcal{W}$	work input	kJ
x	dryness fraction of steam	(—)
$x=(W_H/F)$	(ratio of work output from higher plant to total fuel energy supplied)	(—)
Y	unitised price (price per unit of energy)	£, \$/kW h
$\dot{Z}$	capital cost rate	£, \$ p.a.
A, B, C, D, E, F, K, K'	constants defined in text	various
α	proportions of capital cost	(—)
α, β, γ	areas on T,s diagram; also constants defined in text	kJ/kg; various
β	capitalised cost factor	(—)
δ	fractional irreversibility	(—)
ε_B	energy effectiveness defined in text	(—)
ε	heat exchanger effectiveness	(—)
ζ	cost of fuel per unit of energy	£, \$/kW h
η	efficiency—see note below	(—)
θ	ratio of maximum to minimum temperature, usually	(—)
κ	ratio of specific heat capacities	(—)

λ	ratio of mass flows in cycles (lower to upper); also fractional exergy change across component	(—)
$\dot{\Lambda}$	cost of exergy losses	£, $ p.a.
μ	scaling factor on steam entropy; also ratio of H.P. to total steam flow	(—); (—)
ν	non-dimensional heat supplied (ν_S) or heat unused (ν_{UN})	(—)
μ, ξ, σ, τ	parameters in cycle analysis	(—)
$\dot{\Pi} = \dot{E}c$	total exergy cost	£, $ p.a.
ρ	isentropic temperature ratio; also density	(—), kg/m^3
τ	corporate tax rate	(—)
τ_{CR}	investment tax credit	(—)
ψ	ratio $(\Delta G_0/\Delta H_0)$	(—)
ϕ	temperature function, $\int_0^T \frac{c_p dT}{T}$	kJ/kg K

Subscripts

$a, a', b, b', c, d,$ $e, e', f, f' \ldots$	states in lower cycle (usually steam)
A	air; relating to heat rejection
AUX	auxiliary
B	boiler; relating to heat supply
C	compressor (isentropic efficiency)
CARNOT	Carnot cycle
CL	closed cycle
COMB	combustion (efficiency)
CON	condenser
CP	combined plant (general)
CPP	parallel combined plants (efficiency)
CPS	series combined plant (efficiency)
CPSP	series parallel combined plant (efficiency), i.e. with supplementary heating, no intermediate heat loss
CR	creation—referring to entropy
CS	control surface
CV	control volume
d	debt
e	equity
e	exergetical (efficiency)
E	electrical (price)
F	fuel
FP	feed pump
G	gas

GT	gas turbine
H	higher (upper, topping); relating to heat supply, work output
HL	between higher and lower plants
HR	rejection from higher plant
ht	heat transfer (efficiency)
i	$=$A, B, C . . . component
k	product gas component; also year number ($k=1, 2, \ldots$)
L	lower (bottoming), relating to heat supply, work output
LM	log mean temperature difference
LR	rejection from lower plant
max	maximum
min	minimum
M	mixture
O	overall (efficiency)
OP	open plant
p	polytropic (efficiency)
P	product of combustion
P′	product of supplementary combustion
PH	pre-heater
R	rational; also reactants; also radiation heat loss; also Rankine cycle mass flow
REV	reversible
s	state after isentropic compression or expansion
S	state at entry to stack; also supplementary heating
ST	steam (usually in lower cycle); also steam turbine
str	streams
td	thermodynamic (efficiency)
T	turbine (isentropic efficiency)
UN	rejection of unused heat
w	saturated water
x, y, z, z′	referring to mass flows in Kalina cycle
X, Y	states leaving heat exchanger; also states at entry and exit from component
1, 1′, 2, 2′, 3, 3′, 4, 4′, 5, 6 . . .	miscellaneous, but usually referring to gas states in higher plant
0	conceptual environment (ambient state); also year "zero"

Superscripts

CR	referring to internal irreversibility
Q	referring to thermal exergy (associated with heat transfer); also to lost work due to external irreversibility associated with heat transfer to the atmosphere
$\dot{}$ (e.g. $\dot{M}$, $\dot{Q}$, $\dot{W}$	*rate* of (mass flow, heat supply, work output, etc.)
$'$ (e.g. η', i$'$)	new or changed value of (e.g. efficiency, interest rate, etc.)
$'$ (e.g. a$'$, b$'$, 3$'$, 4$'$)	states in feed heating train, in reheating or intercooling
$-$ (e.g. $\bar{T}$, $\bar{A}$)	mean or average (e.g. temperature); also levelised (cash flow)
$\sim$ (e.g. $\tilde{\eta}$)	relating to series combined plant with no temperature drop between higher and lower plants ($T_{HR} = T_L$)
*	maximal efficiency

Note on Efficiencies

η without brackets is used for thermal efficiency of a closed cycle, but often with a subscript (e.g. η_H for thermal efficiency of a higher cycle, η_{CPS} for thermal efficiency of a "series" combined plant). For other efficiencies the symbol is used with brackets [e.g. (η_O) for overall efficiency of a plant, (η_C) for compressor isentropic efficiency, (η_B) for boiler efficiency]. Subscripts may then appear outside the brackets [e.g. $(\eta_O)_H$ is the overall efficiency of a higher plant].

A list of efficiencies is given below.

Thermal Efficiencies η

η_H	higher cycle
η_L	lower cycle
η_{CP}	combined cycle (general)
η_{CPS}	series combined cycle (no intermediate heat loss)
η_{CPP}	parallel combined cycle
η_{CPSP}	series parallel combined cycle (no intermediate heat loss)
η_{GT}	gas turbine cycle
η_M	mercury cycle
$\tilde{\eta}_{CPS}$	series cycles with no temperature drop between plants
η_{CARNOT}	Carnot cycle

Overall Efficiencies (η_O)

$(\eta_O)_H$	higher plant
$(\eta_O)_{CP}$	combined plant

$(\eta_O)_{CPS}$	series combined plant (no intermediate heat loss)
$(\eta_O)_{CPSP}$	series parallel combined plant (no intermediate heat loss)
$(\eta_O)_{GT}$	gas turbine plant

Rational Efficiencies (η_R)

$(\eta_R)_{CP}$	combined plant
$(\eta_R)_{GT}$	gas turbine plant
$(\eta_R)_i$	component *i*

Other Efficiencies

(η_e)	exergetical
(η_{td})	thermodynamic
η^*_{GT}, η^{**}_{GT}	relating to Seippel and Bereuter analysis (defined in text)
(η_B)	boiler
(η_C)	compressor, isentropic
(η_T)	turbine, isentropic
(η_{COMB})	combustion
(η_p)	polytropic
ε_B	energy effectiveness (defined in text)

CHAPTER 1

Elementary Thermodynamics of Power Plants

1.1 Introduction

In studying the thermodynamics of thermal power plants our objective is the determination and maximisation of plant efficiency, i.e. the most efficient production of work from a supply of fuel with chemical energy. Figure 1.1 shows a block diagram of a conventional power plant receiving fuel energy (F), producing work (W) and rejecting heat (Q_A) to a sink at low temperature. We wish to achieve the least fuel input for a given work output because this will clearly give economic benefit in the operation of the plant, minimising fuel costs. However, the capital cost of achieving high efficiency has to be assessed and balanced against resulting savings in fuel costs.

We first briefly review the thermodynamics of conventional power

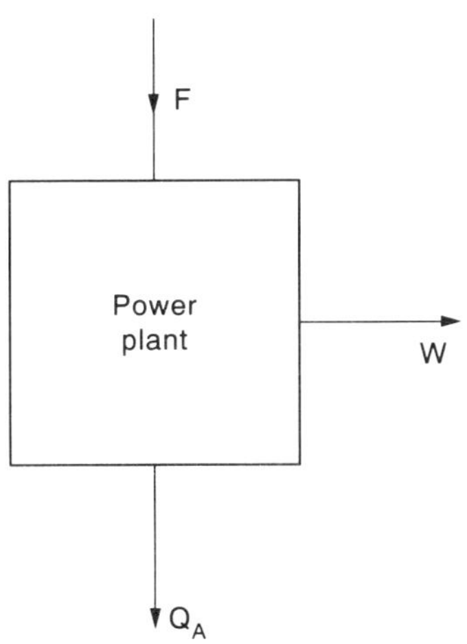

FIGURE 1.1 Basic power plant

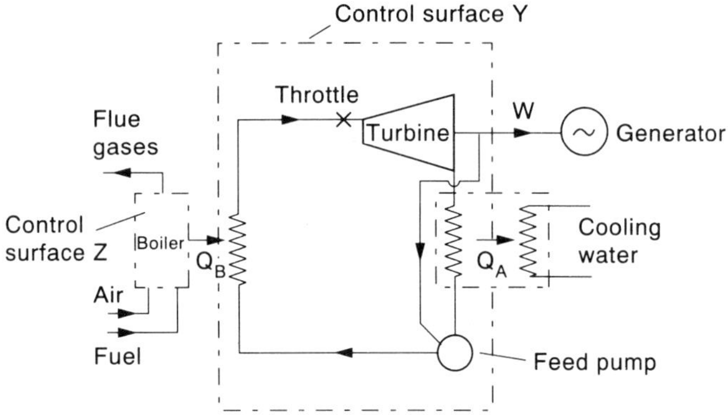

FIGURE 1.2 Steam plant (after Haywood [1])

plants following Haywood [1] and Horlock [2]; we shall restrict our discussion to plants in which the flow is steady, since virtually all the combined plants (and their components) with which this book is concerned have steady flow.

It is important first to distinguish between a cyclic power plant (or heat engine) and an open circuit power plant. In the former, fluid passes continuously round a closed circuit, through a thermodynamic cycle in which heat (Q_B) is received from a source at high temperature, heat (Q_A) is rejected to a sink at low temperature and a work output (W) is delivered, usually to drive an electric generator. Thus in Fig. 1.2, which shows a simple steam power plant, the part within the chain dotted control surface (Y) is a cyclic power plant. Heat is transferred to the boiler tubes, which are within the cyclic power plant, but a second "open" control surface (Z) surrounds the boiler furnace. This receives air and fuel, and discharges flue gases; they are obviously not recirculated, so the boiler furnace is not a cyclic plant. The two control volumes within the control surfaces—one closed, the other open—together form a steam power plant.

A gas turbine power plant may similarly operate on a closed circuit (see Fig. 1.3). Again the chain dotted control surface (Y) surrounds a cyclic heat power plant (or cyclic heat engine) through which air or gas circulates, and the combustion chamber is located within the second, open control surface (Z). Heat Q_B is transferred from Z to Y, and heat Q_A is rejected from Y. The two control volumes form a gas turbine power plant.

More usually, a gas turbine plant operates on "open circuit", with internal combustion (Fig. 1.4). Air and fuel pass across the single control surface into the compressor and combustion chamber respectively, and combustion products leave the control surface after expansion through the turbine. This open circuit plant cannot be said to operate on a thermodynamic cycle; however, its performance is often assessed by treating it as equivalent to a closed cyclic power plant, but care must be used in such an approach.

The classical cycles for power production in steam and gas turbine plant are those associated with the names of Rankine and Joule-Brayton respectively, and the temperature-entropy diagrams for these cycles are shown in Figs. 1.5 and 1.6. The Rankine cycle is the basis of a cyclic steam power plant, with steady flow through a boiler, a turbine driving a generator delivering electrical power, a condenser and a feed-pump. The Joule-Brayton constant pressure cycle is the basis of the cyclic gas turbine power plant, with steady flow of air (or gas) through a compressor, heater, turbine, cooler within a closed circuit. The turbine drives both the compressor and a generator delivering electrical power.

An increasingly important field of study for conventional power plant is that of the "combined plant" (see Wood [3], for example) and this is the subject of the present book. A broad definition of the conventional combined power plant is one in which a "higher" (upper or topping) thermodynamic cycle produces power, but part or all of its heat rejection goes to supply heat to a "lower" (or bottoming) cycle. (In practice, the "upper" plant is often open circuit, not cyclic.) The objective is to achieve a greater work output for a given heat (or fuel energy) supply, by

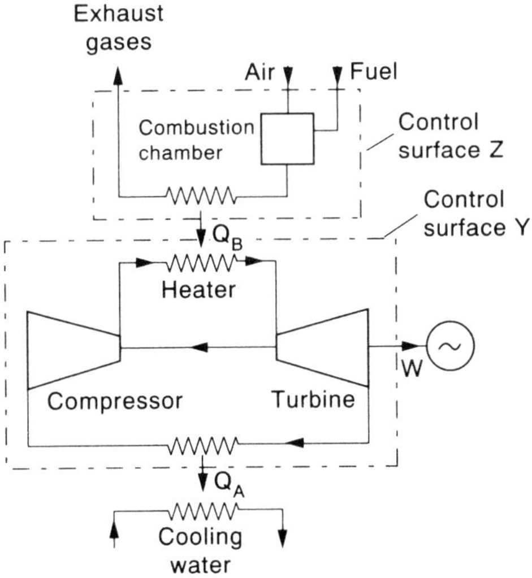

FIGURE 1.3 Closed circuit gas turbine plant (after Haywood [1])

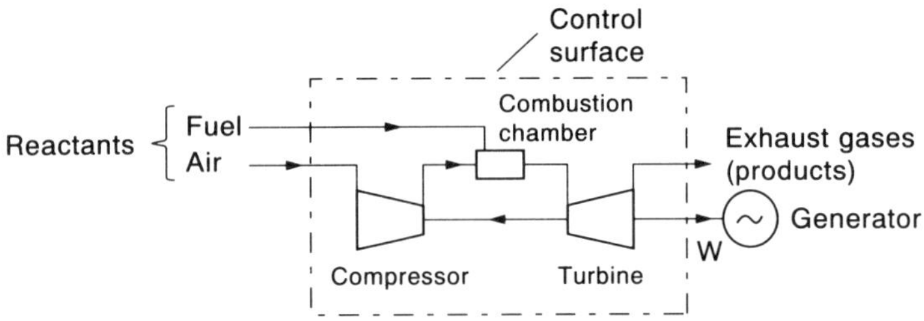

FIGURE 1.4 Open circuit gas turbine plant (after Haywood [1])

approaching more closely the best possible power plant, based on a Carnot cycle (see Section 1.3 below).

The term "cogeneration" is sometimes used to describe combined power plant, but we shall not use it in that way here. "Cogeneration plant" is better used only for a combined *heat and power* (CHP) plant (the subject of an early companion volume in this series, Horlock [2]).

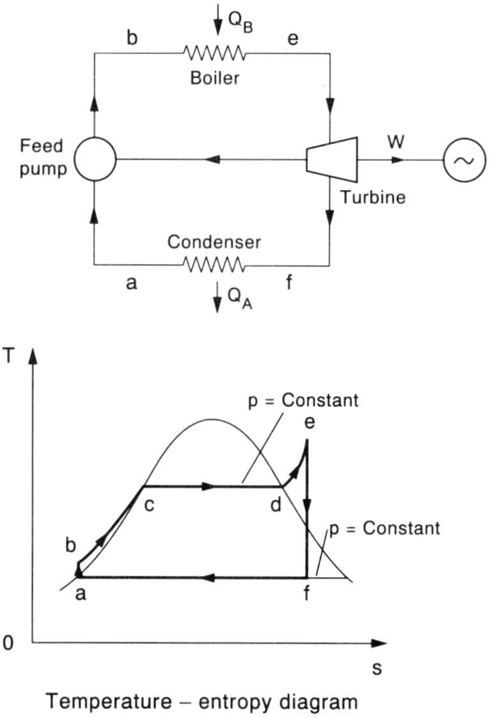

FIGURE 1.5 Rankine cyclic power plant

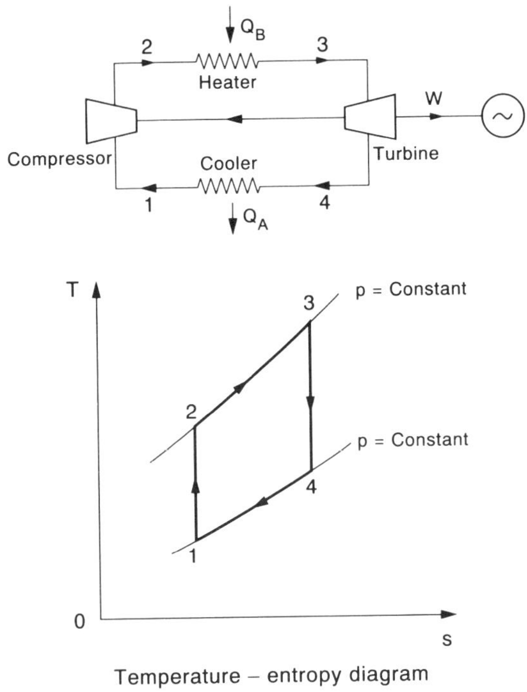

FIGURE 1.6 Joule–Brayton cyclic power plant

1.2 Criteria for Performance of Power Plants

1.2.1 Efficiency of a Closed Circuit Plant

For a cyclic power plant in which fluid circulates continuously within the plant (e.g. the plant enclosed within the control surface Y in Fig. 1.2), one criterion of performance* is simply the thermal or cycle efficiency,

$$\eta \equiv \frac{W}{Q_B}, \tag{1.1}$$

where W is the net work output and Q_B is the heat suplied. W and Q_B may be measured over a given period of time (or per unit mass of the fluid circulating). The efficiency may thus also be expressed in terms of the *power* output ($\dot{W}$) and the *rate* of heat transfer ($\dot{Q}_B$),

*As in *Cogeneration: Combined Heat and Power* [2] we use the words "criterion of performance" to embrace definitions of efficiency. We reserve the word parameter for a "quantity constant in the case considered but varying in different cases" (*Oxford English Dictionary*), such as pressure ratio, which we shall vary in the determination of efficiency.

$$\eta = \frac{\dot{W}}{\dot{Q}_B}, \tag{1.2}$$

and this formulation is sometimes more convenient for a steady flow cycle. In most of the thermodynamic analyses in this book we shall work in terms of W, Q_B and mass flow M (all measured over a given period of time), rather than in terms of the rates $\dot{W}$, $\dot{Q}_B$ and $\dot{M}$ (we shall call M a mass flow and $\dot{M}$ a mass flow rate). Note also the use of η without parentheses around it implies thermal efficiency in this text, although a subscript may be used (e.g. η_{ST} for the thermal efficiency of a cyclic steam power plant).

The heat supply to the cyclic power plant of Fig. 1.2 comes from the control surface Z. Within this second control surface, a steady flow heating device is supplied with reactants (fuel and air) and discharges products of combustion, and we may define a second efficiency for the "heating device" (or boiler) efficiency,

$$(\eta_B) \equiv \frac{Q_B}{F} = \frac{Q_B}{M_F(CV)_0}. \tag{1.3}$$

Q_B is the heat transfer from Z to the closed cycle within control surface Y, which occurs during the time interval that M_F, the mass of fuel, is supplied; and $(CV)_0$ is its calorific value per unit mass for the ambient temperature (T_0) at which the reactants enter. $F = M_F(CV)_0$ is equal to the heat that would be transferred from Z if the products were to leave the control surface at the entry temperature of the reactants, taken as the temperature of the environment, T_0. Figure 1.7 illustrates the definition of calorific value.

The *overall* efficiency of the entire power plant, including the cyclic power plant (within Y) and the heating device (within Z), is given by

$$(\eta_O) \equiv \frac{W}{F} = \left(\frac{W}{Q_B}\right)\left(\frac{Q_B}{F}\right) = \eta(\eta_B). \tag{1.4}$$

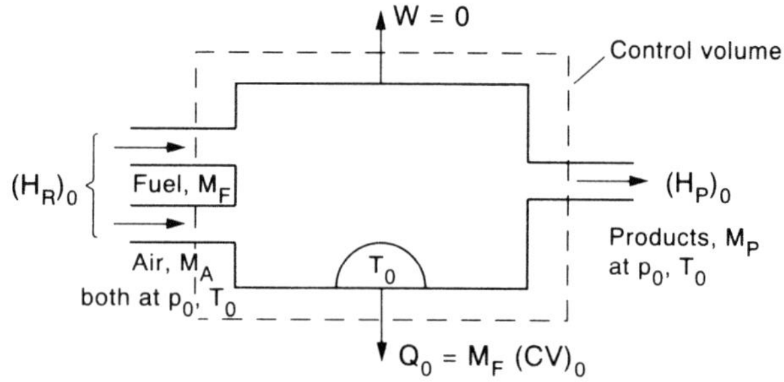

FIGURE 1.7 Determination of calorific value $(CV)_0$

1.2.2 Efficiency of an Open Circuit Plant

For an open circuit (non-cyclic) plant (Fig. 1.4) a different criterion of performance is used—the *rational* efficiency (η_R). This is defined as the ratio of the actual work output to the maximum (reversible) work output that could be achieved between reactants, each at the pressure (p_0) and temperature (T_0) of the environment, and products each at the same p_0, T_0. Thus

$$(\eta_R) \equiv \frac{W}{W_{REV}} \tag{1.5 a}$$

$$= \frac{W}{(-\Delta G_0)}, \tag{1.5 b}$$

where $\Delta G_0 = (G_P)_0 - (G_R)_0$ is the change in Gibbs function (from reactants to products). (The Gibbs function is $G \equiv H - TS$, where H is the enthalpy and S the entropy.) We shall discuss later, in Section 1.6, how further work could be obtained from the products, before entry to the atmosphere, but here we take the maximum available work as $W_{REV} = (-\Delta G)_0$.

ΔG_0 is not readily determinable, and frequently an (*arbitrary*) *overall* efficiency is defined as

$$(\eta_O) \equiv \frac{W}{(-\Delta H_0)} = \frac{W}{M_F(CV)_0} = \frac{W}{F}, \tag{1.6}$$

where $\Delta H_0 = (H_P)_0 - (H_R)_0$ is the change in enthalpy from reactants to products, at the temperature of the environment (Fig. 1.7). This is sometimes unfortunately also referred to as a thermal efficiency, it being implied that there is a heat supply F to an "equivalent" closed circuit plant. For many rections ΔH_0 is numerically almost the same as ΔG_0 and, following Haywood [1], we therefore refer to (η_O) as the (*arbitrary*) *overall* efficiency.

In this book we shall use the criteria of *thermal* efficiency, *heating device* (*or "boiler"*) efficiency (and their product, *overall* efficiency) or *arbitrary overall* efficiency for power plants, basing our analyses of performance upon the assumption of steady flow. We shall also discuss rational efficiency.

1.2.3 Heat Rate

As an alternative to the thermal or cycle efficiency of equation (1.1), the cycle heat rate (the ratio of heat supply rate to power output) is sometimes used:

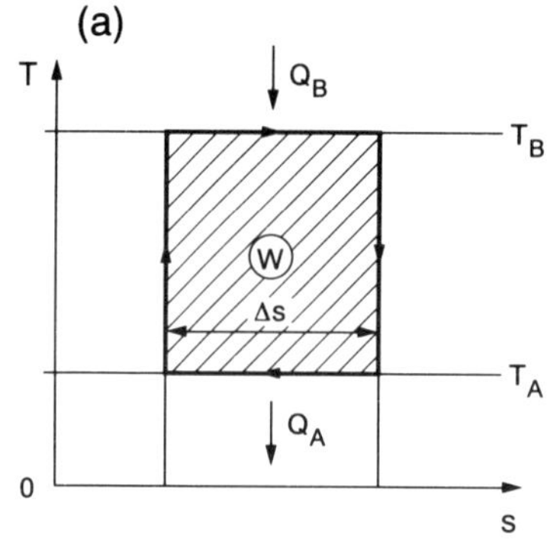

FIGURE 1.8a Temperature-entropy diagram for Carnot power plant

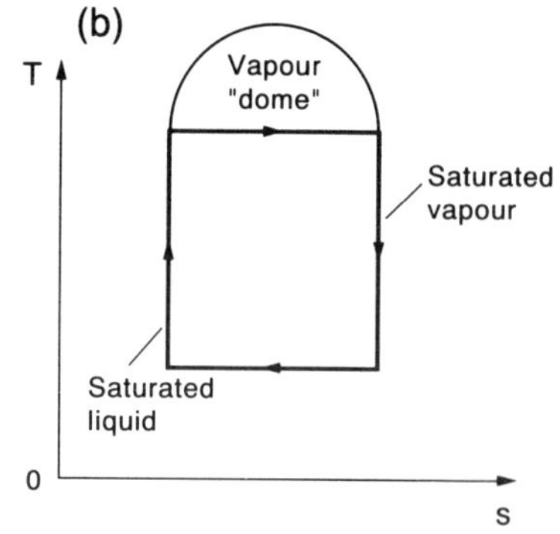

FIGURE 1.8b Carnot cycle using hypothetical fluid

$$\text{Heat rate} \equiv \frac{\dot{Q}_B}{\dot{W}} = \frac{Q_B}{W}. \qquad (1.7)$$

This is the inverse of the thermal or cycle efficiency, when $\dot{Q}_B$ and $\dot{W}$ are expressed in the same units.

Occasionally a "heat" rate is used which is based on energy supplied in the fuel. It is then defined as

$$\text{"Heat rate"} \equiv \frac{M_F(CV)_0}{W} = \frac{F}{W}, \qquad (1.8)$$

which is the inverse of the overall efficiency of the closed circuit plant, as defined in equation (1.4), or the inverse of the (arbitrary) overall efficiency of the open circuit plant, as defined in equation (1.6).

1.3 Ideal (Carnot) Power Plant Performance

The Second Law of Thermodynamics may be used to show that a cyclic heat power plant (or cyclic heat engine) achieves maximum efficiency by operating on a reversible cycle called the Carnot cycle, for given (maximum) temperature of supply (T_{max}), and given (minimum) temperature of heat rejection (T_{min}). Such a Carnot power plant receives all its heat (Q_B) at the top temperature (i.e. $T_B = T_{max}$) and rejects all its heat (Q_A) at the minimum temperature (i.e. $T_A = T_{min}$); the other processes are reversible and adiabatic, and therefore isentropic (see the temperature–entropy diagram of Fig. 1.8a). Its thermal efficiency is

$$\eta_{\text{CARNOT}} = \frac{W}{Q_B} = \frac{Q_B - Q_A}{Q_B} = \frac{T_{max}\Delta s - T_{min}\Delta s}{T_{max}\Delta s} = (T_{max} - T_{min})/T_{max}. \quad (1.9)$$

Clearly raising T_{max} and lowering T_{min} both lead to higher efficiency.

The Carnot engine (or cyclic power plant) is a useful hypothetical device in the study of the thermodynamics of power-producing cycles, for it provides a measure of the best performance that can be achieved under the given boundary conditions on temperature. It has three features which give it maximum thermal efficiency:

(i) all processes involved are reversible;
(ii) all heat is supplied at the maximum (specified) temperature (T_{max});
(iii) all heat is rejected at the lowest (specified) temperature (T_{min}).

In his search for high efficiency, the designer of a conventional power plant attempts to emulate these features of the Carnot cycle.

A fluid with a saturated liquid line and a saturated vapour line as shown in Fig. 1.8 b could in theory be used to obtain a Carnot cycle. In practice no fluids have these properties.

1.4 Limitations of Other Cycles

Other closed cycles do not achieve Carnot efficiency because they do not match these features, i.e.

(i) the actual (variable) temperature of heat supply is less than T_{max};
(ii) the actual (variable) temperature of heat rejection is greater than T_{min};
(iii) irreversibility occurs within the cycle.

An enlightening analysis by Caputo [4] (quoted by Cerri [5]) illustrates these failures to achieve Carnot efficiency. Caputo first defines mean temperatures of heat supply and rejection,

$$\bar{T}_B = \frac{Q_B}{\int \frac{dQ_B}{T}}, \quad \bar{T}_A = \frac{Q_A}{\int \frac{dQ_A}{T}}. \quad (1.10)$$

He then defines parameters ξ_B and ξ_A which measure failure to achieve the maximum and minimum temperatures, T_{max} and T_{min},

$$\xi_B \equiv \frac{\overline{T}_B}{T_{max}}, \quad \xi_A \equiv \frac{\overline{T}_A}{T_{min}}, \tag{1.11}$$

where ξ_B is less than unity and ξ_A is greater than unity. The combined parameter

$$\xi \equiv \frac{\xi_B}{\xi_A} = \left(\frac{T_{min}}{T_{max}}\right)\left(\frac{\overline{T}_B}{\overline{T}_A}\right) = \tau\left(\frac{\overline{T}_B}{\overline{T}_A}\right) \tag{1.12}$$

[where $\tau = (T_{min}/T_{max})$] is an overall measure of the failure of the real cycle to achieve the maximum and minimum temperatures and is always less than unity. (Note that for the Carnot cycle, $\xi_{CARNOT} = 1$.)

Caputo then introduces a parameter (σ) which is a measure of the irreversibilities within the real cycle. He first defines

$$\sigma_B \equiv \frac{Q_B}{\overline{T}_B}, \quad \sigma_A \equiv \frac{Q_A}{\overline{T}_A}, \tag{1.13}$$

which, from the definitions of $\overline{T}$, can be seen to be simply the entropy changes in heat supply and heat rejection respectively. The parameter σ is then defined as

$$\sigma \equiv \frac{\sigma_B}{\sigma_A}, \tag{1.14}$$

the ratio of entropy change in heat supply to entropy change in heat rejection. For the Carnot cycle σ_{CARNOT} is unity, but for other (irreversible) cycles, a value of σ less than unity indicates a "widening" of the cycle on the (T,s) diagram due to irreversibilities (e.g. in compression and/or expansion) and a resulting loss in thermal efficiency.

The overall effect of these failures to achieve Carnot efficiency is then encompassed in a new parameter, μ, where

$$\mu \equiv \xi\sigma. \tag{1.15}$$

The efficiency of the real cycle may then be expressed in terms of τ (the ratio of minimum to maximum temperature) and μ. For

$$\begin{aligned}\eta &= 1 - \left(\frac{Q_A}{Q_B}\right) \\ &= 1 - \frac{\sigma_A \overline{T}_A}{\sigma_B \overline{T}_B} \\ &= 1 - \frac{\sigma_A \xi_A T_{min}}{\sigma_B \xi_B T_{max}} \\ &= 1 - \frac{\tau}{\sigma\xi} \\ &= 1 - \frac{\tau}{\mu}.\end{aligned} \tag{1.16}$$

For the Carnot cycle, $\sigma_{CARNOT}=1$, $\xi_{CARNOT}=1$ and $\mu_{CARNOT}=1$, so $\eta_{CARNOT}=1-\tau$.

For a reversible cycle, such as the Rankine cycle or the Joule-Brayton cycle, there is no "widening" of the cycle due to irreversibilities, so that $\sigma_A=\sigma_B$ and $\sigma=1$. The efficiency is then

$$\eta=1-\frac{\tau}{\xi}<\eta_{CARNOT}, \tag{1.17}$$

and the failure to reach Carnot cycle efficiency is entirely due to non-achievement of T_{max} and/or T_{min}. ξ is less than unity and $\eta<\eta_{CARNOT}$.

For an irreversible cycle (e.g. a ("modified" Carnot cycle with heat supply at T_{max} and heat rejection at T_{min}, but irreversible compression and/or expansion), $\sigma_A>\sigma_B$ and σ is less than unity. For such a "modified" Carnot cycle

$$\eta=1-\frac{\tau}{\sigma}<\eta_{CARNOT}, \tag{1.18}$$

and the non-achievement of Carnot efficiency is entirely due to the irreversibilities.

Calculation of ξ and σ for real cycles illustrates the thermodynamic weaknesses of those cycles, i.e. the reasons for failure to achieve Carnot cycle efficiency.

1.5 Modifications of Other Cycles to Achieve Higher Thermal Efficiency

There are several modifications to basic cycles which may be introduced to raise thermal efficiency.

Two objectives are immediately clear. If top temperature can be raised and bottom temperature lowered, then the ratio $\tau=(T_{min}/T_{max})$ is decreased and, as with a Carnot cycle, thermal efficiency will be increased (for given μ). The limit on top temperature is likely to be metallurgical, and that on the bottom temperature that of the surrounding atmosphere.

A third objective is similarly obvious. If compression and expansion processes can attain more nearly isentropic conditions, then the cycle "widening" due to irreversibility is decreased, σ moves nearer to unity and the thermal efficiency increases (for a given τ).

Other modifications are aimed at increasing ξ (by increasing ξ_B or decreasing ξ_A). For example, for a given T_{max}, raising boiler pressure and using reheat in a steam turbine cycle increases the mean temperature of supply (and hence ξ_B) as does the introduction of feed heating. In a gas turbine cycle, reheat alone has the same effect (increasing ξ_B) but also increases ξ_A so that ξ decreases and efficiency drops. Similarly, whereas intercooling alone lowers the mean temperature of heat rejected (decreasing ξ_A), it also decreases ξ_B, so that ξ decreases and efficiency drops. However, when reheating and intercooling are coupled with the use of a heat exchanger, there is a marked increase in ξ and hence in overall thermal efficiency.

We discuss these various modifications separately, for gas turbine plant (modification of the Joule cycle) and for steam turbine plant (modification of the Rankine cycle).

1.5.1 Gas Turbine Power Plants

We first discuss the way thermal efficiency of the gas turbine plant depends on some major parameters (e.g. pressure ratio, the ratio of maximum to minimum temperature), and how the basic plant may be modified to increase efficiency. To simplify our discussion we develop the basic analyses for closed cycles operating with a single perfect gas of constant specific heat capacities (c_p, c_v), but allow for some irreversibilities (turbine and compressor efficiency, pressure losses, etc.). For a fuller discussion the reader is referred to the books by Cohen, Rogers and Saravanamuttoo [6] and Haywood [1].

1.5.1.1 THE BASIC CYCLE (CBT)

For the basic compressor, heater, turbine cycle (CBT, Fig. 1.9), the thermal efficiency is given by

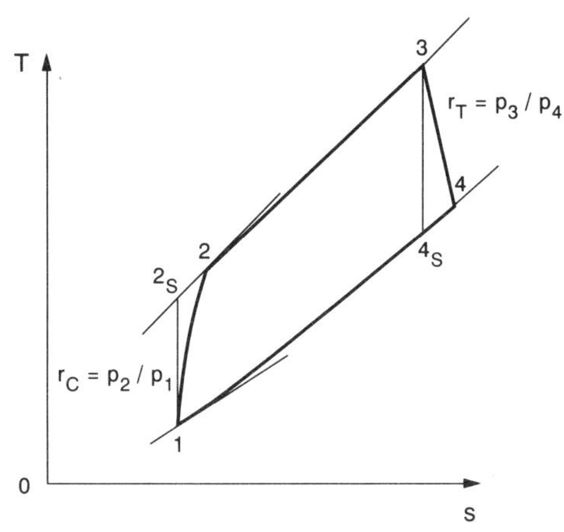

FIGURE 1.9 CBT cycle

$$\eta = 1 - \frac{\text{Heat rejected}}{\text{Heat supplied}}$$
$$= 1 - \frac{c_p(T_4 - T_1)}{c_p(T_3 - T_2)}$$
$$= 1 - \frac{\left[\left(\frac{T_4}{T_3}\right)\left(\frac{T_3}{T_1}\right) - 1\right]}{\left[\left(\frac{T_3}{T_1}\right) - \left(\frac{T_2}{T_1}\right)\right]}.$$

The temperature ratios (T_4/T_3) and (T_2/T_1) may be related to the compressor and turbine pressure ratios r_C and r_T and isentropic efficiencies (η_C) and (η_T):

$$\frac{T_4}{T_3} = 1 - \frac{(T_3 - T_4)}{T_3}$$
$$= 1 - \frac{(\eta_T)(T_3 - T_{4_s})}{T_3}$$
$$= 1 - (\eta_T)\left(1 - \frac{1}{\rho_T}\right),$$

where $\rho_T = (r_T)^{(\kappa - 1)/\kappa}$;

$$\frac{T_2}{T_1} = 1 + \frac{(T_2 - T_1)}{T_1}$$
$$= 1 + \frac{(T_{2_s} - T_1)}{(\eta_C)T_1}$$
$$= 1 + \frac{(\rho_C - 1)}{(\eta_C)},$$

where $\rho_C = (r_C)^{(\kappa - 1)/\kappa}$.

Alternatively, the temperature ratios may be written directly in terms of pressure ratios and polytropic efficiency, η_p,

$$\frac{T_4}{T_3} = \left(\frac{1}{r_T}\right)^{(\eta_p)_T(\kappa - 1)/\kappa}, \quad \frac{T_2}{T_1} = (r_C)^{(\kappa - 1)/(\eta_p)_C\kappa}.$$

Ideally $r_C = r_T = r$, but pressure losses occur in entry and exhaust ducts, and in the heater (or the combustion chamber in the open cycle). An expression for the thermal efficiency is thus

$$\eta = 1 - \frac{\left\{\theta\left[1 - (\eta_T)\left(1 - \frac{1}{\rho_T}\right)\right] - 1\right\}}{\left[(\theta - 1) - \frac{(\rho_C - 1)}{(\eta_C)}\right]}, \tag{1.19}$$

where $\theta \equiv (T_3/T_1) = (T_{max}/T_{min})$.

If $r_C = r_T$, and $\rho_C = \rho_T = \rho$, then

$$\eta = \frac{\left(\frac{\rho - 1}{\rho}\right)[(\eta_T)(\eta_C)\theta - \rho]}{(\eta_C)(\theta - 1) - (\rho - 1)}. \tag{1.20}$$

For maximum efficiency

$$\rho = \frac{\alpha - [\alpha^2 - \alpha\beta(1 + \alpha - \beta)]^{1/2}}{1 + \alpha - \beta}, \tag{1.21}$$

where $\alpha = (\eta_T)(\eta_C)\theta$, $\beta = 1 + (\eta_C)(\theta - 1)$ (see Woods *et al.* [7])

The specific work (the work output per unit flow) is also an important parameter which we shall use later in our studies of combined power plant. For a closed cycle it is given by

$$w = \frac{W}{M} = (h_3 - h_4) - (h_2 - h_1)$$

$$= c_p T_3(\eta_T)\left(1 - \frac{1}{\rho_T}\right) - \frac{c_p T_1}{(\eta_C)}(\rho_C - 1),$$

so that

$$\frac{w}{c_p T_1} = \frac{[(\eta_T)(\eta_C)\theta - \rho](\rho - 1)}{\rho(\eta_C)}, \tag{1.22}$$

if $\rho_T = \rho_C = \rho$.

For maximum specific work, $\rho = (\eta_T \eta_C \theta)^{1/2}$, which gives a pressure ratio less than that for maximum efficiency.

For an open circuit gas turbine, fuel calorific values are required together with tables of gas properties (e.g. Keenan and Kaye [8]). A useful practical plot of overall efficiency (for an open circuit gas turbine) against specific work (per unit air flow) for various r and (T_3/T_1) is given by Rice [9] and is shown in Fig. 1.10, together with a list of his assumptions. The optimum pressure ratio for maximum efficiency increases with maximum temperature, and is larger than that for maximum output (specific work). Figure 1.11 shows turbine exit temperature plotted against the same parameters; this is a vitally important factor in the choice of a bottom cycle to the gas turbine in a combined power plant.

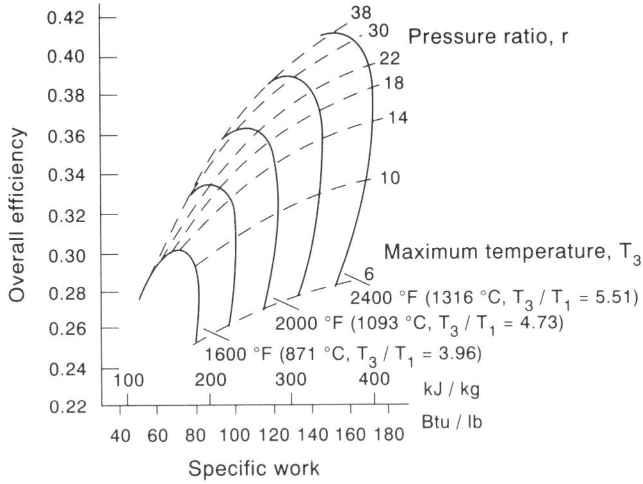

FIGURE 1.10 Overall efficiency (LHV) and specific work for CBT plant. $\eta_C=0.85$, $\eta_T=0.87$, $(p_3-p_2)/p_2=0.04$, fuel $(CH_2)_n$, $(CV)_0=42.8$ MJ/kg, $T_0=15.6°$ C, $p_1=1.0136$ bar, $p_4=1.0274$ bar (after Rice [9])

1.5.1.2 THE REGENERATIVE CYCLE (CBTX)

Introduction of a heat exchanger increases overall thermal efficiency by raising the mean temperature of heat supply and lowering the mean temperature of heat rejection (Fig. 1.12). The effectiveness of the heat exchanger is defined as

$$\varepsilon \equiv \frac{\text{Cold side temperature rise}}{\text{Maximum temperature difference}} = \frac{(T_X - T_2)}{(T_4 - T_2)}. \tag{1.23}$$

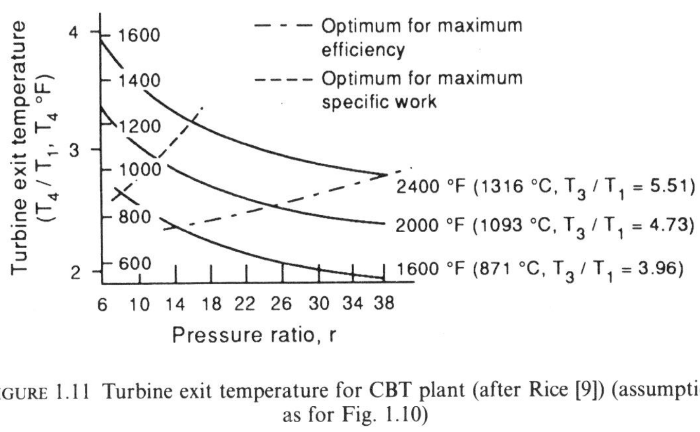

FIGURE 1.11 Turbine exit temperature for CBT plant (after Rice [9]) (assumptions as for Fig. 1.10)

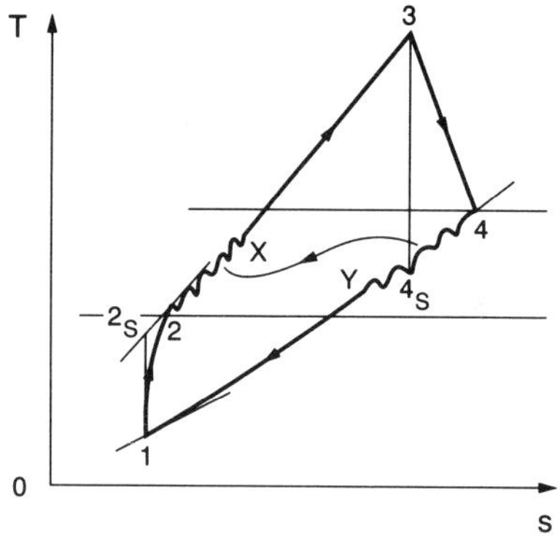

FIGURE 1.12 CBTX cycle

The thermal efficiency of the plant is

$$\eta = 1 - \left(\frac{\text{Heat rejected}}{\text{Heat supplied}}\right)$$
$$= 1 - \frac{c_p(T_Y - T_1)}{c_p(T_3 - T_X)}$$
$$= 1 - \frac{c_p\left(\frac{T_Y}{T_1} - 1\right)}{c_p\left(\frac{T_3}{T_1} - \frac{T_X}{T_1}\right)}. \tag{1.24}$$

The expressions for (T_2/T_1) and (T_4/T_3) given in the last section remain unchanged, and with the heat exchanger effectiveness defined the temperature ratio (T_X/T_1) follows immediately. (T_Y/T_1) is derived from the heat balance in the exchanger. The thermal efficiency may then be expressed in the form

$$\eta = f[\theta, r_C, r_T, (\eta_C), (\eta_T), \varepsilon], \tag{1.25}$$

but the algebraic expression is complicated and is not reproduced here.

Pfenninger [10] gives a practical diagram (Fig. 1.13) of how η varies with pressure ratio for a CBTX open circuit plant with $T_3 = 880\,°C$ and $\varepsilon = 0.8$. The efficiency is higher (at low pressure ratio) than that of the CBT plant (also shown) (and will again increase with $\theta = T_3/T_1$). There is a limiting pressure ratio at which heat exchange is not possible (the turbine exhaust temperature becomes less than that at compressor outlet). Pfenninger also shows the variation of the steady flow availability of the gas at the turbine exhaust ($b=h-T_0s$) in Fig. 1.13, emphasising that the increased efficiency of the CBTX plant means that the availability (b) of the exhaust gas (its usefulness in supplying heat to a lower (bottoming) cycle) is less, and the benefit of adding a lower (steam) cycle is reduced.

1.5.1.3 REHEATING AND INTERCOOLING

Further increases in efficiency can be obtained by adding reheat between an H.P. turbine and a L.P. turbine, and intercooling between a L.P. compressor and a H.P. compressor, provided that a heat exchanger is also used. (In the absence of a heat exchanger, the effect of reheating and intercooling is to reduce efficiency, although specific work output is increased).

Referring to Fig. 1.14, the thermal efficiency is

$$\eta = 1 - \frac{\text{Heat rejected}}{\text{Heat supplied}}$$

$$= 1 - \frac{(T_Y - T_1) + (T_{2'} - T_{1'})}{(T_3 - T_X) + (T_{3'} - T_{4'})}. \tag{1.26}$$

Again the temperature ratios $(T_3/T_{4'})$, $(T_{3'}/T_4)$ may be related to H.P. and L.P. turbine pressure ratios and efficiencies; the temperature ratios

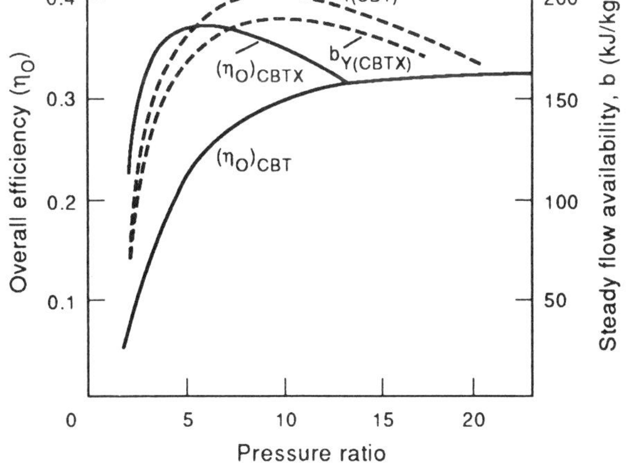

FIGURE 1.13 Overall efficiency and exhaust availability of CBT and CBTX plants (after Pfenninger [10]) $T_3=880°$ C, $\varepsilon=0.80$

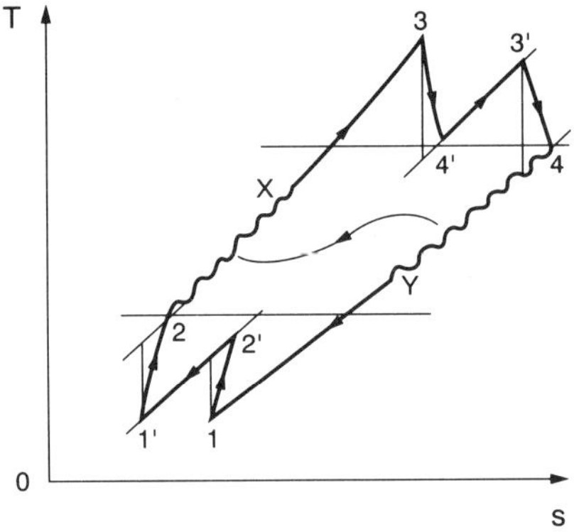

FIGURE 1.14 Temperature-entropy diagram for CICBTBTX cycle

$(T_{2'}/T_1)$, $(T_2/T_{1'})$ may be related to L.P. and H.P. compressor ratios and efficiencies. The effectiveness of the heat exchanger is defined as before,

$$\varepsilon \equiv \frac{(T_X - T_2)}{(T_4 - T_2)}, \tag{1.23}$$

and the overall efficiency may be expressed in the form

$$\eta = f[(T_3/T_1), (T_3'/T_1), (\eta_C)_{12'}, (\eta_C)_{1'2}, (\eta_T)_{34'}, (\eta_T)_{3'4}, (r_C)_{12'}, (r_C)_{1'2}, (r_T)_{34'}, (r_T)_{3'4}, \varepsilon]. \tag{1.27}$$

Obviously the calculation of η is complex, and the establishment of the optimum split in pressure ratio between compressors and between turbines is even more complicated. In fact the optimum pressure ratio across each of the two compressors is almost equal to $[(r_C)_{12'}(r_C)_{1'2}]^{1/2}$ and that between the two turbines is close to $[(r_T)_{34'}(r_T)_{3'4}]^{1/2}$. A full analysis is given by Hawthorne and Davis [11].

Finally, we may note that if many stages of reheat and intercooling were employed and all processes were reversible, then Carnot efficiency would be obtained, for all the heat supplied would be at the top temperature and all the heat rejected would be at the lowest temperature.

1.5.2 Steam Turbine Power Plants

1.5.2.1 THE PRACTICAL CYCLE

For the practical steam turbine cycle the reversible Rankine cycle is modified by the presence of irreversibilities in the components (pump, boiler, turbine, condenser), as indicated in Fig. 1.15a. The thermal efficiency of the plant is

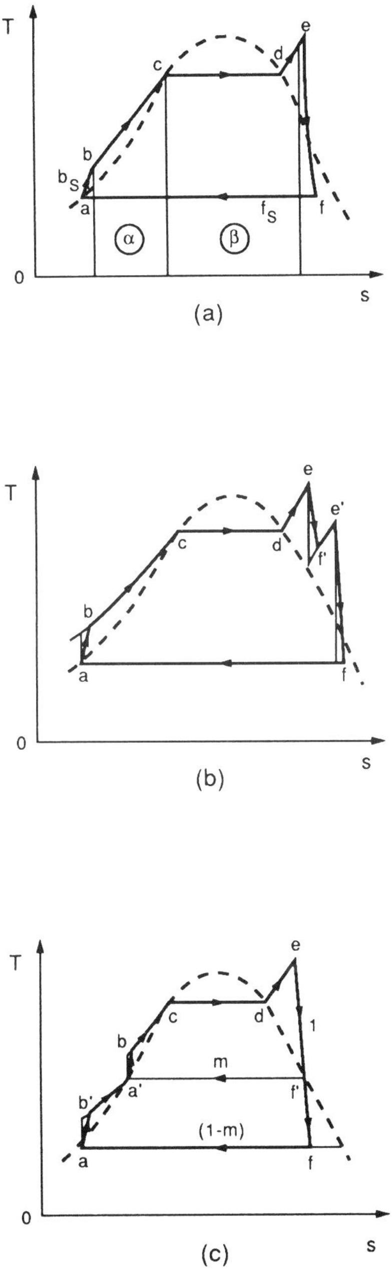

FIGURE 1.15 Temperature-entropy diagrams for steam turbine power plants
(a) Basic cycle
(b) Cycle with reheat
(c) Regenerative cycle

$$\eta = 1 - \left(\frac{\text{Heat rejected}}{\text{Heat supplied}}\right)$$

$$= 1 - \frac{(h_f - h_a)}{(h_e - h_b)}$$

$$= \frac{(h_e - h_f) - (h_b - h_a)}{(h_e - h_b)}. \qquad (1.28)$$

If the pump work $(h_b - h_a)$ is small, as it is for low boiler pressure, the efficiency may be written approximately

$$\eta = 1 - \frac{(h_f - h_a)}{(h_e - h_a)} = \frac{(h_e - h_f)}{(h_e - h_a)}. \qquad (1.29)$$

Unlike the closed cycle gas turbine, the steam turbine cycle is not amenable to analytical treatment and the values of enthalpy in the expression for thermal efficiency have to be determined from tables of steam properties. However, it is important to relate the enthalpy differences $(h_e - h_f)$ and $(h_b - h_a)$, the work output in the turbine and the work input to the pump, to isentropic enthalpy changes, through turbine and pump efficiencies, (η_T) and (η_{FP}), so that

$$\eta = \frac{(\eta_T)(h_e - h_{f_s}) - \dfrac{1}{(\eta_{FP})}(h_{b_s} - h_a)}{(h_e - h_b)}$$

$$\approx \frac{(\eta_T)(h_e - h_{f_s}) - [(p_b - p_a)/\rho_w(\eta_{FP})]}{(h_e - h_a) - [(p_b - p_a)/\rho_w(\eta_{FP})]}. \qquad (1.30)$$

Parametric computer studies showing efficiency variation with boiler and condenser pressure, turbine efficiency, turbine inlet temperatures, etc., may be undertaken to give

$$\eta = f[p_c, T_e, p_a, (\eta_T), (\eta_{FP})]. \qquad (1.31)$$

However, it is worthwhile mentioning an approximate analytical approach suggested by Salisbury [12], and used by Horlock [13] and Haywood [14], in which it is assumed that the turbine expansion line follows a path on the diagram such that $(h - h_w) = \text{constant} = \beta$, where h is the (local) enthalpy on the expansion line at a given pressure, and h_w is the enthalpy of saturated water at that pressure.

The approximation $\beta = \text{constant}$ is valid over but a limited range of steam conditions (see Weir [15]), but fortuitously is quite good for the operating range of steam turbines used in combined power plant. Figure 1.16, given by Weir, shows how β varies along expansion lines of constant

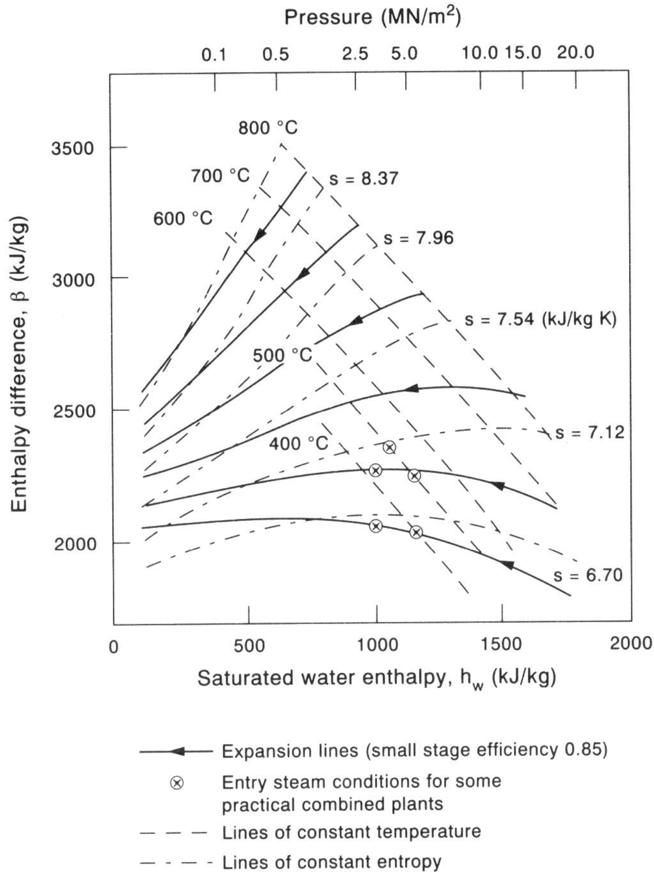

FIGURE 1.16 Variation in $\beta=(h-h_w)$ along expansion lines of constant small stage efficiency (after Weir [15])

small stage efficiency (0.85). The entry steam conditions for several practical combined plants are also shown, and they lie in the region where β remains substantially unchanged.

Accepting this simplification, therefore, the approximate expression for thermal efficiency (equation (1.29)) may then be written

$$\eta=1-\frac{(h_f-h_a)}{(h_e-h_c)+(h_c-h_a)}$$

$$=1-\frac{\beta}{(\alpha+\beta)}$$

$$=\frac{\alpha}{(\alpha+\beta)}, \tag{1.32}$$

where

$$\alpha = (h_c - h_a)$$

$$\approx \int_b^c c_p \, dT.$$

The areas on the T, s diagram corresponding to α and β are shown on Fig. 1.15a; α is approximately the work output, and β the heat rejected.

Calculations using this approximate expression for η for varying boiler pressure (p_c) are shown in Fig. 1.17, for a condenser pressure $p_a = 4\,\text{kN/m}^2$, $T_e = 300\,°\text{C}$, $400\,°\text{C}$, $500\,°\text{C}$, $600\,°\text{C}$, $700\,°\text{C}$. (The turbine isentropic efficiency (η_T) is implicit in the assumption β = constant for the expansion line, and is approximately 80–83.5% for the range of pressures shown.)

There are two major improvements which can be made to the basic steam turbine cycle—reheating and regenerative feed heating.

1.5.2.2 REHEATING

Whereas in the gas turbine cycle the addition of reheat (without a heat exchanger) lowered the efficiency (the mean heat rejection temperature being also raised), in the steam turbine cycle the efficiency is increased as long as the reheat pressure is sufficiently high to raise the mean heat

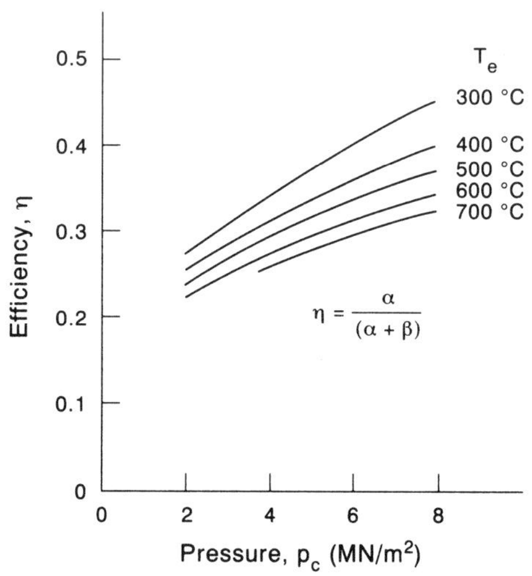

FIGURE 1.17 "Rankine" cycle thermal efficiency for constant β expansion in steam turbine

supply temperature. Detailed calculations show the optimum reheat pressure is about a quarter of the boiler pressure for reversible, non-regenerative cycles, but for actual plant there may be advantage in going to lower reheat pressures because of increased turbine efficiency from decreasing wetness in the L.P. stages.

The calculation of efficiency is straightforward and is given by

$$\eta = 1 - \frac{\text{Heat rejected}}{\text{Heat supplied}}$$

$$= 1 - \frac{(h_f - h_a)}{(h_e - h_b) + (h_{e'} - h_{f'})}, \tag{1.33}$$

where the various states are as indicated in Fig. 1.15b.

1.5.2.3 REGENERATIVE FEED HEATING

The second major improvement in steam turbine cycle efficiency involves regenerative feed heating—bleeding steam from the turbine to heat the feed water before it enters the boiler. Figure 1.15c shows a single abstraction of steam from the turbine at point f', which in a direct contact feed heater raises the feed water temperature from $T_{b'} (\approx T_a)$ to $T_{a'}$. We neglect feed pump work and the heat balance in the heater is simply

$$m(h_{f'} - h_{a'}) = (1 - m)(h_{a'} - h_a),$$

so that the bled steam fraction is

$$m = \frac{(h_{a'} - h_a)}{(h_{f'} - h_a)}. \tag{1.34}$$

The thermal efficiency of the plant is

$$\eta = 1 - \frac{\text{Heat rejected}}{\text{Heat supplied}}$$

$$\approx 1 - \frac{(1 - m)(h_f - h_a)}{(h_e - h_{a'})}. \tag{1.35}$$

If we again make the assumption of Salisbury about the form of the full turbine expansion line, $\beta = (h - h_w) = \text{constant}$, then equations (1.34) and (1.35) may be written

$$m = \frac{\gamma}{\beta + \gamma},$$

where

$$\gamma=(h_{a'}-h_a), \quad \beta=(h_{f'}-h_{a'})$$

and

$$\eta=1-\left[\frac{(1-m)\beta}{(\alpha+\beta-\gamma)}\right]. \tag{1.36}$$

Substituting for m it follows that

$$\eta=1-\frac{\left[1-\dfrac{\gamma}{(\beta+\gamma)}\right]\beta}{(\alpha+\beta-\gamma)}$$

$$=1-\frac{\beta^2}{(\alpha+\beta-\gamma)(\beta+\gamma)}. \tag{1.37}$$

With α and β held fixed, the maximum efficiency results when the amount of bled steam is varied (and γ with it) to give

$$\frac{d\eta}{d\gamma}=0=\beta^2[(\alpha+\beta-\gamma)-(\beta+\gamma)],$$

$$\gamma=\frac{\alpha}{2}. \tag{1.38}$$

This is a widely recognised result, that optimum efficiency occurs when the steam bled is sufficient to divide the water heating equally between the feed water heater and the economiser. More generally, equal divisions between several feed water heaters (and the economiser) give optimum performance (see Haywood [14]).

The resulting efficiency of the single bleed cycle is

$$\eta=1-\frac{\beta^2}{\left(\beta+\dfrac{\alpha}{2}\right)^2}$$

$$=\frac{\alpha^2+4\alpha\beta}{(\alpha+2\beta)^2}. \tag{1.39}$$

The efficiency of the basic plant has been increased by the increase in the mean temperature of supply, but some further irreversibility has been introduced in the mixing process in the direct contact heater. The efficiency of this simple regenerative cycle is compared with the basic "Rankine" cycle in Fig. 1.18 for $T_e=500\,°\text{C}$, $p_a=4\,\text{kN/m}^2$ and various boiler pressures.

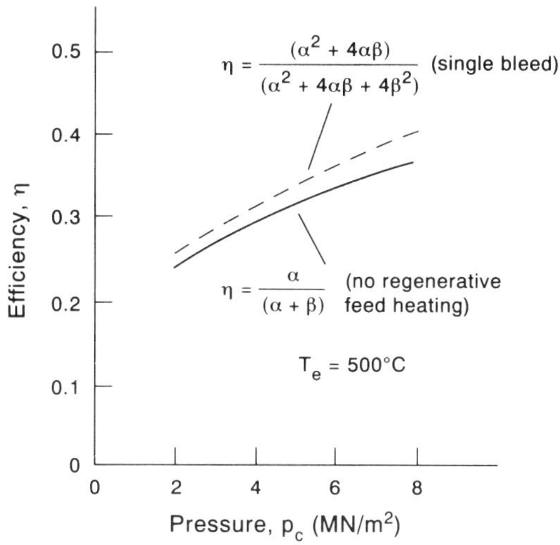

FIGURE 1.18 Effect of feed heating on thermal efficiency

Complex analysis is involved in the study of the performance of a steam cycle plant with many feed heaters, of the direct contact or surface heating type. The introduction or deletion of some limited feed heating in the closed steam cycle is important in our later studies of combined plant and we shall return to a discussion of whether feed heating should be used in such a plant.

1.5.2.4 DISCUSSION

In modern steam plant, both reheating and feed heating are vital parts of the cycle, and it is of course important to raise boiler pressure and lower condenser pressure in order to achieve high mean temperature of heat supply and low heat rejection temperature.

We have indicated trends in efficiency by using the simple assumption β=constant. For complex, advanced steam cycles calculation of overall thermal efficiency becomes extremely complicated. Here we show the results of one of many practical calculations undertaken by Haywood [16] assisted by the author. Figure 1.19 shows the variation of thermal efficiency η with final feed temperature for a plant with double reheat at various high boiler pressures (above and below critical pressure). Seven feed heaters were assumed, the top heating tapping being at the reheat point.

The introduction of many stages of reheat and reversible regenerative feed heating can in theory lead to an approach to Carnot efficiency.

1.6 Reversibility, Availability and Exergy

The concepts of reversibility and irreversibility are important in the analysis of conventional and combined plants. Reduction of irreversibility in the components of a power plant leads to greater overall plant efficiency, and the concepts of thermodynamic availability may be used to demonstrate this. A summary of the results of availability theory for steady flow processes, leading to the definition and use of exergy, is given here.

1.6.1 Flow in the Presence of an Environment at T_0 (not involving chemical reaction)

We first consider the steady flow of a fluid through a control volume CV between prescribed stable states, X and Y (Fig. 1.20), in the presence of a conceptual environment at temperature T_0 (i.e. with heat transfer to that environment only). The maximum work which is obtained in reversible flow between X and Y is given by

$$[(W_{CV})_{REV}]_X^Y = B_X - B_Y, \tag{1.40}$$

where B is the steady flow availability function

$$B = H - T_0 S \tag{1.41}$$

(see Haywood [18]). The reversible (outward) heat transfer between X and Y is

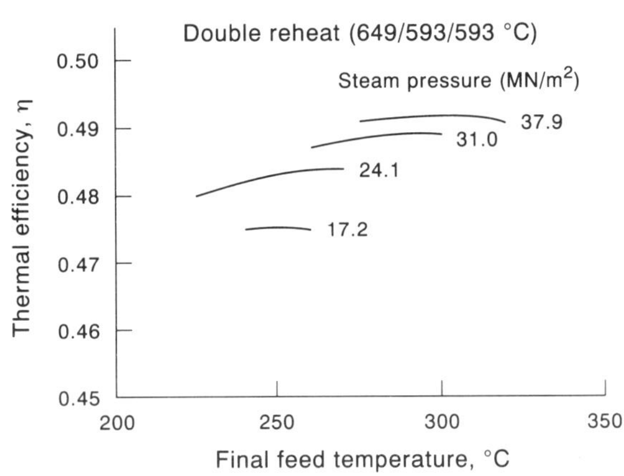

FIGURE 1.19 Thermal efficiency of steam cycles with high pressures, double reheat and seven feed heaters. Variation with final feed temperature (after Haywood [16])

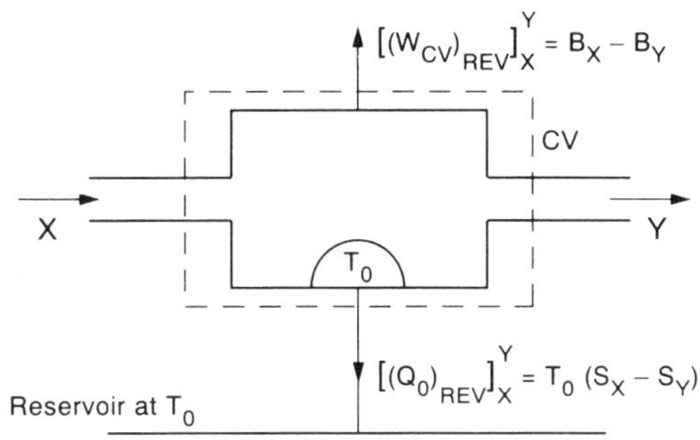

FIGURE 1.20 Reversible process with heat transfer at T_0

$$[(Q_0)_{\mathrm{REV}}]_{\mathrm{X}}^{\mathrm{Y}} = T_0(S_{\mathrm{X}} - S_{\mathrm{Y}}). \tag{1.42}$$

A corollary of this theorem is that the maximum work that can be extracted from fluid at prescribed state X is the exergy

$$E = B_{\mathrm{X}} - B_0. \tag{1.43}$$

Here B_0 is the steady flow availability function at the so-called "dead state", where the fluid is in (restricted) equilibrium with the environment, at state (p_0, T_0). The maximum work obtainable between state X and Y may thus be written as

$$[(W_{\mathrm{CV}})_{\mathrm{REV}}]_{\mathrm{X}}^{\mathrm{Y}} = (B_{\mathrm{X}} - B_0) - (B_{\mathrm{Y}} - B_0) = (E_{\mathrm{X}} - E_{\mathrm{Y}}). \tag{1.44}$$

From the steady flow energy equation the work output in an *actual* (irrcvcrsiblc) flow through a control volume CV, between states X and Y in the presence of an environment at T_0, is

$$[W_{\mathrm{CV}}]_{\mathrm{X}}^{\mathrm{Y}} = (H_{\mathrm{X}} - H_{\mathrm{Y}}) - [Q_0]_{\mathrm{X}}^{\mathrm{Y}}, \tag{1.45}$$

where $[Q_0]_{\mathrm{X}}^{\mathrm{Y}}$ is the heat transferred to the environment from the control volume (Fig. 1.21). $[W_{\mathrm{CV}}]_{\mathrm{X}}^{\mathrm{Y}}$ is less than $[(W_{\mathrm{CV}})_{\mathrm{REV}}]_{\mathrm{X}}^{\mathrm{Y}}$, and $[Q_0]_{\mathrm{X}}^{\mathrm{Y}}$ is greater

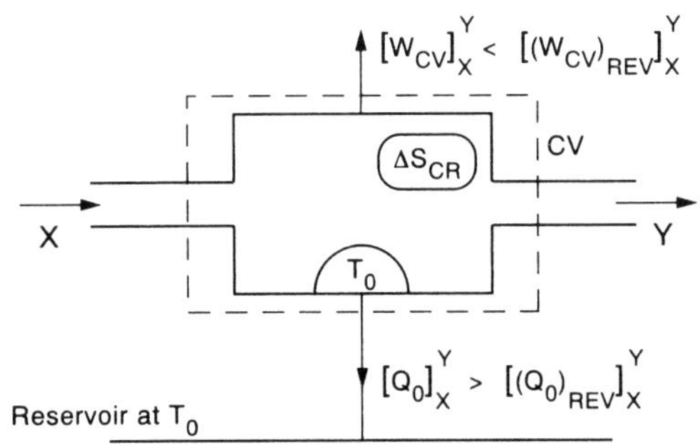

FIGURE 1.21 Actual process with heat transfer at T_0

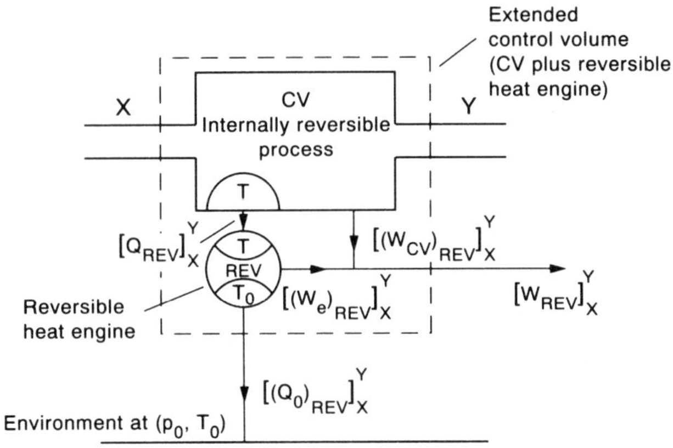

FIGURE 1.22 Fully reversible steady flow process, with heat transfer from CV at temperature T to reversible heat engine (after Haywood [18])

than $[(Q_0)_{REV}]_X^Y$. The leaving entropy flux associated with this outward heat transfer is $[Q_0]_X^Y/T_0$ so that the increase in entropy across the control volume is

$$S_Y - S_X = \Delta S_{CR} - [Q_0]_X^Y/T_0, \tag{1.46}$$

where ΔS_{CR} is the entropy created within the control volume. The lost work due to irreversibility (sometimes simply called the irreversibility) is therefore

$$\begin{aligned} I^{CR} &= [(W_{CV})_{REV}]_X^Y - [W_{CV}]_X^Y \\ &= (B_X - B_Y) - (H_X - H_Y - [Q_0]_X^Y) \\ &= T_0(S_Y - S_X) + [Q_0]_X^Y \\ &= [Q_0]_X^Y - [(Q_0)_{REV}]_X^Y \\ &= T_0 \Delta S_{CR}. \end{aligned} \tag{1.47}$$

1.6.2 *Flow with Heat Transfer at Temperature* T

We consider next the case where heat $[Q_{REV}]_X^Y = \int_X^Y dQ_{REV}$ is transferred from the control volume CV at temperature T, in a reversible steady-flow process between states X and Y (but again in the presence of an environment at T_0).

Figure 1.22 shows such a fully reversible steady flow through the control volume CV. The heat transferred, $[Q_{REV}]_X^Y$, supplies a reversible heat engine which delivers external work $[(W_e)_{REV}]_X^Y$ and rejects heat $[(Q_0)_{REV}]_X^Y$ to the environment.

The total work output from the extended (dotted) control volume is $(B_X - B_Y)$, if the flow is again between states X and Y. But the work from the reversible external engine is

$$[(W_e)_{REV}]_X^Y = \int_X^Y \left(\frac{T-T_0}{T}\right) dQ_{REV}. \tag{1.48}$$

The maximum (reversible) work obtained from the "inner" control volume is therefore equal to

$$\begin{aligned}[(W_{CV})_{REV}]_X^Y &= B_X - B_Y - [(W_e)_{REV}]_X^Y \\ &= B_X - B_Y - \int_X^Y \left(\frac{T-T_0}{T}\right) dQ_{REV} \\ &= E_X - E_Y - \int_X^Y \left(\frac{T-T_0}{T}\right) dQ_{REV}. \end{aligned} \tag{1.49}$$

(We may note that if additional heat were transferred from CV *directly* to the environment, this expression for $[(W_{CV})_{REV}]$ would not change. Part of the last term in equation (1.49) would become zero.)

For a real (irreversible) flow process through the control volume CV, between fluid states X and Y (Fig. 1.23), with the *same* heat rejected at temperature T ($[Q]_X^Y = [Q_{REV}]_X^Y$) the work output is $[W_{CV}]_X^Y$. Heat $[Q_0]_X^Y$ may also be transferred from CV to the environment at T_0.

From the steady-flow energy equation,

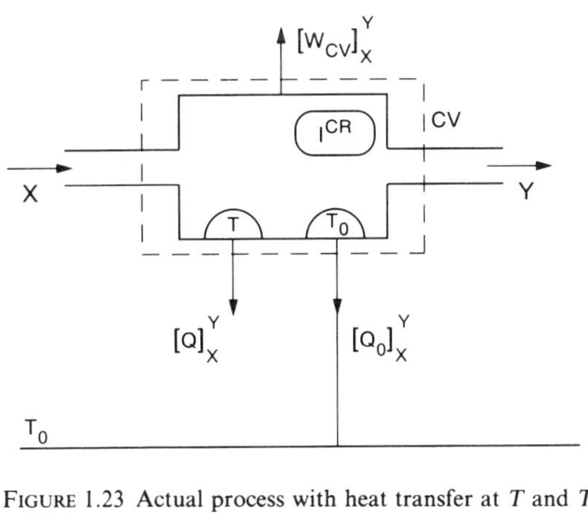

FIGURE 1.23 Actual process with heat transfer at T and T_0

$$[W_{CV}]_X^Y = H_X - H_Y - [Q]_X^Y - [Q_0]_X^Y. \tag{1.50}$$

The entropy flux from the control volume associated with the heat transfer is

$$\int_X^Y \frac{dQ}{T} + \frac{[Q_0]_X^Y}{T_0},$$

so the entropy increase across it is

$$S_Y - S_X = \Delta S_{CR} - \int_X^Y \frac{dQ}{T} - \frac{[Q_0]_X^Y}{T_0}. \tag{1.51}$$

The lost work due to irreversibility I^{CR} within the control volume CV (with $dQ = dQ_{REV}$) is

$$\begin{aligned} I^{CR} &= [(W_{CV})_{REV}]_X^Y - [W_{CV}]_X^Y \\ &= B_X - B_Y - \int_X^Y \left(\frac{T - T_0}{T}\right) dQ_{REV} - \{(H_X - H_Y) - [Q]_X^Y - [Q_0]_X^Y\} \\ &= T_0(S_Y - S_X) + T_0 \int_X^Y \frac{dQ}{T} + [Q_0]_X^Y \\ &= T_0 \Delta S_{CR}. \end{aligned} \tag{1.52}$$

Thus the lost work due to internal irreversibility within the control volume when heat transfer takes place is still $T_0 \Delta S_{CR}$, as when the heat transfer is limited to exchange with the environment.

The actual work output is a *real* irreversible process between stable states X and Y is therefore

$$\begin{aligned} [W_{CV}]_X^Y &= [(W_{CV})_{REV}]_X^Y - \int_X^Y \left(\frac{T - T_0}{T}\right) dQ - I^{CR} \\ &= B_X - B_Y - \int_X^Y \left(\frac{T - T_0}{T}\right) dQ - I^{CR} \\ &= E_X - E_Y - E^Q - I^{CR}, \end{aligned} \tag{1.53}$$

where

$$E^Q = \int \left(\frac{T - T_0}{T}\right) dQ.$$

This is an important equation which we shall use subsequently.

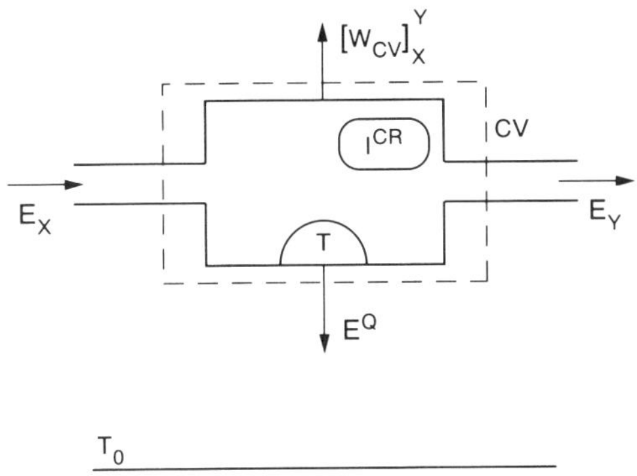

FIGURE 1.24 Exergy flux

1.6.3 Exergy Flux

Equation (1.53) may be interpreted in terms of exergy flows, work output and work producing potential (Fig. 1.24). The equation may be rewritten as

$$E_X = [W_{CV}]_X^Y + E^Q + I^{CR} + E_Y. \tag{1.54}$$

Thus the exergy E_X of the entering flow (its total capacity for producing work) is translated into

(i) the actual work output, $[W_{CV}]_X^Y$,
(ii) the work potential, sometimes called the thermal exergy, of the heat transferred,

$$E^Q = \int_X^Y \left(\frac{T - T_0}{T}\right) dQ,$$

(iii) the lost work due to internal irreversibility, $I^{CR} = T_0 \Delta S_{CR}$,
(iv) the leaving energy, E_Y.

If the heat transferred from the control volume is not used externally to create work, but is simply lost to the atmosphere (in which further entropy is created) then E^Q is numerically equal to I^Q, the lost work due to external irreversibility.

Then equation (1.54) becomes

$$E_X = [W_{CV}]_X^Y + I^Q + I^{CR} + E_Y. \tag{1.55}$$

1.6.4 Maximum Work Output in a Chemical Reaction at T_0

The (maximum) reversible work in steady flow between reactants at an entry state $R_0(p_0, T_0)$ and products at a leaving state $P_0(p_0, T_0)$ is

$$[(W_{CV})_{REV}]_{R_0}^{P_0} = (B_R)_0 - (B_P)_0. \tag{1.56}$$

It is supposed here that the various reactants are separated at p_0, T_0; similarly for the various products. The maximum work may then be written

$$(B_R)_0 - (B_P)_0 = (G_R)_0 - (G_P)_0 = -\Delta G_0, \tag{1.57}$$

where G is the Gibbs function, $G = H - TS$. This is the maximum work obtainable from such a combustion process, and is usually used in defining the rational efficiency of open circuit plant.

However, it should be noted that if reactants and/or products are not at pressure p_0, but mixed at a total pressure of p_0, then work of delivery or extraction has to be allowed for in obtaining the maximum possible work from reactants and products drawn from and delivered to the atmosphere, and the expression for maximum work has to be modified.

Kotas [17] draws a distinction between the "environmental" state (but called the dead state by Haywood [18]) in which reactants and products are in (restricted) thermal and mechanical equilibrium with the environment (each at p_0, T_0) and the truly or completely dead state in which they are also in chemical equilibrium, finally attaining the corresponding partial pressures in the atmosphere. He defines the chemical exergy as the sum of the maximum work obtained from the reaction with components at p_0, T_0 (i.e. $(-\Delta G_0)$) plus the work extraction and delivery terms $\Sigma_k RT_0 \ln(p_0/p_k)$, where p_k is a partial pressure, for the case of gaseous components.

We shall not subsequently consider these extraction and delivery work terms, but shall use $(-\Delta G_0)$ as an approximation to the maximum work output obtainable from a chemical reaction.

1.6.5 The Adiabatic Combustion Process

Returning to the general availability equation, we may write for an adiabatic combustion process between states R_X (reactants) at temperature T_X and P_Y (products) at temperature T_Y (see Fig. 1.25),

$$(B_R)_X = (B_P)_Y + T_0 \Delta S_{CR}, \tag{1.58}$$

since there is no heat or work transfer. The lost work due to internal irreversibility is

$$I^{CR} = T_0 \Delta S_{CR} = (B_R)_X - (B_P)_Y. \tag{1.59}$$

In forming the exergy at the stations X and Y we must be careful to subtract the steady flow availability function in the final equilibrium state, which we take here as the product (environmental) state at (p_0, T_0). Then equation (1.58) may be written, using equation (1.57),

$$(B_R)_X - (G_P)_0 = (B_P)_Y - (G_P)_0 + T_0 \Delta S_{CR}$$

or

$$(B_R)_X - (B_R)_0 + (G_R)_0 - (G_P)_0 = (B_P)_Y - (B_P)_0 + T_0 \Delta S_{CR}. \tag{1.60}$$

It is convenient for exergy tabulations to associate the term $(-\Delta G_0) = (G_R)_0 - (G_P)_0$ with the fuel supplied, (of mass M_F), i.e. $(E_F)_0 = (-\Delta G_0)$. For a combustion process burning liquid or solid fuel (at temperature T_0) with air at temperature T_1, the left-hand side of the equation may be written

$$E_X = (B_R)_X - (B_R)_0 + (G_R)_0 - (G_P)_0 \tag{1.61}$$

$$= (B_A)_1 - (B_A)_0 + (-\Delta G_0)$$

$$= (E_A)_1 + (E_F)_0, \tag{1.62}$$

where subscript A refers to the air supply. Hence the exergy equation becomes

$$E_X = E_Y + I^{CR}, \text{ or } (E_A)_1 + (E_F)_0 = (E_P)_2 + I^{CR}, \tag{1.63}$$

where $E_Y = (E_P)_2$, and

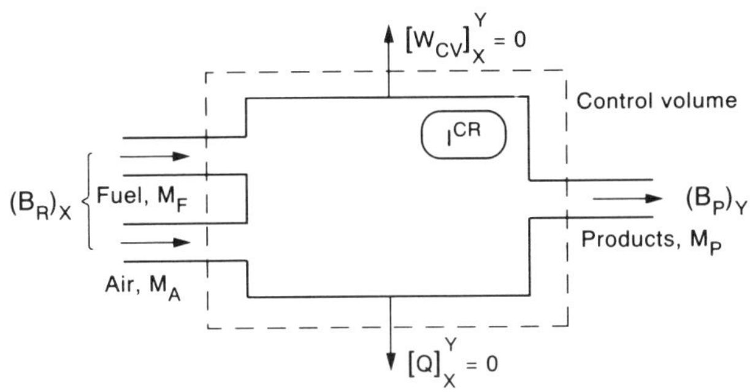

FIGURE 1.25 Adiabatic combustion

$$I^{CR} = T_0 \Delta S_{CR}$$
$$= (E_A)_1 + (E_F)_0 - (E_P)_2$$

For a combustion process burning gaseous fuel (which may have been compressed from state 0 to state 1′), the left-hand side of the exergy equation may be written

$$\begin{aligned} E_X &= (B_A)_1 - (B_A)_0 + (B_F)_{1'} - (B_F)_0 + (-\Delta G_0) \\ &= (E_A)_1 + [(B_F)_{1'} - (B_F)_0] + (-\Delta G_0) \\ &= (E_A)_1 + [(B_F)_{1'} - (B_F)_0] + (E_F)_0, \end{aligned} \tag{1.64}$$

and the exergy equation becomes

$$(E_A)_1 + (E_F)_{1'} + (E_F)_0 = (E_P)_2 + I^{CR}$$
$$I^{CR} = (E_A)_1 + (E_F)_{1'} + (E_F)_0 - (E_P)_2. \tag{1.65}$$

In general for any gas of mass M we may write

$$\begin{aligned} E &= B - B_0 \\ &= M(b - b_0) \\ &= M(h - h_0) - MT_0(s - s_0), \end{aligned} \tag{1.66}$$

where h_0 and s_0 are the specific enthalpy and specific entropy respectively at the ambient pressure p_0 and the temperature T_0. For a semi-perfect gas with $p = \rho RT$ and $c_p = c_p(T)$

$$h - h_0 = \int_{T_0}^{T} c_p \, dT, \tag{1.67}$$

$$s - s_0 = \phi - R \ln (p/p_0), \tag{1.68}$$

and

$$b - b_0 = \int_{T_0}^{T} c_p \, dT - T_0 \phi + RT_0 \ln (p/p_0), \tag{1.69}$$

where

$$\phi \equiv \int_{T_0}^{T} \frac{c_p \, dT}{T}.$$

We shall use these expressions subsequently, neglecting the exergy changes associated with the differences in pressure between unmixed gaseous components (in the ideal work-producing combustion process) and the partial pressures of the gases (in the real combustion processes).

1.6.6 Work Output from a Power Plant

The statements on work output made for a real process and for the ideal chemical reaction or combustion process at (p_0, T_0) may be compared:

$$B_X = B_Y + [W_{CV}]_X^Y + \int_X^Y \frac{(T-T_0)}{T}\, dQ + T_0 \Delta S_{CR} \qquad (1.70)$$

$$(G_R)_0 = (G_P)_0 + [(W_{CV})_{REV}]_{R_0}^{P_0}. \qquad (1.71)$$

The first equation may be thought of as applied to a complete power plant receiving reactants at state R_X, and discharging products at state P_Y; either

(i) a boiler (or heating device) plus a closed cycle power plant as in Fig. 1.3; or
(ii) an open circuit power plant (as in Fig. 1.4).

As for the combustion process we may subtract the steady flow availability function for the equilibrium product state from each side of equation (1.70) to give

$$[(B_R)_X - (G_R)_0] - [(G_P)_0 - (G_R)_0] = [(B_P)_Y - (G_P)_0] + [W_{CV}]_X^Y + \int_X^Y \left(\frac{T-T_0}{T}\right) dQ + T_0 \Delta S_{CR}, \qquad (1.72)$$

or rearranging, and introducing equation (1.71)

$$\begin{aligned}(-\Delta G_0) &= (G_R)_0 - (G_P)_0 \\ &= [(W_{CV})_{REV}]_{R_0}^{P_0} \\ &= \underset{①}{[W_{CV}]_X^Y} + \underset{②}{\int_X^Y \left(\frac{T-T_0}{T}\right) dQ} + \underset{③}{T_0 \Delta S_{CR}} \\ &\quad + \underset{④}{[(B_P)_Y - (G_P)_0]} - \underset{⑤}{[(B_R)_X - (G_R)_0]}.\end{aligned} \qquad (1.73)$$

This equation shows clearly how the maximum possible work output, from the ideal combustion process, splits down into the various terms on the right-hand side:

(1) the actual work output from the power plant;
(2) the work potential of any heat transferred out from various components, which if transferred to the atmosphere at T_0, becomes the lost work due to external irreversibility, I^Q;
(3) the lost work due to internal irreversibility, I^{CR} (which may occur in various components);
(4) the work potential of the discharged exhaust gases.

The last term (5) will usually be zero, since reactants will enter at ambient temperature and $(B_R)_X = (G_R)_0$ (but not in the case of compressed gaseous fuel).

The rational efficiency of either

(i) a closed cycle plant plus boiler, or
(ii) an open circuit plant

may be defined as the ratio of the actual work output $[W_{CV}]_X^Y$ to the maximum possible work output, approximately $(-\Delta G_0)$

$$(\eta_R) = \frac{[W_{CV}]_X^Y}{(-\Delta G_0)}$$

$$= 1 - \frac{\int_X^Y \left(\frac{T - T_0}{T}\right) dQ}{(-\Delta G_0)} - \frac{T_0 \Delta S_{CR}}{(-\Delta G_0)} - \frac{(B_P)_Y - (G_P)_0}{(-\Delta G_0)}. \qquad (1.74)$$

We shall discuss briefly later, in Chapter 5, the rational efficiency of a component, and how it can be related to the rational efficiency of a combined plant (a fuller discussion is given by Horlock [19]).

1.6.7 Tracing Exergy Fluxes through a Power Plant

More recent analyses have been presented in terms of exergy flows through the various components of a power plant. Thus equation (1.54) is used, in the form

$$E_Y = E_X - E^Q - I^{CR} - [W_{CV}]_X^Y, \qquad (1.75)$$

where

$$I^{CR} = T_0 \Delta S_{CR},$$

and

$$E^Q = \int_X^Y \left(\frac{T - T_0}{T}\right) dQ$$

$$= I^Q,$$

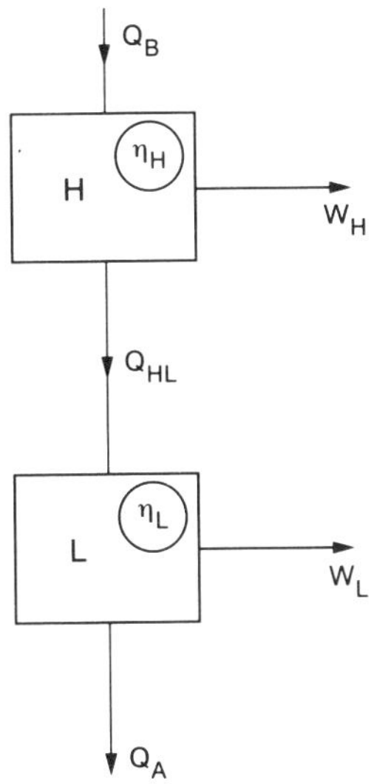

FIGURE 1.26 Combined power plant (CPS)

if the heat transfer is to the atmosphere at T. This equation is valid for any component, between states X and Y. The approach usually followed is to tabulate the exergy flows into and out of each component, together with the component work output and the "lost work" terms I^{CR} and I^{Q}. We shall describe such an exergy analysis of complete plants, giving examples, in Chapter 5.

1.7 The Incentive for Using Combined Plant to Increase Efficiency

The modifications to single cycles described in Section 1.5 may not achieve a high enough overall efficiency, or may involve excessive capital expenditure in order to do so. The plant designer therefore explores the possibility of using a combined plant—essentially one plant thermodynamically on top of the other, the lower plant receiving some or all of the heat rejected from the upper plant. If a higher mean temperature of heat supply, and/or a lower temperature of heat rejection, can be achieved in this way [an increase in ξ, equation (1.12)] then a higher overall plant efficiency will be achieved, as long as substantial irreversibilities are not introduced.

This is most simply illustrated by considering a combined power plant (CPS) (Fig. 1.26) made up of two cyclic plants (H, L) in series. Heat rejected from the higher (topping) plant (H), of thermal efficiency η_H, is used as a supply to the lower (bottoming) plant L (of thermal efficiency η_L), with no heat loss and no supplementary heating.

The work output from the lower cycle is

$$W_L = \eta_L Q_{HL}. \tag{1.76}$$

But

$$Q_{HL} = Q_B(1-\eta_H), \tag{1.77}$$

where Q_B is the heat supplied to the upper plant, which delivers work

$$W_H = \eta_H Q_B. \tag{1.78}$$

Thus the total work output is

$$\begin{aligned} W_{CPS} &= W_H + W_L \\ &= \eta_H Q_B + \eta_L(1-\eta_H)Q_B \\ &= Q_B(\eta_H + \eta_L - \eta_H\eta_L). \end{aligned} \tag{1.79}$$

The thermal efficiency of the combined plant is therefore

$$\eta_{CPS} = \frac{W}{Q_B} = \eta_H + \eta_L - \eta_H\eta_L. \tag{1.80}$$

The thermal efficiency of the combined plant is greater than that of the upper cycle alone, by the amount $\eta_L(1-\eta_H)$, as a result of adding the bottoming cycle receiving heat rejected from the upper cycle. An alternative interpretation is that the efficiency of the combined cycle is greater than that of the lower cycle alone, by the amount $\eta_H(1-\eta_L)$, as a result of the addition of a topping cycle.

This simple analysis is based on two cyclic plants, with all the heat rejected from the upper cycle being supplied to the lower cycle. We shall find that there are many variations of this "series" combined plant. The component plants may be open circuit rather than cyclic, heat may be lost and/or supplementary heat may be supplied between the plants; there may be joint heating of the two plants from a common "boiler" heating device. We shall develop the simple analysis above to allow for these variations in Chapter 3, but first, in Chapter 2, we provide a classification of many possible combined plants, with brief descriptions of each.

References

[1] Haywood, R. W. *Analysis of Engineering Cycles*, 4th edition. Pergamon Press, Oxford, 1991.

[2] Horlock, J. H. *Cogeneration; Combined Heat and Power*. Pergamon Press, Oxford, 1987.

[3] Wood, B. Combined Cycles—A General Review of Achievements. Modern Steam Practice. *Proc. I.Mech.E.*, 75–86, April 1971.

[4] Caputo, C. Una Cifra di Merito Dei Cicli Termodinamici Directti. *Il Calore*, **7,** 291–300, 1967.
[5] Cerri, G. Parametric Analysis of Combined Gas-Steam Cycles. *Trans. ASME Journal of Engineering for Gas Turbines and Power*, **109,** 1, 46–55, 1987.
[6] Cohen, H., Rogers, G. F. C. and Saravanamuttoo, H. I. H. *Gas Turbine Theory*, 2nd edition. Longman, London, 1972.
[7] Woods, W. A., Bevan, P. J. and Bevan, D. I. Output and Efficiency of the Closed Cycle Gas Turbine. *Proc. I.Mech.E.*, **205,** 59–66, 1991.
[8] Keenan, J. H. and Kaye, J. *Gas Tables*. Wiley, New York, 1941.
[9] Rice, I. G. The Combined Reheat Gas Turbine/Steam Turbine Cycle, Part I. *Trans. ASME Journal of Engineering for Power*, **102,** 1, 35–41, 1980.
[10] Pfenninger, H. Combined Steam and Gas Turbine Power Stations. *Brown Boveri Review*, **60,** 9, 389–397, 1973.
[11] Hawthorne, W. R. and Davis, G. de V. Calculating Gas Turbine Performance. *Engng.*, **181,** 361, 1956.
[12] Salisbury, J. K. The Steam Turbine Regenerative Cycle—an Analytical Approach. *Trans. ASME*, **64,** 231, 1942.
[13] Horlock, J. H. Approximate Analysis of Feed and District Heating Cycles for Steam Combined Heat and Power Plant. *Proc. I.Mech.E.*, **201,** A3, 1987.
[14] Haywood, R. W. A Generalised Analysis of the Regenerative Steam Cycle for a Finite Number of Heaters. *Proc. I.Mech.E.*, **161,** 157, 1949.
[15] Weir, C. D. An Analytical Study of the Regenerative Multiple-Reheat Steam Turbine Cycle. *I.Mech.E., Engineering Science Monograph*, **5,** 1967.
[16] Haywood, R. W. Steam Cycle Calculations for Supercritical Pressures. [Unpublished] 1957.
[17] Kotas, T. J. *The Exergy Method of Thermal Plant Analysis*. Butterworths, London, 1985.
[18] Haywood, R. W. *Equilibrium Thermodynamics for Engineers and Scientists*. Wiley, Chichester, 1980.
[19] Horlock, J. H. The Rational Efficiency of Power Plants and Their Components. *ASME COGEN V*, 103–114, 1991.

CHAPTER 2

Classification of Combined Power Plants

2.1 Introduction

In this chapter we shall briefly review the range of combined power plant options that are available. We shall not include combined *heat* and power plants [which have been the subject of another volume (see reference 1)], restricting ourselves to plants which produce power only. We shall not discuss the detailed performance of the combined power plants at this stage, but simply attempt to classify them.

Some of the plants which we shall consider involve theoretical concepts only in that they have not yet been developed for practical use. However, we include a great variety of options discussed in the literature in order to show the way thinking about combined plants has developed. We develop classifications suggested by Buxmann [2] and Haywood [3], and first introduce some new definitions.

We first define a *doubly cyclic plant* as one in which a single working fluid is used, but two closed cycles are employed. There is *internal coupling* between them, through the use of a single fluid.

Secondly, we define a *binary cycle plant* as one in which *two* working fluids are used, each operating on a separate closed cycle, with heat transfer (*external coupling*) between them. A natural extension is to a *ternary cycle plant*, with three fluids operating on three closed cycles and with external coupling between them, but such plants have not been developed as yet.

Thirdly, we consider *open circuit/closed cycle* combined plants. Two working fluids are involved; fuel is burned with air in the open circuit plant (e.g. a gas turbine) and water/steam is usually used in the closed cycle plant. External coupling is involved when the steam turbine cycle is heated from the exhaust of the gas turbine (with or without supplementary heating), but internal coupling is implied if fuel is supplied to a pressurised boiler, which supplies "heat" to both plants (joint "heating"). Combined power plants using a magnetohydrodynamic or a thermionic device as the open circuit plant have been proposed, but we shall not discuss these here, as they have not been developed.

Fourthly, combined power plants involving *doubly open circuit plants*, internally coupled, are considered. An example of this type is an open gas turbine plant with steam injection in the combustion chamber. Essentially, the latter operates as an open circuit plant without condensation of the steam.

A summary of the combined plants we shall discuss in the chapter, based on this classification, is given in Table 2.1. Note that as a shorthand we refer to gas turbine and steam turbine closed cycles as "Joule" and "Rankine" cycles. Strictly a Joule cycle is a reversible constant pressure gas cycle, with the isentropic efficiencies of compressor and turbine both

TABLE 2.1 *Classification of Combined Power Plants*

1. *Doubly Cyclic Plants*
   (using one working fluid with "internal" coupling between closed cycles)
   (a) "Joule"/"Rankine" Cycles
      (i) Ideal, super-regenerative steam cycle
      (ii) Field cycle
      (iii) Sonnenfeld cycle
   (b) "Rankine"/"Rankine" Cycles
      Split "Rankine" (two-level) cycle (approximates to dual pressure cycle)
   (c) "Joule"/"Joule" Cycles
      Sulzer half-closed gas turbine cycle

2. *Binary and Ternary Closed Cycle Plants*
   (using two or three working fluids, with "external" coupling between cycles)
   (a) "Rankine"/"Rankine" Cycles
      (i) "Topping" mercury/steam or potassium/steam cycles
      (ii) "Bottoming" steam/ammonia or steam/organic fluid cycles
   (b) "Rankine"/"Rankine"/"Rankine" Cycles
      Mercury, steam, ammonia cycles
   (c) "Joule"/"Rankine" Cycles
      Gas (e.g. helium), steam cycles

3. *Open Circuit/Closed Cycle Plants*
   (using two working fluids, usually in an open "Joule"/closed "Rankine" combination)
   (a) Heating of steam turbine cycle by gas turbine exhaust (external coupling)
      (i) No supplementary firing of heat recovery steam generator (HRSG)
      (ii) Supplementary firing of heat recovery steam generator
      (iii) Fully (or maximally) fired exhaust boiler
   (b) Pressurised boiler (internal coupling through joint "heating")

4. *Doubly Open Circuit Plants*
   (using two working fluids with internal coupling)
   Open "Joule", open "Rankine" plant—open gas turbine with steam injection.

Notes

External coupling—both working fluids flow separately in their own cycles, and heat transfer takes place from one cycle to another.

Internal coupling—either
(i) the fluid being common to both cycles, or
(ii) two working fluids being heated together.

100%, and no pressure losses in heat supply and rejection processes. Similarly, a Rankine cycle is defined as a liquid/vapour cycle, with constant pressure heating of the fluid (from liquid state to vapour state), constant pressure condensation, and with a perfect adiabatic turbine and a perfect adiabatic feed-pump. However, for convenience we shall refer to real cycles (including irreversibilities) of these types using inverted commas, i.e. as "Joule" and "Rankine" cycles; but we also refer to open "Joule" and open "Rankine" plants. For most plants we shall show the various component processes on temperature-entropy (T, s) diagrams.

2.2 Doubly Cyclic Plants (Single Fluid)

There are three main types of this plant (see Table 2.1)—"Joule"/"Rankine", "Rankine"/"Rankine" and "Joule"/"Joule".

2.2.1 "The Super-regenerative Cycle ("Joule"/"Rankine")

The most ambitious double cycle that has been proposed is the super-regenerative steam cycle, which is illustrated in Fig. 2.1 (see Field [4] and [5]). Essentially, the higher cycle is a modified Joule (CICI BTRTRTX) cycle using a large number of turbines, with intermediate reheat, the exhaust from the last turbine (state 5) passing through an ideal heat exchanger and heating higher pressure steam from its saturated state (3) to a superheated state (4). The steam flow at exit from the heat exchanger (state 6) splits, part (M_R) passing through a lower cycle [an ideal "regenerative Rankine" cycle (6712)], rejecting heat at condenser

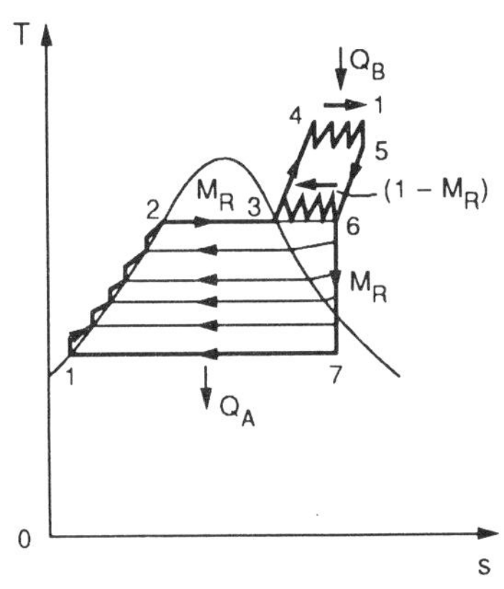

FIGURE 2.1 Temperature-entropy diagram for super-regenerative cycle

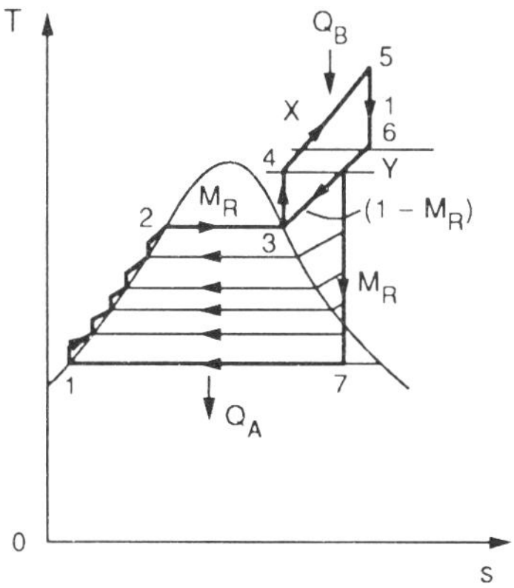

FIGURE 2.2 Temperature-entropy diagram for Field cycle

temperature $T_1 = T_7$. The remaining part of the steam flow $(1 - M_R)$ passes through a series of compressors between 6 and 3, with intermediate cooling. This internally rejected heat is used to evaporate the saturated feed water at boiler pressure (state 2) so that the two streams meet again as two saturated steam flows (state 3), and together enter the high pressure side of the heat exchanger. The higher cycle is thus internally coupled to the lower cycle.

This ideal cycle is fully reversible, with all the heat supplied externally at the top temperature $T_4 = T_5$ (in the reheaters), and all the heat rejected externally at condenser temperature $T_1 = T_7$. It therefore has Carnot efficiency; this statement depends not only on processes within the cycles being reversible, but also upon steam approximating to a semi-perfect gas (with specific heat a function of temperature only) if the temperature rise of the high pressure steam in the heat exchanger is to be equal to the temperature drop of the low pressure steam.

This internally coupled double cycle is indeed impracticable, but it provides a good example of the design objective for combined cycle plant, in that it achieves the highest attainable efficiency for given $T_{max}(= T_5 = T_4)$ and $T_{min}(= T_1 = T_7)$.

2.2.2 The Field Cycle ("Joule"/"Rankine")

The mixing of the two streams in the super-regenerative cycle is reversible, but the multi-stage reversible compression/intercooling process is complex and impracticable. Field [5] subsequently proposed an irreversible mixing process to overcome this complexity; a particular version of the Field double cycle, involving compression of superheated steam, is illustrated on the (T, s) diagram shown in Fig. 2.2.

Again the top cycle is a modified Joule cycle with an ideal heat exchanger. [A single stage of expansion (56) is shown in the figure, but reheating could be employed.] Again the steam flow is divided at exit from the ideal heat exchanger (state Y); part of the flow (M_R) passes through a regenerative Rankine cycle (Y712) and the remaining flow ($1-M_R$) is mixed with the saturated water at boiler pressure (at state 2, if fully reversible regenerative feed heating is used in the lower cycle). The mixing process ($Y+2=3$) is irreversible; saturated steam in state 3 results, and is compressed to state 4 before entry into the ideal heat exchanger. External heat (Q_B) is supplied between states X and 5.

Again, the efficiency of this cycle is high, because external heat is supplied near the top temperature (between states X and 5). All external heat rejection takes place at condenser temperature.

Variations on this cycle include one in which mixing produces wet steam before compression. Theoretically, this involves less compressor work, but efficient wet steam compression has not been achieved in practice. In spite of the irreversibility introduced in the mixing process, the Field cycle has a potential for high efficiency. Although it was subject to detailed study and some developmental work in the fifties, it has not been developed to a practical state.

An analysis of the efficiency of the Field cycle has been given by Horlock [6]. In spite of the internal irreversibility introduced by the mixing process, the efficiency is still given by

$$\eta_{CP}=\eta_H+\eta_L-\eta_H\eta_L, \tag{2.1}$$

where η_H is the efficiency of the higher (modified "Joule") cycle and η_L is the efficiency of the lower regenerative ("Rankine") cycle. This is simply illustrated by regarding the higher cycle (34X56Y3) as having a steam flow circulation of unity, and the lower cycle (123Y71) a steam flow circulation of M_R (see Fig. 2.2). The heat transferred between the two cycles is

$$Q_{HL}=(h_Y-h_3)=M_R(h_Y-h_2) \tag{2.2 a}$$

[which is equivalent to an equation describing the mixing process

$$M_R(h_3-h_2)=(1-M_R)(h_Y-h_3)]. \tag{2.2 b}$$

The relation derived in Section 1.7 then applies directly, since all the heat rejected from the higher cycle is received by the lower cycle.

Equation (2.1) may also be derived as follows:

$$1-\eta_{H}=\frac{h_{Y}-h_{3}}{h_{5}-h_{X}}, \tag{2.3}$$

$$1-\eta_{L}=\frac{h_{7}-h_{1}}{h_{Y}-h_{2}}, \tag{2.4}$$

$$\begin{aligned}\eta_{CP}&=1-\frac{M_{R}(h_{7}-h_{1})}{h_{5}-h_{X}}=1-\frac{(h_{Y}-h_{3})(h_{7}-h_{1})}{(h_{Y}-h_{2})(h_{5}-h_{X})}\\&=1-(1-\eta_{H})(1-\eta_{L})\\&=\eta_{H}+\eta_{L}-\eta_{H}\eta_{L},\end{aligned} \tag{2.1}$$

identical to the series plant efficiency, η_{CPS}.

2.2.3 The Sonnenfeld Cycle ("Joule"/"Rankine")

Another double steam cycle which approaches the ideal super-regenerative cycle is the Sonnenfeld process, an example of which (taken from reference 7) is illustrated in Fig. 2.3. The basis is

(i) a supercritical modified "Rankine" cycle with three stages of reheat, and complex feed heating in two main stages, the first at low pressures (by bleeding steam from the low pressure turbines),

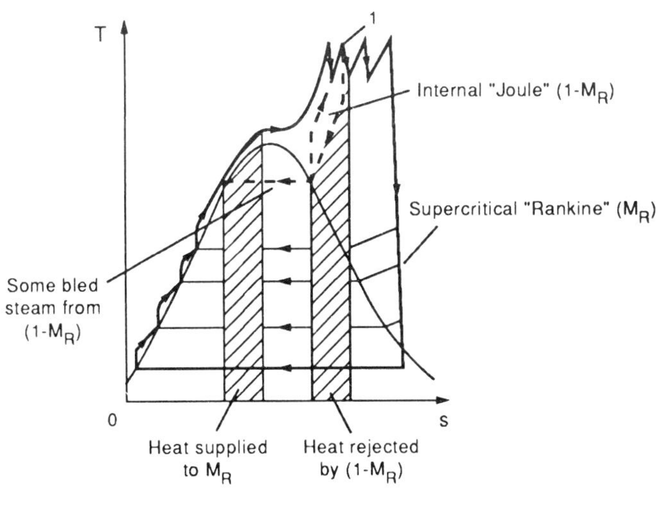

FIGURE 2.3 Temperature-entropy diagram for Sonnenfeld cycle

the second at boiler pressure, partially by bled steam but mainly by heating with rejected heat from
(ii) an "inner" "Joule" cycle.

After expansion of the combined flow in the second turbine, the steam flow is split into two streams. The first (M_R) proceeds through the modified "Rankine" cycle (second reheater, third turbine, third reheater and fourth (final) turbine, condenser and the feed heating train). The second stream ($1-M_R$) passes through the inner "Joule" cycle rejecting heat at (almost) constant pressure down to a dry saturated state, to raise the temperature of the feed water of the first stream (after feed heating by bled steam) to above the critical temperature. This second stream is then compressed to the state reached by the first stream after (i) heating in the supercritical boiler and superheater and (ii) expansion through the first high pressure turbine. The two streams then mix and are jointly superheated.

Referring to Fig. 2.3, the external heat supplies are therefore

(i) for the first stream (M_R): in the boiler and superheater, and three reheaters
(ii) for the second stream ($1-M_R$): in the first reheater.

External heat rejection to the condenser is by the first stream only.

The Sonnenfeld process approaches Carnot efficiency in that the external heat supplies are close to the top temperature in the cycle, and external heat rejection is at the lowest (ambient) temperature.

The process is primarily an example of "joint" or parallel heating in that heat is supplied to both cycles in the second reheater, but heat is also transferred from the inner "Joule" cycle to partially heat the parallel modified "Rankine" cycle (series heating of the lower cycle from the higher cycle). The overall efficiency cannot therefore be described by the simple relationship for η_{CPS} [equation (2.1)]. We shall consider such combined plants with series/parallel heating in Chapter 3.

2.2.4 The Split "Rankine" Cycle and the Dual Pressure Cycle ("Rankine"/"Rankine")

Several proposals have been made for split Rankine cycles—"Rankine"/"Rankine" (see Colosimo, Rose and Decker [8]). Figure 2.4 shows the simplest resulting double cycle.

The purpose of splitting the Rankine cycle in this way is to reduce the irreversibility involved in the heat transfer between a hot gas source of heat and the heating of water in the closed cycles, and its evaporation and superheating. If such heating is divided between two pressure levels, then

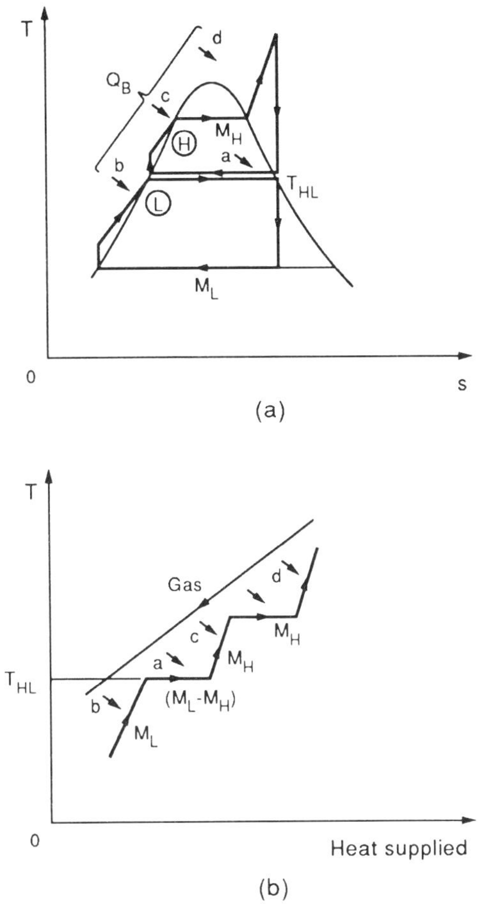

FIGURE 2.4 Split "Rankine" cycles
(a) Overall temperature-entropy diagram
(b) Heating process

this irreversibility is reduced, the temperature difference between the gas and the water/steam being decreased.

Figure 2.4a shows a hypothetical double cycle "Rankine" plant, made up of a higher Rankine cycle, with a mass flow M_H, and a lower Rankine cycle, with a mass flow M_L. Heat rejection from the higher cycle at the intermediate temperature T_{HL} is $M_H l_{HL}$, where l_{HL} is the latent heat. But the heat required to evaporate M_L is $M_L l_{HL}$ and if $M_L > M_H$ then the external heat supply required (a) at the intermediate pressure and temperature is $(M_L - M_H) l_{HL}$. This in addition to the external heat required (b) to heat the low pressure feed water, (c) to heat the high pressure feed water, and (d) to evaporate the high pressure water and to superheat the steam produced. The total heat supply ($Q_B = a + b + c + d$) can be provided by the hot gas source with small irreversibility, as temperature differences are reduced (Fig. 2.4b). Colosimo *et al.* have also considered a combined cycle in which the lower cycle is also superheated.

Thermodynamically, the basic double Rankine cycle is equivalent to a dual pressure Rankine cycle as is used in a nuclear power plant. It is also adopted in the steam plant of an open circuit (gas turbine)/closed cycle (steam turbine) combined plant. In such a dual pressure cycle, M_L passes through the condenser and the low pressure water heater. The flow is then split, with M_H flowing into the high pressure water heater, evaporator and H.P. turbine, and $(M_L - M_H)$ flowing into the low pressure evaporator after which it mixes with the exhaust from the H.P. turbine, before the combined flow $M_H + (M_L - M_H) = M_L$ expands through the L.P. turbine.

Other variants of the dual pressure Rankine cycle are illustrated in Fig. 2.5. In the first (proposed by Mangan and Pettit [9], Fig. 2.5a), superheated steam at exit from the H.P. turbine is mixed with dry saturated steam from the L.P. evaporator, before expansion in the L.P. turbine. In the second (proposed by Wardall and Doorly [10], Fig. 2.5b), superheated steam at exit from the H.P. turbine is mixed with superheated steam from the L.P. boiler/superheater, before expansion in the L.P. turbine. In both proposals all the heat rejected by the higher cycle is effectively transferred to the lower cycle, but with supplementary heating. The thermal efficiencies of all these combined plants are *not* then given by the simple series expression (2.1).

Further, as explained in Chapter 1, the overall performance is subject to a thermodynamic penalty because of the remaining temperature difference between the heating gas and the evaporating steam/water, and lower efficiency η_L.

2.2.5 The Sulzer Half-Closed Plant ("Joule"/"Joule")

A plant which approaches our definition of a doubly cyclic plant (single fluid with external coupling) is the Sulzer half-closed gas turbine illustrated in Fig. 2.6a, in which we have shown isentropic compression and expansion processes and heat exchange. Air from the atmosphere (1) is compressed in an L.P. compressor to state 2 and then mixes with air leaving the heat exchanger of the "closed" cycle (state Y) to form state 3. The mixed stream is cooled in an intercooler to state 4 before the second stage of compression to state 5 ($T_5 = T_3$). After compression it is then heated in the "cold" side of the heat exchanger from T_5 to T_X. The combined stream of air is then split into an "open" stream and a "closed" stream. Fuel is burnt with the "open" stream in a combustion chamber, reaching a higher temperature (T_8). The "closed" air stream is heated by the combustion gases, both leaving the joint "heater" at $T_6 = T_9$. The "closed" air stream is expanded in an air turbine from temperature T_6 to T_7, and then passes through the hot side of the heat exchanger to state Y before mixing with the air delivered by the L.P. compressor. The

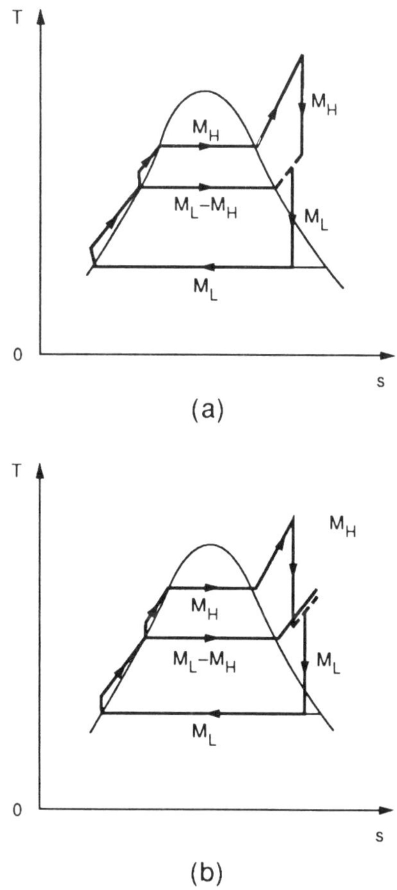

FIGURE 2.5 "Split Rankine" cycles
after (a) Mangan and Pettit [9]
(b) Wardall and Doorly [10]

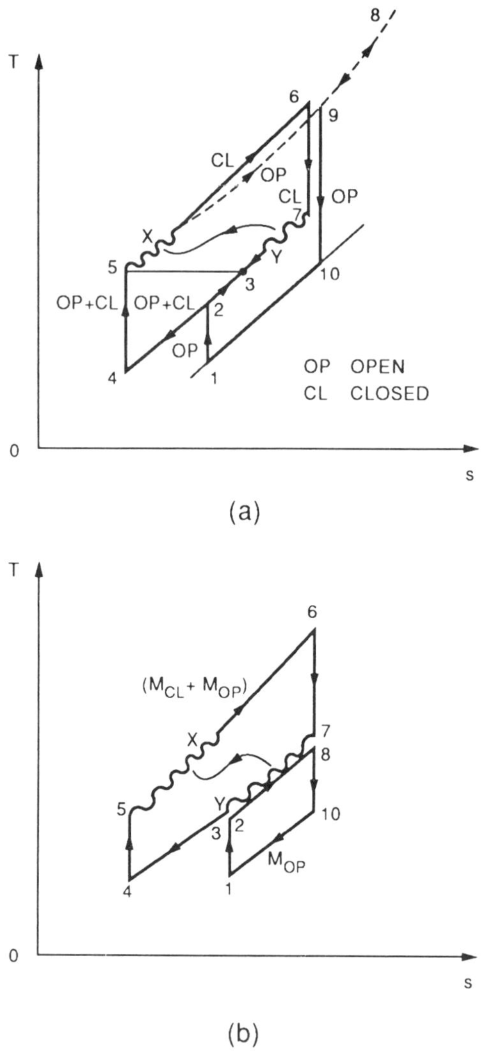

FIGURE 2.6 Sulzer half-closed plant
(a) Practical plant
(b) Idealised limiting Sulzer cycles ($M_{OP}/M_{CL} \rightarrow 0$)
(after Buxmann [2])

combustion gases are expanded through the full pressure ratio, from temperature T_9 to temperature T_{10} and then discharged at atmospheric pressure.

If the air mass flow (in unit time) through the closed cycle is M_{CL}, and that entering the open circuit is M_{OP}, then, neglecting the fuel mass flow and assuming heating processes instead of combustion, the thermal efficiency of the combined plant is

$$\eta_{CP}=1-\frac{\text{Heat rejected}}{\text{Heat supplied}}$$

$$=1-\frac{(M_{CL}+M_{OP})(h_3-h_4)+M_{OP}(h_{10}-h_1)}{(M_{CL}+M_{OP})(h_6-h_X)} \tag{2.5}$$

But in the heat exchanger

$$M_{CL}(h_7-h_Y)=(M_{CL}+M_{OP})(h_X-h_5), \tag{2.6}$$

so that $h_X=h_3+\left[\frac{M_{CL}(h_7-h_Y)}{(M_{CL}+M_{OP})}\right]$, since $h_3=h_5$; from which the denominator in equation (2.5) may be written

$$(M_{CL}+M_{OP})(h_6-h_7)+M_{OP}(h_7-h_3)+M_{CL}(h_Y-h_3). \tag{2.7}$$

For a very large heat exchanger, and if the thermal capacity of the two streams is similar, states 2, Y and 3 come together and the last term vanishes, $M_{CL}(h_Y-h_3)\rightarrow 0$.

For such a limiting case, with (M_{OP}/M_{CL}) small, this analysis shows that the Sulzer cycle can be represented ideally by two "Joule" cycles as sketched in Fig. 2.6b. The upper "Joule" cycle (4567—a CBTX cycle) has a mass flow $(M_{CL}+M_{OP})$ and the lower "Joule" cycle (12810— a CBT cycle) has a mass flow M_{OP}. These two cycles may be regarded as operating in parallel, with joint heating and no heat transfer between the two; the states Y, 2 and 3 are identical. The efficiency of the combined plant is given by

$$\eta_{CP}=1-\left(\frac{\text{Heat rejected from the two cycles}}{\text{Heat supplied to the two cycles}}\right)$$

$$=1-\frac{[(M_{CL}+M_{OP})(h_2-h_4)+M_{OP}(h_{10}-h_1)]}{[(M_{CL}+M_{OP})(h_6-h_7)+M_{OP}(h_7-h_3)]} \tag{2.8}$$

which is the same as the efficiency given in equation (2.5), with $h_Y=h_3$, $h_X=h_7$. This efficiency is less than that of the upper (CBTX) cycle which is

$$\eta_H=1-\frac{(h_2-h_4)}{(h_6-h_7)}, \tag{2.9}$$

but the work output is increased. We shall discuss the efficiency of parallel, or jointly heated, plants more generally in Chapter 3.

2.3. Binary and Ternary Closed Cycle Plants (Two or Three Condensing Fluids)

We have defined a combined plant operating on a binary system as one with two closed cycles, using different working fluids, with heat transfer across the boundaries of the two component plants, from the higher (or topping) cycle to the lower (or bottoming) cycle. There are three types of this combined plant (see Table 2.1)—"Rankine"/"Rankine", "Rankine"/ "Rankine"/"Rankine" and "Joule"/"Rankine".

2.3.1 "Rankine"/"Rankine" Cycles

The earliest binary system to be developed was the mercury/steam binary vapour cycle (Fig. 2.7).). Mercury has a much higher critical temperature than water (1500 °C cf. 374 °C), so that it can be evaporated at a maximum allowable temperature with a large fraction of the heat supply occurring near the top temperature. The critical pressure is also high (106 MN/m^2), so the saturation pressure at a working temperature (say 550° C) is quite modest (1.4 MN/m^2). Further, in a binary cycle the mercury can be condensed at a temperature well above atmospheric, and excessively low mercury condenser pressures avoided.

In Fig. 2.7 the entropy scales of the mercury and steam are in the ratio of the mass flows through the common condenser-boiler. The area representing the heat rejected by the mercury (Q_{HL}) is then the same as the area indicating the heat supplied to evaporate the steam.

In an ideal cycle, with all the heat rejected from the higher cycle being absorbed by the lower cycle, the efficiency would be

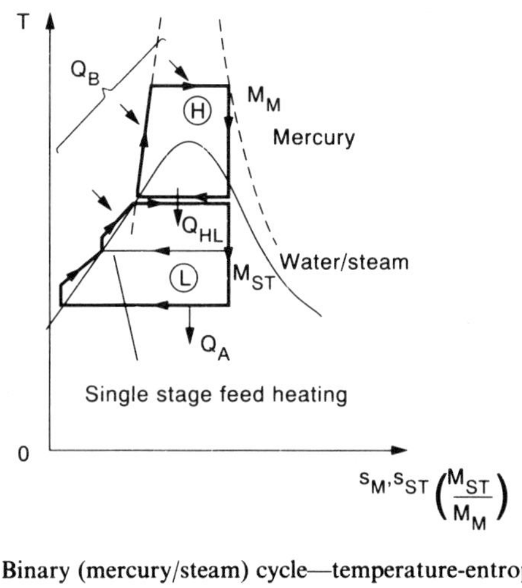

FIGURE 2.7 Binary (mercury/steam) cycle—temperature-entropy diagrams

$$\eta_{CP} = \eta_{CPS} = \eta_H + \eta_L - \eta_H \eta_L. \tag{2.1}$$

In the more practical cycle the external heat supply may be a "joint heating" system with a fraction supplied to the steam cycle (to the economiser and the superheater). We show later that in such a case the efficiency will then be reduced below η_{CPS}, and will in fact be between η_H and η_L.

The property requirements for "higher" cycle working fluids are as follows:

(i) the critical temperature should be well above the upper cycle temperature so that the latent heat of evaporation is high;
(ii) the working medium should be chemically resistant, non-toxic, inert and cheap;
(iii) the saturation pressure at the top temperature should not be too high and at the bottom temperature not too low;
(iv) the saturation line for the saturated liquid should be steep; this reduces the fraction of heat supplied below the top temperature;
(v) the saturation line for the dry saturated vapour should also be steep; this means that the turbine expansion line does not pass through too wet a region.

In addition to the mercury/steam combined plant, other combinations that have been studied include the use of ammonia as a bottoming cycle to an upper steam cycle. Ammonia has a higher vapour pressure than water at ambient temperatures (728 kN/m^2 cf 1.7 kN/m^2 at 15° C) so that its low specific volume enables a smaller low pressure turbine to be built. Organic fluids (such as the freon gases) serve the same purpose and have been considered for small power plants. We consider the choice of possible fluids further in the parametric studies of Chapter 4.

2.3.2 "Joule"/"Rankine" Cycles

A binary plant employing a closed gas turbine cycle, rejecting heat to a closed steam turbine cycle, has been considered in some detail by Buxmann [11]. Heat supply to the topping cycle could come from a nuclear reactor.

For the combined plant shown in Fig. 2.8 (in which the steam entropy has been scaled by a factor μ), Buxmann assumed that the higher gas turbine operated on a closed (CBTX) cycle, but not all the heat rejected by the higher plant was transferred to the steam cycle. The enthalpy drop $(H_Y - H_S)$ is equal to the heat Q_{HL} supplied to the lower cycle, but the enthalpy drop $(H_S - H_1)$ is an unused heat "loss" from the gas turbine cycle. The efficiency is therefore less than that given by equation (2.1) for η_{CPS} because of the heat loss between the cycles.

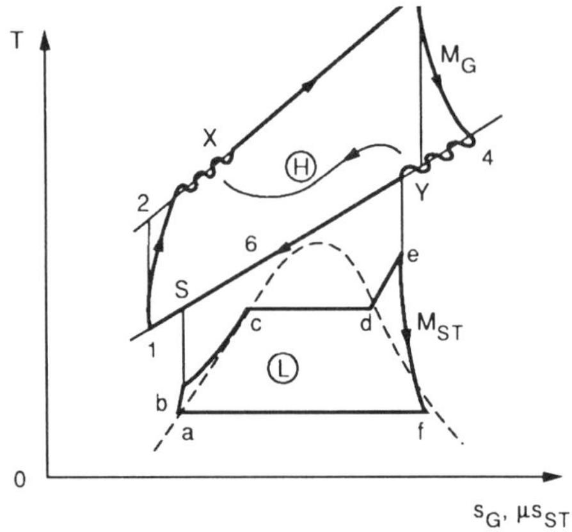

FIGURE 2.8 Binary "Joule"/"Rankine" combined plant (with heat exchanger)

A detailed parametric study of this combined cycle has been given by Buxmann and is described in Chapter 4. Although such a combined plant has not been widely used, this study is useful in that it gives an initial guide to the performance of the open gas turbine/closed steam turbine plant which has become the most popular combined system.

2.3.3 "Rankine"/"Rankine"/"Rankine" Cycles

Ternary cycles have also been proposed, with the heat cascading down from one cycle to a second, and from the second to a third. Mercury/water/ammonia and potassium/mercury/water combinations have been considered, together with an open gas turbine/closed potassium/binary cycle. To the author's knowledge, no ternary plants have been built or are under development.

The work output from an ideal ternary cycle, receiving heat Q_B at the highest level and with complete heat transfer between the higher (H) and intermediate (I) cycles, and between the intermediate and lower cycles, is

$$W=\eta_H Q_B+(1-\eta_H)\eta_I Q_B+(1-\eta_H)(1-\eta_I)\eta_L Q_B,$$

so the efficiency is

$$\eta_{CPS}=\frac{W}{Q_B}=\eta_H+\eta_I+\eta_L-\eta_H\eta_I-\eta_H\eta_L-\eta_I\eta_L \qquad (2.10)$$

2.4 Open Circuit/Closed Cycle Plant (Two Working Fluids)

The most well-developed and commonly used combined power plant involves a combination of an open circuit plant (usually a gas turbine) and a closed cycle (usually a steam turbine) plant. Other systems, such as a magnetohydrodynamic generator, have been proposed for the open circuit plant, but none is yet operational.

Many different combinations of gas turbine and steam turbine plant have been proposed. Seippel and Bereuter [12] provided a wide-ranging review of possible proposed plants. We consider their work in detail later, in Chapter 3, but here we accept their basic point—that essentially there are two types of this combined plant (see Table 2.1).

In the first type, heating of the steam turbine cycle is by the gas turbine exhaust with or without additional firing (there is normally sufficient excess air in the turbine exhaust for additional fuel to be burnt, without an additional air supply). In the second the main combustion chamber is pressurised and joint "heating" of gas turbine and steam turbine plants is involved.

Most major developments have been of the first (*exhaust heated*) system, with and without additional firing of the exhaust. The firing may be

(i) "supplementary"—burning additional fuel in the heat recovery steam generator (HRSG), up to a maximum gas temperature of about 760° C; or
(ii) fully fired, in which exhaust boilers are used (particularly in the repowering of existing steam plant).

Recently coal gasification systems have been introduced into the open circuit/closed cycle type of combined plant:

(i) generation of synthetic gas which is supplied directly to the combustion chamber of the gas turbine in a combined plant, with an unfired HRSG—usually called an integrated gasification combined cycle (IGCC);
(ii) combustion of coal in a fluidised bed, the combustion chamber supplying gas to the gas turbine alone (in an exhaust heated unfired combined system) or also being used as a boiler for the steam plant (i.e. in "joint heating" mode).

Thus a sub-classification of open circuit (gas turbine)/closed cycle (steam turbine) combined plant is as follows:

(a) *Exhaust heated*:
Unfired HRSG including
(i) gas or liquid fuel firing or

(ii) coal firing [IGCC plant or circulating (atmospheric pressure) fluidised bed combustion (CFBC)];

Supplementary fired HRSG, gas or liquid fuel firing;

Maximally fired exhaust boiler, gas or liquid fuel firing.

(b) *Pressurised combustion chamber/boiler*:

including

(i) gas or liquid fuel firing or

(ii) coal firing pressurised fluidised bed (PFB).

We consider these various systems below.

2.4.1 Exhaust Heating Systems

As we indicated above, there are essentially three types of combined gas turbine/steam turbine plant employing exhaust heating—those employing an unfired heat recovery steam generator (HRSG), those using supplementary firing in the HRSG, and those involving full firing (maximally fired).

2.4.1.1 UNFIRED HRSG

In the first system exhaust gases from the gas turbine are used to raise steam in the lower cycle without the burning of additional fuel (Fig. 2.9). A limitation on this application lies in the heat recovery steam generator; the temperatures of the gas and water/steam flows are indicated in the figure. Choice of the pressure for evaporation is related to the temperature difference ($T_6 - T_c$) at the pinch point as shown, and a compromise

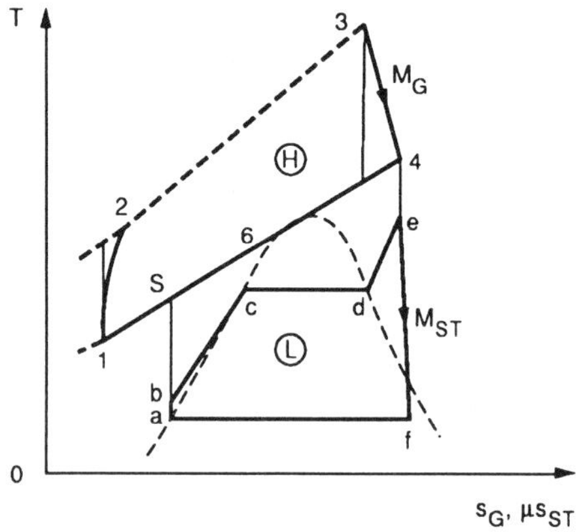

FIGURE 2.9 Open ("Joule")/closed ("Rankine") combined plant (unfired exhaust, no heat exchanger, $T_a \approx T_1$)

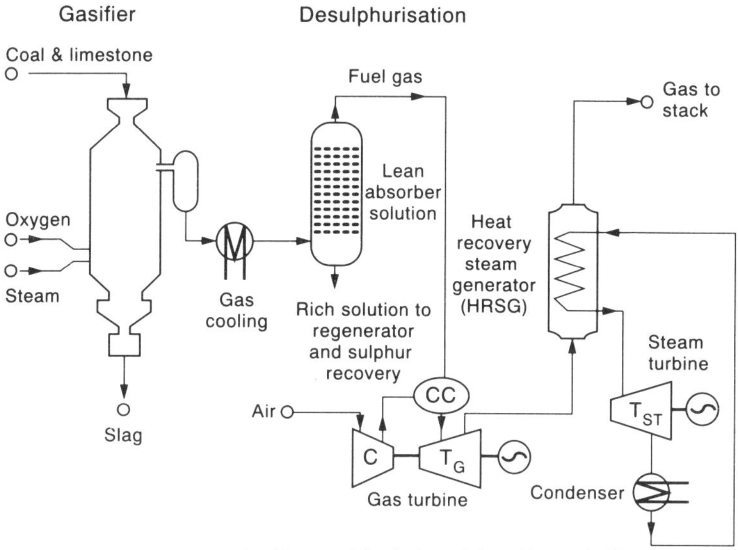

FIGURE 2.10 Gasifier combined plant (after Chester [13])

has to be reached between that pressure and the stack temperature of the gases leaving the exchanger, T_S (and the consequent "heat" loss). A detailed discussion of the thermodynamics of this type of plant is given in Chapter 3, and parametric studies are described in Chapter 4.

Once again, since there is heat unused ("lost") between the component plants, the ideal efficiency of a combined plant (η_{CPS}) cannot be attained. We shall find that there may be little or no advantage in using feed heating in the steam cycle of this combined plant, depending on the steam cycle chosen.

The exhaust heated system is the most widely used combined plant. A current development of this type is the integrated coal gasification combined cycle (IGCC) plant, the basic features of which are shown in Fig. 2.10, taken from the review by Chester *et al.* [13]. Coal and limestone are fed to a pressure vessel, the coal being gasified by oxygen and steam and the synthetic gas produced then being cooled. (Air can be used instead of oxygen in some proposed applications, but this produces a gas of lower calorific value.) The ash and limestone form a molten slag which is discharged, but the gas contains hydrogen sulphide which has to be removed. The resultant clean gas is supplied as fuel to the air from the compressor of a gas turbine plant and combustion takes place. The combustion gases expand through a gas turbine driving the compressor and a generator, and subsequently raise steam in an exhaust heated steam generator.

Several different forms of gasifier have been developed (e.g. the Texaco, Shell, Dow, British Gas/Lurgi gasifiers). The Texaco process has been used in the Cool Water demonstration plant described in Chapter 6.

The circulating (atmospheric) fluidised bed combustion system (CFBC) has also been considered for the simple combined plant with exhaust heating of the steam cycle (see Fig. 2.11 given by Chester *et al.*) but no plant has yet been designed and built. Here the air delivered by the compressor is first heated by solids circulating through the CFBC cyclone, before entry to the air turbine. The main combustor, which operates at low pressure, burns coal with air from the turbine exhaust. The combustion gases raise steam in an HRSG, but the circulating solids from the CFBC also transfer heat to the steam. While it is primarily exhaust heating that is involved, there are also elements of joint and supplementary heating, in that products of combustion transfer heat to the air (before it enters the air turbine) and to superheat the steam.

2.4.1.2 SUPPLEMENTARY FIRED HRSG

The exhaust gases from a gas turbine contain substantial amounts of excess air, since the main combustion process has to be diluted to reduce the combustion temperature to well below that which could be obtained in stoichiometric combustion, because of the metallurgical limits on the gas turbine operating temperature. This excess air enables supplementary firing of the exhaust to take place, and higher steam temperatures may then be obtained in the heat recovery steam generator.

The (T, s) diagram for a combined plant with supplementary firing is

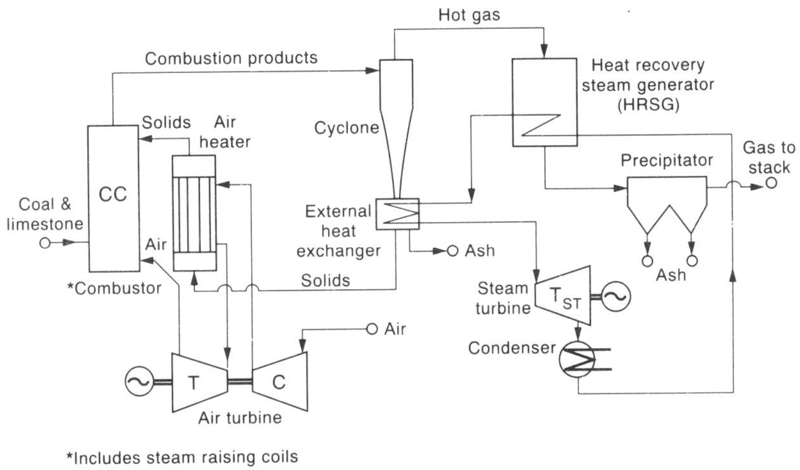

FIGURE 2.11 Circulating (atmospheric) fluidised bed combined plant (after Chester [13])

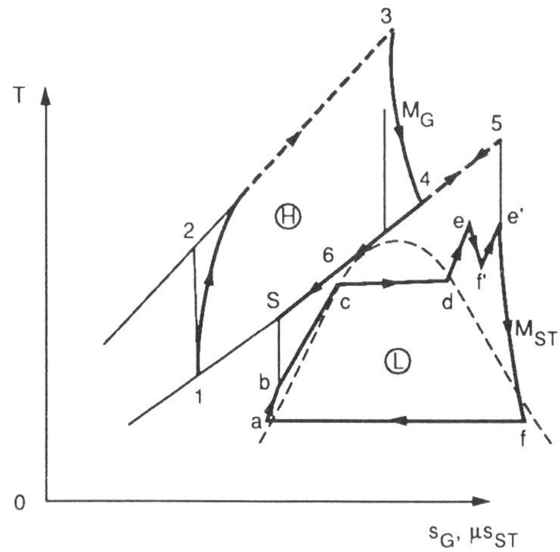

FIGURE 2.12 Open circuit/closed cycle combined plant with supplementary firing ($T_a \approx T_1$)

illustrated in Fig. 2.12. For HRSG temperatures up to about 760° C, the introduction of regenerative feed heating of the water is again of doubtful value, as we shall discuss later. Supplementary heating generally lowers the overall efficiency of the combined plant, below that of a series combined plant (η_{CPS}) in which heat supplied to the lower cycle is entirely the heat rejected by the higher plant (see Chapter 3). Essentially this is because a fraction of the total heat supplied is utilised to produce work in the lower cycle, of lower efficiency than that of the higher cycle. Parametric analysis of this combined plant is given in Chapter 4.

2.4.1.3 FULL FIRING OF THE EXHAUST GASES

If the exhaust gases from the gas turbine are supplied to a conventional boiler, as opposed to a simple heat recovery steam generator, then a "fully fired" or "maximally fired" combined plant results (Fig. 2.13). This system is particularly relevant for repowering a conventional steam power plant. The steam plant usually employs substantial feed heating, but in some cases one (or more) of the higher temperature feed heaters is by-passed, part of the feed water being passed through an economiser to reduce the temperature of the stack gases (as originally proposed by Seippel and Bereuter [12]).

2.4.1.4 A NOTE ON HEAT RECOVERY STEAM GENERATORS

The HRSG is clearly an important component in the exhaust heated combined power plant. It can be of the natural convection type (horizontal design), where space is of minor importance, or of the forced ("controlled") circulation type (vertical design), where space is limited.

A useful comparison between the basic features of an HRSG (unfired, forced circulation) and a conventional boiler (natural circulation) is given by Lugand and Paren [14] and is shown in Fig. 2.14, with illustrative gas, water and steam temperatures. The HRSG is simple, employs convective heat transfer and has a low heat flux, about 150,000 W/m^2 (tube internal surface). The conventional boiler is more complicated, employs radiation and convection about equally, and has a higher heat flux (about 300,000 W/m^2). The table below Fig. 2.14 gives further comparisons between the major features of these two forms of steam generator.

2.4.2 Pressurised Combustion Plants

2.4.2.1 CLEAN COMBUSTION PLANT

The simplest form of pressure charged combustion plant is illustrated in Fig. 2.15 after Seippel and Bereuter [12]. The gas turbine is essentially a (CBTX) open plant, but the combustion of compressed air with distillate oil or natural gas takes place in a boiler, generating steam, the flue gases subsequently expanding through the gas turbine. In the example shown, they then heat the feed water in the steam cycle.

The plant is essentially a "joint heating" plant, although in the example illustrated there is also some use of heat rejected from the upper circuit in the lower cycle (by heating of the feed water by the exhaust from the gas turbine).

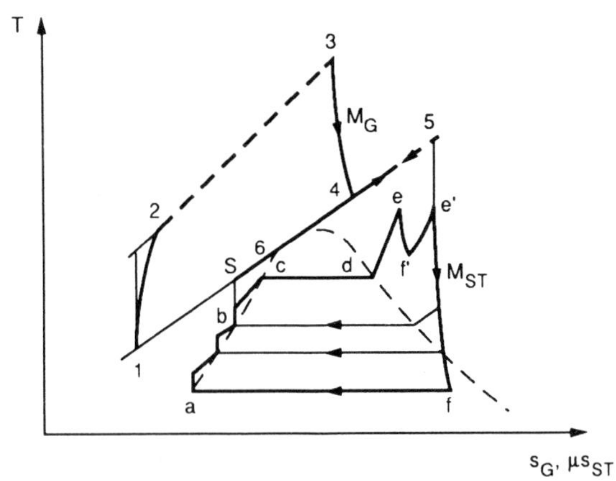

FIGURE 2.13 Maximally fired combined plant with full regenerative feed heating ($T_a \approx T_1$)

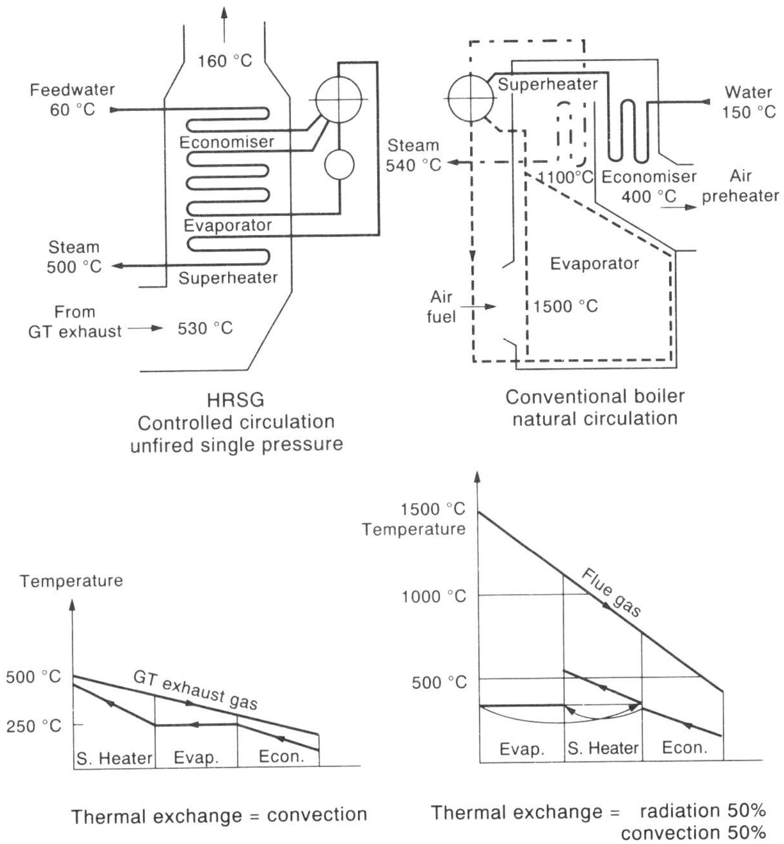

Parameter	HRSG	Conventional boiler
Flue gas flowrate, t/h/MW	17 to 28	7 to 10
Exchange surfaces m^2/MW	800 to 1300	60 to 80 (air preheater excluded)
Boiler volume m^3/MW (MW: steam turbine gross output)	70 to 100	50 to 65 (air preheater excluded)

FIGURE 2.14 Comparison between an HRSG and a conventional boiler (after Lugand and Paren [14])

2.4.2.2 PRESSURISED FLUIDISED BED (COAL COMBUSTION) COMBINED PLANT

While the plant illustrated in Fig. 2.15 can operate with distillate oil or natural gas, it cannot burn coal, since the combustion products cannot be accepted by the gas turbine. This difficulty may be overcome by the development of a pressurised fluidised bed combustion (PFBC) system as illustrated in Fig. 2.16.

Air is compressed and delivered to the fluidised bed boiler which is supplied with dry coal or a coal/water slurry. Full combustion takes place in the fluidised bed, SO_2 emissions being reduced by a solvent (limestone or dolomite) fed into the bed and sulphur removed in the form of calcium sulphate in the ash. Heat is removed from the fluidised bed by a bank of steam raising tubes, and steam is fed to a conventional steam cycle. Hot gas leaving the combustor flows through a gas turbine, driving the compressor and a generator. This hot gas is cleaned before entry to the turbine (in a series of cyclones, or cyclones and a ceramic filter). The turbine exhaust gases may be used to heat the boiler feed water.

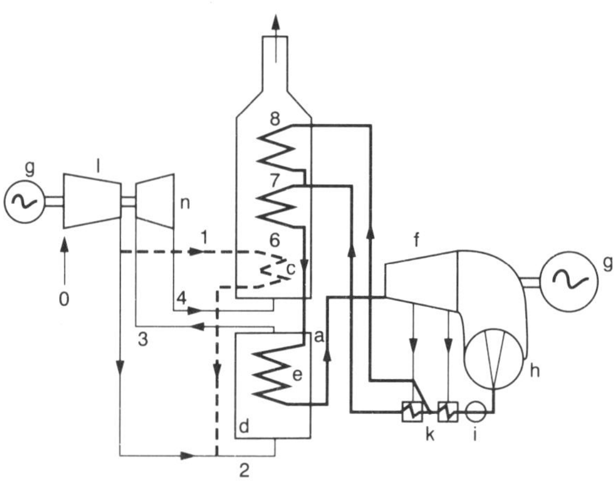

a = Boiler
c = Air heater
d = Combustion chamber with burner
e = Steam part of boiler
f = Steam turbine
g = Generator
h = Condenser
i = Feedwater pump
k = Feedwater heater and de-aerator
l = Compressor
m = Combustion and swirl chamber
n = Gas turbine

FIGURE 2.15 Combined plant with pressure charged boiler (Velox principle) (after Seippel and Bereuter [12])

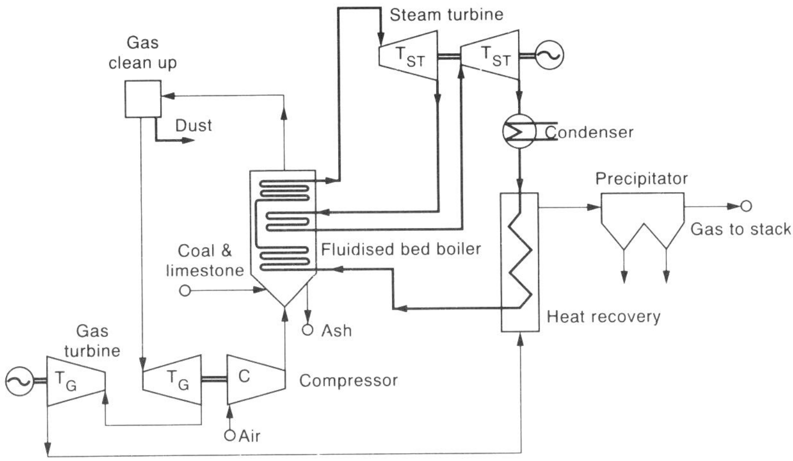

FIGURE 2.16 PFBC combined cycle (after Chester [13])

Essentially the PFBC combined plant is a "joint heating" system (but again a "series" element is introduced if the feed water is exhaust heated).

2.4.3 Some Other Developments

Other possible developments of the open circuit/closed cycle plant include

(i) a proposed modification to the closed cycle to reduce the irreversibility involved in the heat transfer from the higher plant (the Kalina cycle),
(ii) a proposed topping gasifier/fluidised bed combustion plant.

2.4.3.1 THE KALINA CYCLE

An interesting development for the lower liquid/vapour cycle is that proposed by Kalina, in which a mixture of ammonia and water is used. The ammonia begins to boil first, having the lower boiling point. As it boils off, the concentration of the ammonia decreases and the boiling point of the mixture increases progressively. This leads to a better match between the temperature profiles of the exhaust gas from the higher open plant (rejecting heat) and the liquid/vapour of the lower cycle (absorbing heat).

However, the high concentration ammonia solution at the low pressure of the vapour turbine exhaust cannot be condensed at normal cooling water temperatures. In the Kalina cycle this problem is solved by transferring heat from the "working" turbine exhaust to a "basic" solution of higher mass flow, but lower concentration, which achieves most of the heat rejection through condensation at a lower temperature.

This can be a fairly complex process, but Fig. 2.17 shows an early, relatively simple example described by Kalina [15], involving turbines, heater, economiser, flash tank and condenser. A table of concentrations and relative mass flows is given below the figure, and some representative properties are shown.

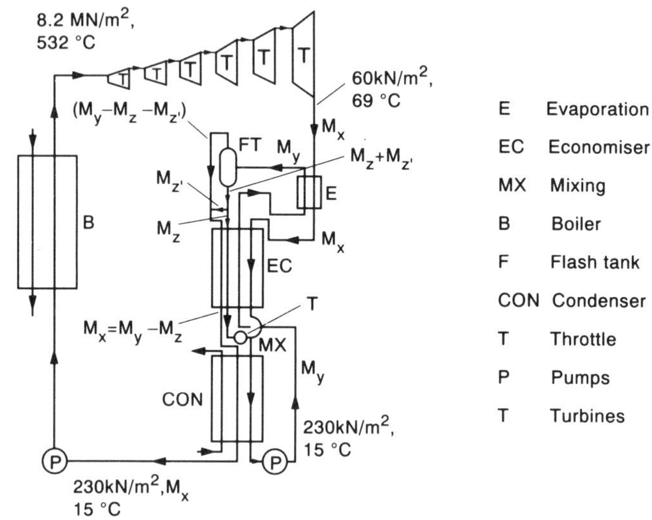

Representative values

	Relative mass flow	Concentration of ammonia
M_x	1.0	0.5
M_y	5.17	0.294
M_z	4.17	0.246
$M_{z'}$	0.63	0.246
$(M_y-M_z-M_{z'})$	0.37	0.931
$M_x=M_y-M_z$	1.0	0.5

FIGURE 2.17 Example of Kalina cycle (after Kalina [15])

The ammonia concentration of the working fluid at exit from the final turbine is 0.5 in this example. This "working" stream (mass flow $M_x = 1.0$) is cooled in an evaporator (E) and an economiser (EC), transferring heat to another "basic" ammonia stream (mass flow $M_y = 5.17$) of higher pressure but lower concentration (0.294), which is thus partially evaporated. Stream M_y is then led to a flash tank (FT) where vapour of high concentration (0.931) is separated from liquid of low concentration (0.246). The major part of this liquid stream ($M_z = 4.17$) is cooled in EC, throttled at T, mixed with the cooled working stream ($M_x = 1.0$) at MX, to reduce the ammonia concentration and form the basic stream ($M_y = 5.17$) of low concentration (0.294). The remaining liquid ($M_{z'} = 0.63$) is mixed with the concentrated vapour ($M_y - M_{z'} - M_z = 5.17 - 4.17 - 0.63 = 0.37$) to form the working, wet vapour stream ($M_x = 1.0$) of concentration 0.5. The basic stream ($M_y = 5.17$) can now be condensed in CON at intermediate pressure (58.6 kN/m^2) but low concentration (0.294) and therefore low temperature (31.6 °C to 15.5 °C). The working stream ($M_x = 1.0$) at a higher pressure (229.6 kN/m^2) and higher concentration (0.5) is cooled in EC and condensed at a slightly higher temperature (46.2 °C to 15.5 °C) before being pumped to the boiler (B) at a pressure of 8274 kN/m^2. Note that the mass flow of the basic stream is over five times that of the working stream, so most of the heat rejection is achieved by that stream at low concentration and low temperature.

Kalina has produced several variations on his cycle (e.g. reference 16) with a higher concentration of ammonia in the working stream (0.7) and involving more components. He has also described variations of the lower cycle, modifying the arrangement of consecutive radial turbines as shown in Fig. 2.17 to involve reheating between an H.P. and an I.P. turbine, and cooling between the I.P. turbine and an L.P. turbine. The increasing complexity needs to be assessed from an economic point of view before the Kalina cycle is widely accepted. It is nevertheless a highly novel proposal meriting careful analytical and experimental study; a demonstration plant is being developed [17].

2.4.3.2 TOPPING GASIFICATION/FLUIDISED BED COMBUSTION PLANT

Developments involving a gasifier "topping" a fluidised bed system (pressurised or circulating) have been proposed by British Coal (Dawes *et al.* [18]). The purpose is to provide high temperature gas for the gas turbine while minimising the energy losses involved in the production of the fuel gas—no oxygen production plant is required.

Figure 2.18 shows the pressurised fluidised bed combustion/topping plant. Compressor air is fed to both the gasifier and the pressurised

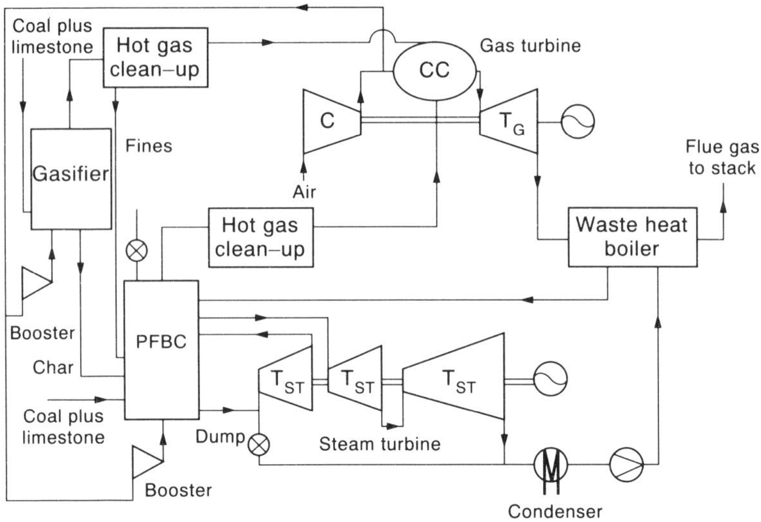

FIGURE 2.18 Outline of pressurised fluidised bed combustion topping cycle (single shaft gas turbine) (after Dawes *et al.* [18])

fluidised bed combustor (PFBC). The former supplies hot gas (after cleaning) to the gas turbine combustion chamber. Char from the gasifier and fines from the clean-up fuel the PFCB which raises steam for the steam turbine. Exhaust gases from the gas turbine transfer heat to the feed water in a waste heat boiler before passing to the flue gas stack. Again this system is a hybrid form of pressurised combustion (joint "heating") and exhaust heating.

Figure 2.19 shows the circulating fluidised bed combustion (CFBC)/ topping plant. (It is not strictly a pressurised combustion system like the PFBC system described above.) The essential difference is that the CFBC, again fuelled with char and fines from the pressurised gasifier, is now fed with low pressure turbine exhaust gases, which have sufficient oxygen for further combustion. The CFBC raises steam for the steam turbine plant, but of course the low pressure flue gas produced is not supplied to the gas turbine combustion chamber in this case. This plant essentially consists of an open circuit gas turbine, an HRSG with maximally fired after "heating" and a steam turbine.

2.5 Doubly Open Circuit Plants (Two Working Fluids)

A combined plant which has been the subject of much recent interest and development is the doubly open circuit plant. In this scheme (Fig. 2.20) water is evaporated by the exhaust from a "gas" turbine, the

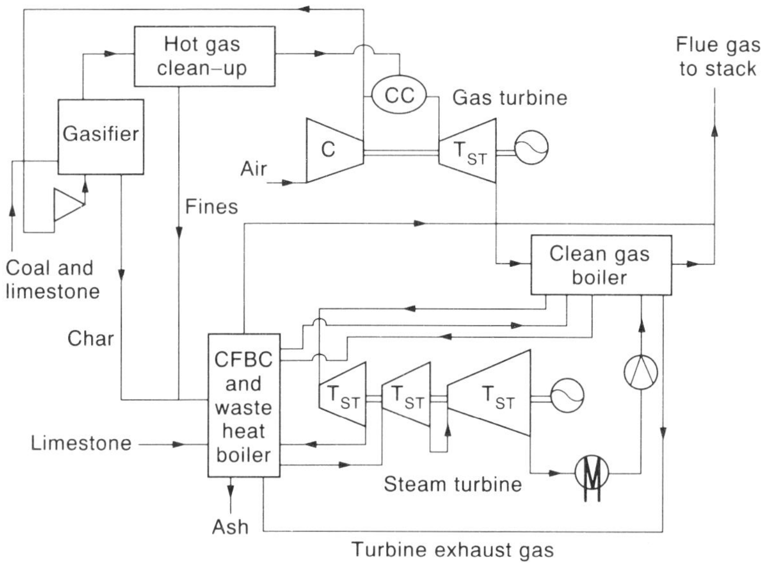

FIGURE 2.19 Outline of circulating fluidised bed combustion topping cycle (after Dawes *et al.* [18])

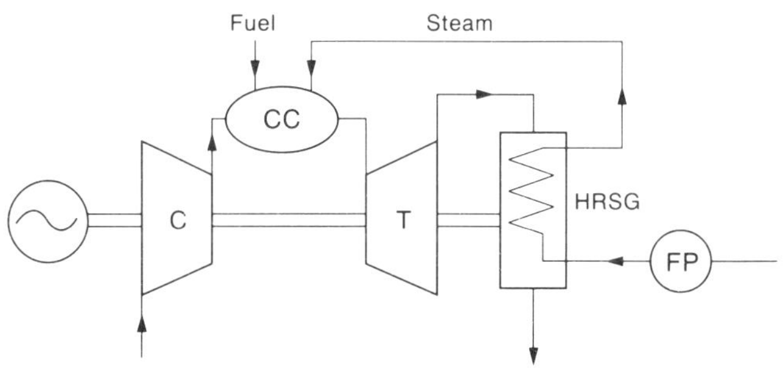

C	Compressor
T	Turbine
FP	Feed pump
HRSG	Heat recovery steam generator
CC	Combustion chamber

FIGURE 2.20 Basic steam injection gas turbine (STIG)

resulting steam is injected into the combustion chamber of the gas turbine plant, and the two fluids expand through the turbine. The combined plant may then be said to be composed of an open circuit gas turbine plant and a "hidden" open circuit steam turbine plant. A combination of series and parallel heating between the two circuits is involved, in that heat rejection from the turbine exhaust is used to heat the feed water and evaporate it, and joint (parallel) "heating" in the combustion chamber "heats" the gas and superheats the steam. Substantial increases in power output are obtained by this process (sometimes called the Cheng cycle). Here we simply classify it as a doubly open circuit combined plant, and discuss its performance in detail later in Chapter 4.

References

[1] Horlock, J. H. *Cogeneration: Combined Heat and Power*. Pergamon Press, Oxford, 1987.
[2] Buxmann, J. *Combined Cycles for Power Generation—Introduction*. Von Karman Institute for Fluid Dynamics, Lecture Series 6, Vol. 1, 1978.
[3] Haywood, R. W. *Analysis of Engineering Cyles*, 4th edition. Pergamon Press, Oxford, 1991.
[4] Field, J. F. Improvements in and Relating to Steam Power Plants. Brit. Pat. Specs. 451,571, 1943 and 395,581, 1946.
[5] Field, J. F. The Application of Gas Turbine Technique to Steam Power. *Proc. I.Mech.E.*, **162,** 209, 1950.
[6] Horlock, J. H. *The Thermodynamic Efficiency of the Field Cycle*. ASME Paper 57-A-44, 1957.
[7] Flatt, F. and Scharfenberg, G. Sonnenfeld-Prozess für Uberkritische Dampfkraftwerke. *Elektrizitatswirtschaft*, **67,** 721–729, 1968.
[8] Colosimo, D. D., Rose, R. K. and Decker, O. *Power Generation with Binary Cycles*. Von Karman Institute for Fluid Dynamics, Lecture Series 6, Vol. 2, 1978.
[9] Mangan, J. L. and Pettit, J. L. *A Highly Efficient Steam Turbine Gas Turbine Cycle*. ASME Paper 63-GT-51, 1963.
[10] Wardall, R. M. and Doorly, E. E. Current Prospects for Efficient Combined Cycles for Small Gas Turbines. *Energy Process Can.*, **68**, 42–51, 1976.
[11] Buxmann, J. and Bormann, T. *The Closed Cycle Gas Turbine-Steam Turbine*. Von Karman Institute for Fluid Dynamics, Lecture Series 6, 1976.
[12] Seippel, C. and Bereuter, R. The Theory of Combined Steam and Gas Turbine Installations. *Brown Boveri Review*, **47,** 783–799, 1960.
[13] Chester, P. F. *et al. Prospects for the Use of Advanced Coal Based Power Generation Plant in the United Kingdom*. H.M.S.O. Energy Paper 56, 1988.
[14] Lugand, P. and Paren, J. *Vega—a Versatile Range of Combined Cycle Power Plants*. Alsthom Technical Report AGTR 8809, 1990.
[15] Kalina, A. I. Combined Cycle System with Novel Bottoming Cycle. *Trans. ASME Journal of Engineering for Gas Turbines and Power*, **106,** 737–742, 1984.
[16] Kalina, A. I. *Applying Kalina Technology to a Bottoming Cycle for Utility Combined Cycles*. ASME Paper 87-GT-35, 1987.
[17] Jurgen, R. J. The Promise of the Kalina Cycle. *IEEE Spectrum*, April, 68–69, 1986.
[18] Dawes, S. G., Brown, D. and Hyde, J. A. C. Optics for Advanced Power Generation from Coal. I.Mech.E. Conference, Paper C410/042, 123–134, 1991.

CHAPTER 3

Combined Power Plants—Some Thermodynamic Concepts

3.1 Introduction

As was emphasised at the end of Chapter 1, the primary purpose of combining power plants is to achieve high efficiency of the combined plant. Other objectives may be to reduce capital cost in comparison with a single power plant, into which a high degree of expensive sophistication would have to be introduced to achieve high efficiency (e.g. by raising boiler pressure in a steam cycle). Yet another objective may be to obtain increased power output from an existing single power plant (e.g. by "repowering" a steam plant, replacing the conventional boiler by a gas turbine plant supplying gas to an exhaust fired boiler).

From the thermodynamic point of view, the primary objective is to obtain higher efficiency by raising the mean temperature of "heat" supply, and/or lowering the mean temperature of heat rejection, i.e. seeking to approach more closely Carnot efficiency. There are many variations on how this can be done and we review some of them in this chapter. Utilising heat rejection from the higher plant as a heat supply to the lower plant is an essential part of all the schemes we shall consider. But the introduction of other practical features (e.g. "heat loss", supplementary heat supply to the lower plant) means that the "series" efficiency of the combined cyclic plant,

$$\eta_{CPS} = \eta_H + \eta_L - \eta_H \eta_L \tag{3.1}$$

(see Section 1.7), is rarely if ever achieved.

We consider first an ideal combined cyclic plant, comprising two plants H and L each operating on Carnot cycles and *in series* (S), i.e. with the heat rejected from the upper plant completely supplied to the lower. We then derive the overall plant efficiency if there is a temperature drop between the two Carnot cycles (but no heat loss) assessing the effect of this internal irreversibility. After considering the effect of temperature drop more generally, again for series plants, we consider how the "series" efficiency, η_{CPS}, is reduced by the effect of heat loss between the upper and lower plants.

We next investigate cyclic plants *in parallel* (P), heat being supplied jointly to two power plants from a single heating device. In this case there is no thermodynamic advantage over the operation of the cyclic plant of higher efficiency, a somewhat obvious conclusion.

But by the introduction of supplementary heating between upper and lower cyclic plants (initially in series), an element of parallelism (of lower thermodynamic efficiency) is introduced (i.e. *series/parallel* operation SP) and that is the reason for loss in overall combined plant efficiency in comparison with the "series" efficiency, η_{CPS}. We therefore explore the thermal efficiency of combined cyclic plants with both heat loss and supplementary heating between the upper and lower plants, approaching most closely some practical plants in which both effects are present.

These preliminary analyses are essentially based on two cyclic component plants. But subsequently we make a study of the thermodynamics of combined plants in which the higher plant is open circuit (usually a gas turbine plant) and the lower is a closed cycle (usually a steam plant). We consider exhaust heated (unfired and fired) heat recovery steam generators (HRSGs) and pressurised boiler systems.

Finally, we review an early but comprehensive study of possible open circuit/closed cycle plants carried out by Seippel and Bereuter [1], who discussed the thermodynamic merits of such combined plants.

3.2 Series Plants (Two Closed Cycles in Series)

3.2.1 The Ideal "Series" Plant

We have already described the operation of an ideal "series" plant in Section 1.7. If all the heat rejected from the higher cycle (of efficiency η_H) is used to supply the lower cycle (of efficiency η_L), then the thermal efficiency of the combined "series" plant is

$$\eta_{CPS} = \eta_H + \eta_L - \eta_H \eta_L. \tag{3.1}$$

We discuss this relationship further for "series" plants, in which heat is cascaded down from one cycle to another.

3.2.2 A Carnot Combined Plant

Let us suppose that the two plants, η_H and η_L, each operate on a Carnot cycle and are combined to form a Carnot combined plant (Fig. 3.1). This may appear to be a hypothetical situation, but various complex combined plants have been proposed which do approach this double Carnot concept, as we have described in Chapter 2.

If the temperatures of supply and rejection for plant H are T_B and T_{HR} respectively, then

$$\eta_H = \frac{T_B - T_{HR}}{T_B} = 1 - \frac{T_{HR}}{T_B}. \tag{3.2}$$

If there is no finite drop in temperature between the two plants, then the temperature of supply to plant L is $T_L = T_{HR}$, and

$$\tilde{\eta}_L = \frac{T_{HR} - T_A}{T_{HR}} = 1 - \frac{T_A}{T_{HR}}, \tag{3.3}$$

where T_A is the temperature of heat rejection from the bottoming plant, and the superscript ˜ now indicates the special case with no temperature drop between plant H and plant L, which we now refer to as $\tilde{L}$.

(In Fig. 3.1 we imply that the specific entropy change of the fluid in Carnot engine H (Δs_H) is equal to that of the fluid in Carnot engine $\tilde{L}$ ($\Delta \tilde{s}_L$); we discuss this point later in Section 3.2.4.1.)

The combined cycle efficiency is, from equation (3.1),

$$\begin{aligned}\tilde{\eta}_{CPS} &= \left(1 - \frac{T_{HR}}{T_B}\right) + \left(-\frac{T_A}{T_{HR}}\right) - \left(1 - \frac{T_{HR}}{T_B}\right)\left(1 - \frac{T_A}{T_{HR}}\right) \\ &= 1 - \left(\frac{T_{HR}}{T_B}\right)\left(\frac{T_A}{T_{HR}}\right) \\ &= 1 - \frac{T_A}{T_B},\end{aligned} \tag{3.4}$$

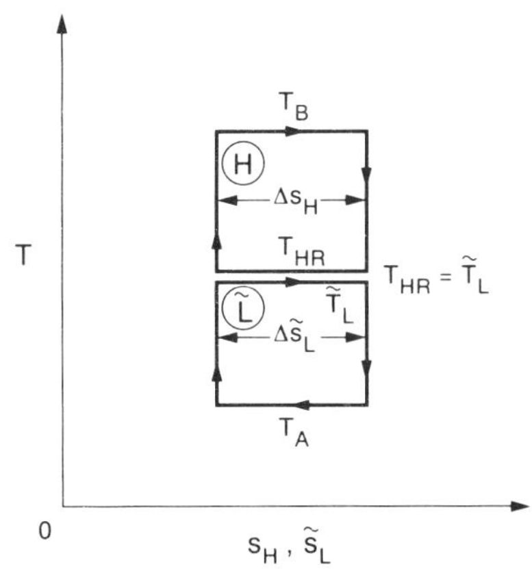

FIGURE 3.1 Carnot cycles (H and $\tilde{L}$) with no intermediate temperature drop ($T_{HR} = \tilde{T}_L$)

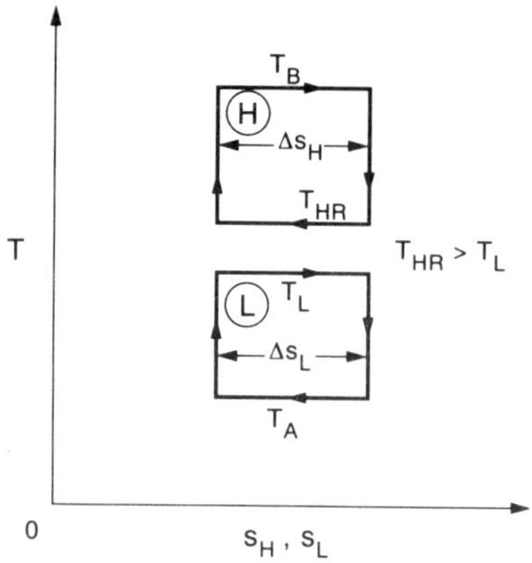

FIGURE 3.2 Carnot cycles (H and L) with intermediate temperature drop ($T_{HR} > T_L$)

which, as to be expected, is the efficiency of a single Carnot cycle (η_{CARNOT}) operating between T_B and T_A. Compared with this single cycle there is no gain or loss in efficiency by combining the two separate Carnot plants, because, with the assumption of no temperature drop between them, no irreversibility is introduced.

This is an obvious conclusion, but the analysis is now developed further to illustrate the general effect of such a temperature drop, for it is of crucial importance in the thermodynamics of combined power plants.

We shall find the existence of the irreversibility associated with the intermediate temperature drop does not invalidate the expression for the overall efficiency; but as this expression includes the efficiency of the lower cycle, and that efficiency is decreased, the overall plant efficiency falls as a result.

3.2.3 Combined Carnot Plant with Intermediate Temperature Drop

We next consider a combined plant based on two Carnot cycles, H and L (Fig. 3.2). As before, the higher upper cycle operates between T_A and T_{HR}, but the lower cycle now operates between T_L and T_A. Heat is now rejected from T_{HR} to T_L across a finite temperature drop ($T_{HR} - T_L$) and the efficiency of the lower plant drops from $\tilde{\eta}_L$ to η_L.

(We again imply in Fig. 3.2 that the specific entropy changes in the two engines are the same, a point later discussed in Section 3.2.4.1.)

The work outputs from H and L are

$$W_H = \eta_H Q_B = \left(\frac{T_B - T_{HR}}{T_B}\right) Q_B, \tag{3.5}$$

(where η_H is the unchanged efficiency of the higher plant) and

$$W_L = \eta_L Q_{HL} = \left(\frac{T_L - T_A}{T_L}\right) Q_{HL} \tag{3.6}$$

(where η_L is the reduced efficiency for the lower plant, $\eta_L < \tilde{\eta}_L$).

But $Q_{HL} = Q_B\left(\frac{T_{HR}}{T_B}\right)$, so that the total work output is

$$\begin{aligned} W_{CP} &= W_H + W_L \\ &= Q_B\left[\left(\frac{T_B - T_{HR}}{T_B}\right) + \left(\frac{T_{HR}}{T_B}\right)\left(\frac{T_L - T_A}{T_L}\right)\right], \end{aligned} \tag{3.7}$$

and the efficiency of the complete plant is

$$\begin{aligned} \eta_{CPS} &= \frac{W_{CP}}{Q_B} \\ &= \left(1 - \frac{T_{HR}}{T_B}\right) + (1 - \eta_H)\left(1 - \frac{T_A}{T_L}\right) \\ &= \eta_H + (1 - \eta_H)\eta_L \qquad (3.1) \\ &= 1 - \left(\frac{T_{HR}\, T_A}{T_L\, T_B}\right), \qquad (3.8) \end{aligned}$$

and $\eta_{CPS} < \tilde{\eta}_{CPS} = \left(1 - \frac{T_A}{T_B}\right)$, since $T_{HR} > T_L$. The general expression for the thermal efficiency of a "series" plant—the sum of the individual efficiencies less their product—is valid for the combined plant with or without an intermediate temperature drop.

The loss in thermal efficiency due to the intermediate temperature drop $(T_{HR} - T_L)$ of the plant of Fig. 3.2 compared with the perfect Carnot engine is

$$\begin{aligned} \Delta\eta &= \tilde{\eta}_{CPS} - \eta_{CPS} \\ &= \left(1 - \frac{T_A}{T_B}\right) - \left(1 - \frac{T_{HR}}{T_L}\frac{T_A}{T_B}\right) \\ &= \frac{T_A}{T_B}\left(\frac{T_{HR} - T_L}{T_L}\right) \\ &= \frac{\Delta W}{Q_B}, \end{aligned} \tag{3.9}$$

where ΔW is the lost work output,

$$\Delta W = Q_B\left(\frac{T_A}{T_B}\right)\left(\frac{T_{HR}-T_L}{T_L}\right). \tag{3.10}$$

3.2.4 The General Effect of Temperature Drop between Higher and Lower Level Plants

There will always be a temperature drop in the heat transfer between the upper and lower plant in any "series" combined power plant. This causes a loss in overall work output and in overall efficiency, as has been illustrated in the analysis for the combined Carnot cycle plant given above. We now consider the effect more generally, with higher and lower cycles H and L, which are not necessarily Carnot cycles.

Consider (Fig. 3.3a) an elementary quantity of heat dQ_{HL} being transferred from the cycle H (at variable rejection temperature T_{HR}) to the lower cycle (at variable receiving temperature T_L). The "lost" work in this process is related to the difference between the potential work output from the heat rejection—the work that could have been obtained in elementary Carnot engines operating between temperatures T_{HR} and $T_A = T_0$, the temperature of the environment, and the work input required by Carnot heat pumps to "pump" the heat back up to temperature T_L from temperature $T_A = T_0$.

Thus the lost work is

$$\begin{aligned}\Delta W &= \int\left(\frac{T_{HR}-T_0}{T_{HR}}\right)dQ_{HL} - \int\left(\frac{T_L-T_0}{T_L}\right)dQ_{HL}\\ &= \int\left(\frac{T_0}{T_L}-\frac{T_0}{T_{HR}}\right)dQ_{HL}\\ &= T_0\int\left(\frac{T_{HR}-T_L}{T_{HR}T_L}\right)dQ_{HL}\\ &= T_0(\Delta S_L - \Delta S_H),\end{aligned} \tag{3.11}$$

where ΔS_L is the entropy gain of plant L, and ΔS_H is the entropy leaving plant H. This is a measure of the irreversibility involved in the heat transfer across the finite temperature drop between T_{HR} and T_L.

An alternative way of assessing this irreversibility is to use the concept of exergy (Fig. 3.4) developed in Section 1.6.3. For the upper cycle, the statement of exergy flux is

$$E_B^Q = W_H + I_H^{CR} + E_{HR}^Q \tag{3.12}$$

where E_{HR}^Q is the work capacity associated with the heat rejection at (variable) temperature T_{HR}, and I_H^{CR} is the internal irreversibility in cycle H.

For the lower cycle,

$$E_L^Q = W_L + I_L^{CR} + E_A^Q, \tag{3.13}$$

where I_L^{CR} is the irreversibility in cycle L.

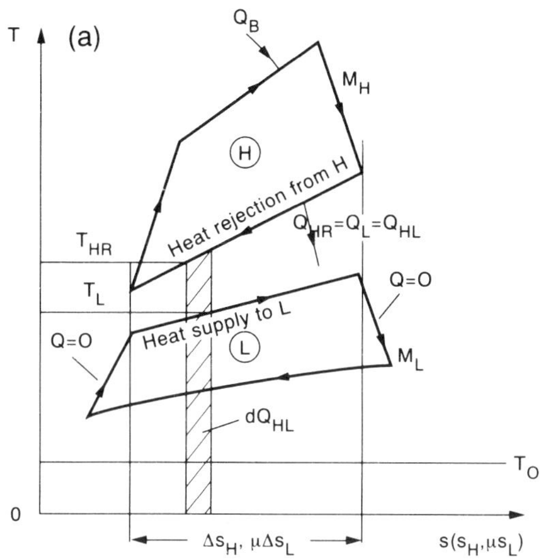

FIGURE 3.3a Heat transfer from higher cyclic plant to lower cyclic plant

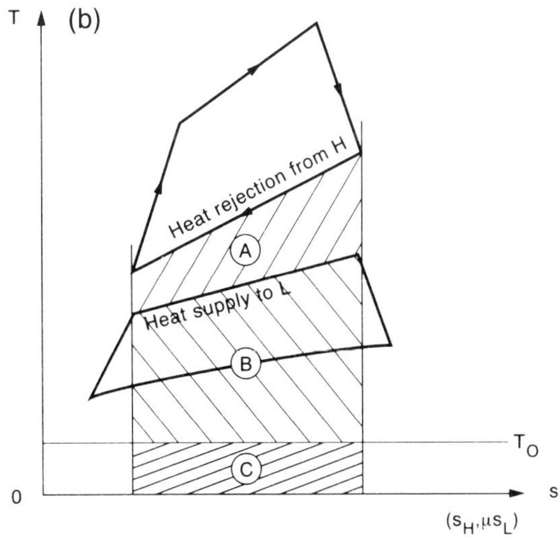

FIGURE 3.3b Areas on T, s diagram

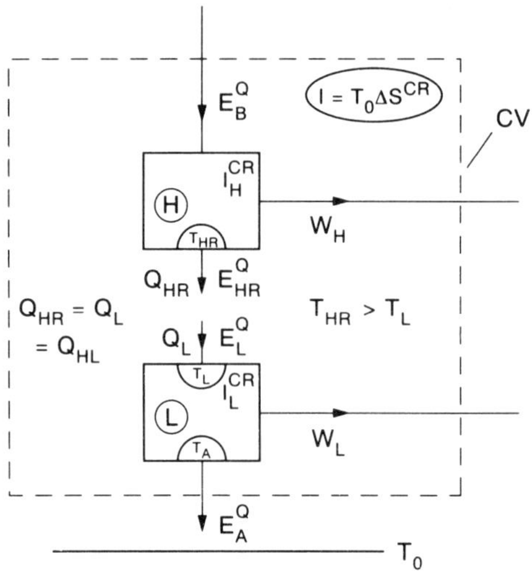

FIGURE 3.4 Exergy fluxes with intermediate temperature drop ($E_{HR}^Q > E_L^Q$)

But if heat is rejected at the ambient temperature, i.e. $T_A = T_0$, then E_A^Q is zero.

Hence from equations (3.12) and (3.13)

$$W_{CP} = W_H + W_L = E_B^Q - I_H^{CR} - I_L^{CR} - I_{HL}^Q, \tag{3.14}$$

where I_{HL}^Q is the "intermediate" irreversibility, external to the two separate plants,

$$\begin{aligned} I_{HL}^Q &= E_{HR}^Q - E_L^Q \\ &= \int \left(\frac{T_{HR} - T_0}{T_{HR}} \right) dQ_{HL} - \int \left(\frac{T_L - T_0}{T_L} \right) dQ_{HL} \\ &= T_0 \int \left(\frac{T_{HR} - T_L}{T_{HR} T_L} \right) dQ_{HL}. \end{aligned} \tag{3.15}$$

Equation (3.14) illustrates how possible work output E_B^Q is not only reduced by internal irreversibilities I_H^{CR} and I_L^{CR} in the two cycles but also by the external irreversibility I_{HL}^Q due to the finite temperature drop involved in the heat transfer between the two cyclic plants. The total irreversibility is $(I_H^{CR} + I_L^{CR} + I_{HL}^Q) = T_0 \Delta S_{CR}$ where ΔS_{CR} is the total entropy created within the control surface surrounding the two plants.

3.2.4.1 GRAPHICAL REPRESENTATION OF THE LOST WORK

The lost work due to the irreversibility in the heat transfer $Q_{HL}(=Q_{HR}=Q_L)$ from the higher plant to the lower plant, across a finite temperature drop, may be related to areas on temperature-entropy (T, s) diagrams.

Figure 3.3a shows such heat transfer between a higher cyclic plant (mass flow M_H) and a lower cyclic plant (mass flow rate M_L). The path line of the heat rejection from the upper cycle is shown on a (T, s) diagram with a true entropy scale $(s=s_H)$. The path line for heat supplied to the lower cycle is shown on the same diagram, but with a different entropy scale, such that $s=\mu s_L$, where μ is a factor to be determined. The width of the two processes in the entropy direction is made the same, so that $\Delta s_H=\mu\Delta s_L$ and $\mu=(\Delta s_H/\Delta s_L)$.

We assume that there is no heat loss or pressure loss in the heat transfer process, so that

$$M_H \int_{HR} T_{HR}\, ds_H = M_L \int_L T_L\, ds_L. \tag{3.16}$$

We show three areas A, B and C on Fig. 3.3b,

$$(A+B+C)=\int_{HR} T_{HR}\, ds_H = Q_{HR}/M_H, \tag{3.17}$$

$$(B+C)=\mu\int_L T_L\, ds_L = \mu Q_L/M_L, \tag{3.18}$$

the area C being related to the (hypothetical) heat rejection Q_0 at temperature T_0 if the lower cycle were an ideal Rankine cycle,

$$C=\mu T_0 \Delta s_L = \mu Q_0/M_L. \tag{3.19}$$

Since $Q_{HR}=Q_L$,

$$\mu=\left(\frac{M_L}{M_H}\right)\left(\frac{B+C}{A+B+C}\right), \tag{3.20}$$

which is the scaling factor on the steam entropy in the diagram.

We should note that this scaling is required to make the end states of the heat rejection and heat supply process line up vertically on the (T, s) diagrams shown in Fig. 3.3. It does not follow that intermediate states line up similarly. With regard to the Carnot cycles shown in Fig. 3.1 the scaling factor is unity, with $M_H=M_L$ and $A=0$. For the Carnot cycles shown in Fig. 3.2 with $A \neq 0$, M_L would have to be greater than M_H to give $\mu = 1$.

The lost work due to irreversibility in this process, shown in Fig. 3.3b, is simply

$$\begin{aligned} I_{HL}^{Q} &= T_0[M_L \Delta s_L - M_H \Delta s_H] \\ &= Q_0\left(1 - \frac{\mu M_H}{M_L}\right) \\ &= Q_0\left[1 - \left(\frac{B+C}{A+B+C}\right)\right] \\ &= \frac{Q_0 A}{(A+B+C)}. \end{aligned} \tag{3.21}$$

The "specific" irreversibility $\left(\frac{I_{HL}^{Q}}{M_H}\right)$ is then

$$\frac{I_{HL}^{Q}}{M_H} = \frac{A Q_0}{Q_{HL}}, \tag{3.22}$$

which is proportional to the area A lying between the gas and steam path lines, for given Q_{HL} and Q_0.

An alternative presentation (Fig. 3.5a) simply involves replacing the abscissae s_H and μs_L by a single horizontal scale representing the heat transferred, since $dQ_{HL} = M_H T_{HR} ds_H = M_L T_L ds_L$. It is not possible to interpret the lost work due to irreversibility on this diagram. However, if the ordinate is changed from T to (T_0/T) (Fig. 3.5b) then the shaded area between the two lines on the diagram is indeed $I_{HL}^{Q} = T_0 \int \left(\frac{1}{T_L} - \frac{1}{T_{HR}}\right) dQ_{HL}$, which is equation (3.15).

In subsequent (T, s) diagrams for combined plants we shall adopt the first practice, showing the entropy scale s_H truly, and scaling that for the lower plant so that the width of the entropy change in the heat transfer is the same for the two plants.

3.2.4.2 SEIPPEL AND BEREUTER'S INTERPRETATION OF THE IRREVERSIBILITY I_{HL}^{Q}

Seippel and Bereuter [1] have drawn particular attention to this limitation of combined cyclic power plants presenting the point in a somewhat different way.

They first give the maximum possible thermal efficiency of the lower cycle $\tilde{L}$ receiving heat at the (variable) temperature T_{HR} and rejecting heat at $T_A = T_0$,

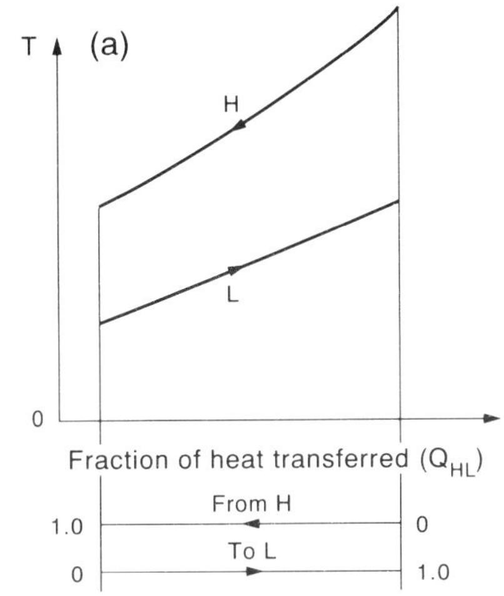

FIGURE 3.5a Temperature-heat transfer diagram

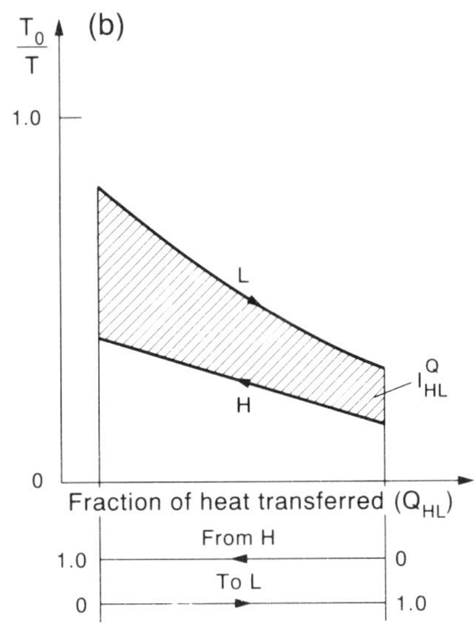

FIGURE 3.5b Graphical representation of irreversibility, I^Q_{HL}

$$\tilde{\eta}_L = \frac{\tilde{W}_L}{Q_{HL}} = \frac{1}{Q_{HL}} \int \left(1 - \frac{T_0}{T_{HR}}\right) dQ_{HL} = \frac{E^Q_{HR}}{Q_{HL}}. \tag{3.23}$$

The best that could be achieved by the lower cyclic plant (L) receiving heat Q_{HL} but at temperature $T_L (< T_{HR})$ would be

$$\eta_{L\,max} = \frac{W_{L\,max}}{Q_{HL}} = \frac{1}{Q_{HL}} \int \left(1 - \frac{T_0}{T_L}\right) dQ_{HL} = \frac{E^Q_L}{Q_{HL}}. \tag{3.24}$$

If the *actual* thermal efficiency of the lower plant receiving heat at temperature T_L is η_L, then

$$\eta_L = \frac{W_L}{Q_{HL}} = \left(\frac{W_L}{W_{L\,max}}\right)\left(\frac{W_{L\,max}}{Q_{HL}}\right) = (\eta_{td})\eta_{L\,max}. \tag{3.25}$$

Seippel and Bereuter call

$$(\eta_{td}) = \frac{W_L}{W_{L\,max}} = \frac{W_L}{E^Q_L}, \tag{3.26a}$$

the "thermodynamic" efficiency of the lower plant; this accounts for the internal irreversibility I^{CR}_L, i.e.

$$1 - (\eta_{td}) = \frac{W_{L\,max} - W_L}{W_{L\,max}} = \left(\frac{I^{CR}_L}{E^Q_L}\right). \tag{3.26b}$$

Further, equation (3.24) for $\eta_{L\,max}$ may be written as

$$\eta_{\mathrm{L\,max}} = \frac{W_{\mathrm{L\,max}}}{Q_{\mathrm{HL}}} = \left(\frac{W_{\mathrm{L\,max}}}{\tilde{W}_{\mathrm{L}}}\right)\left(\frac{\tilde{W}_{\mathrm{L}}}{Q_{\mathrm{HL}}}\right) = \left(\frac{E_{\mathrm{L}}^{\mathrm{Q}}}{E_{\mathrm{HR}}^{\mathrm{Q}}}\right)\left(\frac{E_{\mathrm{HR}}^{\mathrm{Q}}}{Q_{\mathrm{HL}}}\right) = (\eta_{\mathrm{ht}})\tilde{\eta}_{\mathrm{L}}, \tag{3.27}$$

where

$$(\eta_{\mathrm{ht}}) = W_{\mathrm{L\,max}}/\tilde{W}_{\mathrm{L}} = E_{\mathrm{L}}^{\mathrm{Q}}/E_{\mathrm{HR}}^{\mathrm{Q}} = 1 - (I_{\mathrm{HL}}^{\mathrm{Q}}/E_{\mathrm{HR}}^{\mathrm{Q}}) \tag{3.28}$$

is a "heat transfer" efficiency.

Finally, we obtain from equations (3.25), (3.26) and (3.27)

$$\eta_{\mathrm{L}} = \left(\frac{W_{\mathrm{L}}}{E_{\mathrm{L}}^{\mathrm{Q}}}\right)\left(\frac{E_{\mathrm{L}}^{\mathrm{Q}}}{E_{\mathrm{HR}}^{\mathrm{Q}}}\right)\left(\frac{E_{\mathrm{HR}}^{\mathrm{Q}}}{Q_{\mathrm{HL}}}\right)$$

$$= (\eta_{\mathrm{td}})(\eta_{\mathrm{ht}})\tilde{\eta}_{\mathrm{L}} \tag{3.29a}$$

$$= \left(1 - \frac{I_{\mathrm{L}}^{\mathrm{CR}}}{E_{\mathrm{L}}^{\mathrm{Q}}}\right)\left(1 - \frac{I_{\mathrm{HL}}^{\mathrm{Q}}}{E_{\mathrm{HR}}^{\mathrm{Q}}}\right)\frac{E_{\mathrm{HR}}^{\mathrm{Q}}}{Q_{\mathrm{HL}}}. \tag{3.29b}$$

3.2.4.3 SUMMARY

To summarise this discussion, when there is a temperature drop involved in the heat transfer between the two cyclic plants in series:

(i) The expression for thermal efficiency of the combined plant may still be written in terms of the thermal efficiencies of the two cycles, but the efficiency of the lower cycle is reduced (say to η_{L} from $\tilde{\eta}_{\mathrm{L}}$), i.e.

$$\eta_{\mathrm{CPS}} = \eta_{\mathrm{H}} + \eta_{\mathrm{L}} - \eta_{\mathrm{H}}\eta_{\mathrm{L}}. \tag{3.30}$$

(ii) However, the thermal efficiency of the combined plant is reduced (in comparison with that of the combined plant with no temperature drop) by

$$\begin{aligned}\Delta\eta &= \tilde{\eta}_{\mathrm{CPS}} - \eta_{\mathrm{CPS}}\\ &= \Delta W/Q_{\mathrm{B}}\\ &= I^{\mathrm{Q}}_{\mathrm{HL}}/Q_{\mathrm{B}}\\ &= \frac{T_0}{Q_{\mathrm{B}}}\int\left(\frac{T_{\mathrm{HR}}-T_{\mathrm{L}}}{T_{\mathrm{HR}}T_{\mathrm{L}}}\right)dQ_{\mathrm{HL}}\end{aligned} \tag{3.31}$$

(assuming the internal irreversibilities in the lower plant are unchanged).

(iii) In the terms of Seippel and Bereuter the efficiency of the combined series plant may be written

$$\begin{aligned}\eta_{\mathrm{CPS}} &= \frac{W_{\mathrm{H}}+W_{\mathrm{L}}}{Q_{\mathrm{B}}} = \frac{W_{\mathrm{H}}}{Q_{\mathrm{B}}} + \left(\frac{W_{\mathrm{L}}}{Q_{\mathrm{HL}}}\right)\left(\frac{Q_{\mathrm{HL}}}{Q_{\mathrm{B}}}\right)\\ &= \eta_{\mathrm{H}} + (1-\eta_{\mathrm{H}})\left(\frac{W_{\mathrm{L}}}{Q_{\mathrm{HL}}}\right)\\ &= \eta_{\mathrm{H}} + (1-\eta_{\mathrm{H}})\eta_{\mathrm{L}}\\ &= \eta_{\mathrm{H}} + (1-\eta_{\mathrm{H}})(\eta_{\mathrm{td}})(\eta_{\mathrm{ht}})\tilde{\eta}_{\mathrm{L}},\end{aligned} \tag{3.32}$$

where

$$(\eta_{\mathrm{td}}) = \frac{W_{\mathrm{L}}}{(W_{\mathrm{L}})_{\mathrm{max}}} = \frac{W_{\mathrm{L}}}{E^{\mathrm{Q}}_{\mathrm{L}}} = 1 - \left(\frac{I^{\mathrm{CR}}_{\mathrm{L}}}{E^{\mathrm{Q}}_{\mathrm{L}}}\right)$$

$$(\eta_{\mathrm{ht}}) = \frac{(W_{\mathrm{L}})_{\mathrm{max}}}{\tilde{W}_{\mathrm{L}}} = \frac{E^{\mathrm{Q}}_{\mathrm{L}}}{E^{\mathrm{Q}}_{\mathrm{HR}}} = 1 - \left(\frac{I^{\mathrm{Q}}_{\mathrm{HL}}}{E^{\mathrm{Q}}_{\mathrm{HR}}}\right)$$

$$\tilde{\eta}_{\mathrm{L}} = \frac{\tilde{W}_{\mathrm{L}}}{Q_{\mathrm{HL}}} = \frac{E^{\mathrm{Q}}_{\mathrm{HR}}}{Q_{\mathrm{HL}}} = \frac{1}{Q_{\mathrm{HL}}}\int\left(1-\frac{T_0}{T_{\mathrm{HR}}}\right)dQ_{\mathrm{HL}}.$$

3.2.5 Heat "Loss" Between Plants in Series

We have considered the effect of a temperature drop between two plants operating in series, but without a heat "loss" ($Q_{\mathrm{HR}} = Q_{\mathrm{L}} = Q_{\mathrm{HL}}$). We now consider the case in which there is unused heat Q_{UN} (or a heat "loss") between the two plants as in Fig. 3.6, $Q_{\mathrm{HR}} = Q_{\mathrm{L}} + Q_{\mathrm{UN}}$.

The overall thermal efficiency of the combined plant is by definition

$$\eta_{\mathrm{CP}} = \frac{W_{\mathrm{H}}+W_{\mathrm{L}}}{Q_{\mathrm{B}}},$$

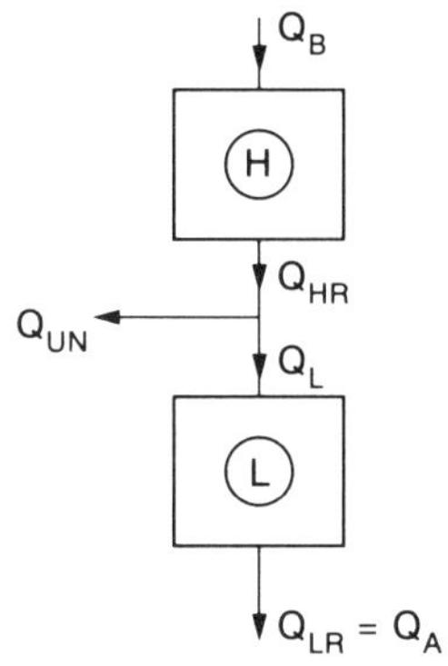

FIGURE 3.6 Heat loss between series plants

and the efficiencies of the higher and lower plants are still

$$\eta_H = \frac{W_H}{Q_B}, \quad \eta_L = \frac{W_L}{Q_L}.$$

However, the heat supplied to the lower cycle is now

$$Q_L = Q_{HR} - Q_{UN} = Q_B(1 - \eta_H) - Q_{UN},$$

so that

$$\eta_{CP} = \frac{\eta_H Q_B + \eta_L [Q_B(1 - \eta_H) - Q_{UN}]}{Q_B}$$

$$= \eta_H + \eta_L - \eta_H \eta_L - \nu_{UN} \eta_L, \qquad (3.33a)$$

where $\nu_{UN} = Q_{UN}/Q_B$. There is thus a loss in efficiency of $\nu_{UN}\eta_L$ in comparison with the "series" cycle with no heat loss between plants H and L.

Haywood [2] has written this result in another form, defining

$$(\eta_B) \equiv \frac{Q_L}{Q_{HR}} = 1 - \frac{Q_{UN}}{Q_{HR}} = 1 - \left[\frac{\nu_{UN}}{(1 - \eta_H)}\right] \qquad (3.34)$$

as the fraction of the total heat rejected by the higher cycle which is supplied to the lower cycle, a form of "boiler efficiency" for the heat transfer process. The combined plant efficiency may then be written

$$\eta_{CP}=\eta_H+\frac{\eta_L Q_L}{Q_B}$$
$$=\eta_H+\eta_L(\eta_B)\frac{Q_{HR}}{Q_B}$$
$$=\eta_H+(\eta_B)\eta_L-(\eta_B)\eta_H\eta_L, \qquad (3.33b)$$

or

$$\eta_{CP}=\eta_H+(\eta_O)_L-\eta_H(\eta_O)_L, \qquad (3.33c)$$

where $(\eta_O)_L$ is the overall efficiency for the lower cycle, the product of thermal efficiency and "boiler efficiency".

The effect of a finite temperature drop between the two series cyclic plants was to modify the lower plant efficiency (from $\tilde{\eta}_L$ to η_L) as described in Section 3.2 and hence the terms in the "series" efficiency expression as a result. The additional effect of a heat "loss" (i.e. not using all the heat rejection from the higher plant in the lower) is to modify the form of the expression for series efficiency to

$$\eta_{CP}=\eta_H+(\eta_O)_L-\eta_H(\eta_O)_L, \qquad (3.33c)$$

or

$$\eta_{CP}=\eta_H+\eta_L-\eta_H\eta_L-\nu_{UN}\eta_L, \qquad (3.33d)$$

if the lower efficiency (η_L) remains unchanged.

3.3 Parallel Plants (Joint Heating of Two Closed Cycles)

So far we have considered two cycles in series, with "external" heat supplied to the higher cycle only; heat is transferred internally to the lower cycle from the higher cycle, with a possible "loss" of heat between the two. There is a clear gain in efficiency of the combined cycle in comparison with either of the individual cycle efficiencies, although that gain decreases with the intermediate heat loss.

Here we examine supplying the "external" heat to two cyclic power plants in parallel. Figure 3.7 shows such a scheme diagrammatically, with heat supplies Q_M and Q_N to the two plants M and N, of efficiency η_M and η_N respectively and delivering work $W_M=\eta_M Q_M$, $W_N=\eta_N Q_N$. We suppose initially that $\eta_M>\eta_N$.

The thermal efficiency of the combined parallel plant (CPP) is

$$\eta_{CPP} = \frac{W_M + W_N}{Q_M + Q_N}$$
$$= \frac{\eta_M Q_M + \eta_N Q_N}{Q_M + Q_N},$$

i.e.

$$\eta_{CPP} = \eta_M - (\eta_M - \eta_N)\nu_N, \tag{3.35a}$$

where

$$\nu_N = Q_N/(Q_M + Q_N);$$

or

$$\eta_{CPP} = \eta_N + (\eta_M - \eta_N)\nu_M, \tag{3.35b}$$

where

$$\nu_M = Q_M/(Q_M + Q_N).$$

The combined plant efficiency thus lies between η_M and η_N. Further interpretation of these equations depends on the comparison that is made. There is no advantage to the parallel system in comparison with the higher efficiency η_M of plant M; on the other hand, in comparison with plant N, the combined plant shows an increase in efficiency.

Clearly if cyclic plant M were able to absorb an increased supply of heat $(Q_M + Q_N)$, then it would be advantageous to use that plant alone. However, we must conclude that heating of the two cycles in parallel gives no thermodynamic advantage, unless there is some further coupling between the cycles.

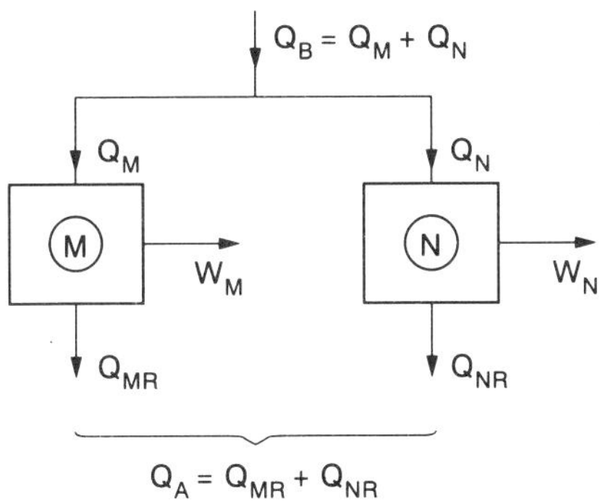

FIGURE 3.7 Joint heating of parallel plants

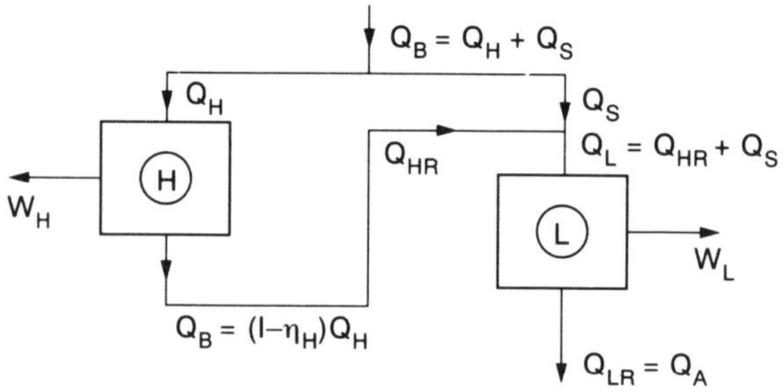

FIGURE 3.8 Series/parallel plants (supplementary heating, no intermediate heat loss)

3.4 Series/Parallel Plants (Two Closed Cycles)

3.4.1 Intermediate Heat Transfer and Supplementary Heat Supply Between Plants

Figure 3.8 shows coupling between two cyclic plants with rejected heat from the higher plant $Q_{HR} = Q_H(1 - \eta_H)$ being transferred to the lower plant (L). Supplementary heat Q_S is also supplied to the latter, so the total heat supply to the combined plant $Q_B = Q_H + Q_S$ may be regarded as joint heating, or as primary heating Q_H and supplementary heating Q_S. We define $\nu_S = Q_S/Q_B$, so that $(1 - \nu_S) = Q_H/Q_B$.

The efficiency of the combined "series/parallel" plant (CPSP) is now

$$\eta_{CPSP} = \frac{W_H + W_L}{Q_B} = \frac{\eta_H Q_H + \eta_L Q_L}{Q_H + Q_S}, \tag{3.36}$$

where

$$Q_L = Q_H(1 - \eta_H) + Q_S,$$

so that

$$\begin{aligned} \eta_{CPSP} &= \eta_H(1 - \nu_S) + \eta_L(1 - \eta_H)(1 - \nu_S) + \eta_L \nu_S \\ &= \eta_H + \eta_L - \eta_H \eta_L - \eta_H(1 - \eta_L)\nu_S \\ &= \eta_{CPS} - \eta_H(1 - \eta_L)\nu_S \end{aligned} \tag{3.37}$$

This gives an increase of $\eta_L(1 - \eta_H)(1 - \nu_S)$ over the efficiency of the parallel heating plant, equations (3.35), but in comparison with the series plant the effect of an element of parallel heating (through Q_S) is a reduction of efficiency of $\eta_H(1 - \eta_L)\nu_S$.

Again equation (3.37) has ben derived in another form by Haywood [2]. He defines $q = 1 - \nu_S = Q_H/Q_B = Q_H/(Q_H + Q_S)$. Substituting for ν_S in equation (3.37) yields

$$\eta_{CPSP} = \eta_H + \eta_L - \eta_H\eta_L - \eta_H(1-\eta_L)(1-q)$$
$$= \eta_L + q\eta_H - q\eta_H\eta_L, \quad (3.38a)$$

or

$$1 - \eta_{CPSP} = (1 - q\eta_H)(1-\eta_L). \quad (3.38b)$$

If q is unity ($v_S = 0$), this equation reduces to that for the ideal series plant.

3.4.2 Intermediate Heat Transfer, Supplementary Heat Supply and Heat Loss Between Plants

We now consider the most general case of intermediate heat transfer, supplementary heat supply and heat loss between two closed cycles (H and L, Fig. 3.9). A simple analysis of combined cycle performance under these conditions has been given by Buxmann [3].

The overall thermal efficiency of the combined plant is

$$\eta_{CP} = \frac{W}{Q_B} = \frac{W_H + W_L}{Q_H + Q_S}. \quad (3.39)$$

Q_S may be regarded as an additional or supplementary heat supply between the two plants, or the total heat Q_B may be viewed as joint heating, so there is an element of parallel heating involved.

The efficiency of the upper cycle is still

$$\eta_H = \frac{W_H}{Q_H}$$

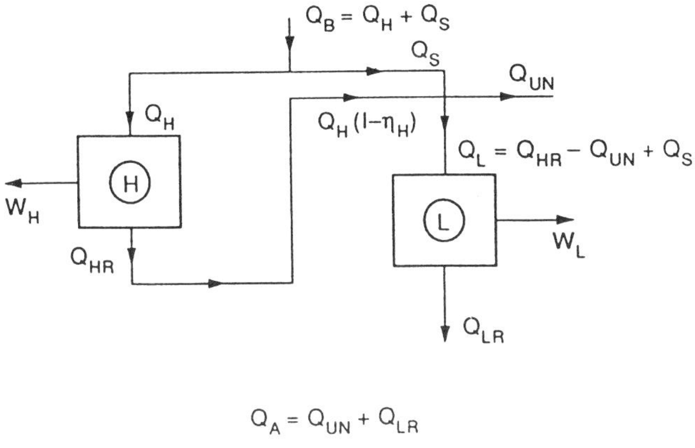

FIGURE 3.9 Series/parallel plants (supplementary heating, but with intermediate heat loss)

and the efficiency of the lower cycle

$$\eta_L = \frac{W_L}{Q_L}.$$

But Q_L, the heat supplied to the lower cycle, is now made up of the heat transferred from the upper cycle $Q_H(1-\eta_H)$, plus the supplementary heat supplied Q_S less the unused heat Q_{UN}, i.e.

$$Q_L = Q_H(1-\eta_H) + Q_S - Q_{UN}. \tag{3.40}$$

Equations (3.39) and (3.40) yield

$$\eta_{CP} = \frac{\eta_H Q_H}{Q_H + Q_S} + \frac{\eta_L Q_L}{Q_H + Q_S}$$

$$= \frac{\eta_H Q_H}{Q_H + Q_S} + \frac{\eta_L [Q_H(1-\eta_H) + Q_S - Q_{UN}]}{Q_H + Q_S}$$

However, if we again non-dimensionalise Q_S and Q_{UN} by $Q_B = Q_H + Q_S$, i.e. as

$$\nu_S \equiv \frac{Q_S}{Q_H + Q_S}, \quad \nu_{UN} \equiv \frac{Q_{UN}}{Q_H + Q_S},$$

then it follows that

$$\eta_{CP} = \eta_H(1-\nu_S) + \eta_L[(1-\eta_H)(1-\nu_S) + \nu_S - \nu_{UN}]$$

$$= \eta_H + \eta_L - \eta_H \eta_L - \nu_{UN}\eta_L - \nu_S \eta_H(1-\eta_L)$$

$$= \eta_{CPS} - \nu_{UN}\eta_L - \nu_S \eta_H(1-\eta_L). \tag{3.41}$$

As for the series plant without supplementary heating, a form of boiler efficiency may be defined as

$$(\eta_B) \equiv \frac{Q_L}{(Q_L + Q_{UN})} = \frac{Q_H(1-\eta_H) + Q_S - Q_{UN}}{Q_H(1-\eta_H) + Q_S}, \tag{3.42a}$$

or

$$(\eta_B) = 1 - \frac{\nu_{UN}}{(1-\eta_H)(1-\nu_S) + \nu_S}. \tag{3.42b}$$

Equation (3.41) may then be rewritten as

$$\eta_{CP} = \eta_H(1-\nu_S) + \eta_L(\eta_B)[\nu_S + (1-\nu_S)(1-\eta_H)] \tag{3.43a}$$

$$= \eta_H(1-\nu_S) + (\eta_O)_L[\nu_S + (1-\nu_S)(1-\eta_H)], \tag{3.43b}$$

where $(\eta_O)_L = \eta_L(\eta_B)$ is an overall efficiency for the lower cycle. A shorter form for η_{CP}, introduced by Haywood [2], is obtained by writing $x = W_H/Q_B$, so that $\nu_S = 1-(x/\eta_H)$. It follows that

$$\eta_{CP} = x + (1-\nu_{UN})\eta_L - x\eta_L, \tag{3.43c}$$

and we shall refer to this form later.

From equation (3.41) it can be seen that this more practical combined power plant does not attain the "series" efficiency, $\eta_{CPS} = \eta_H + \eta_L - \eta_H\eta_L$. The deficit in efficiency $\nu_{UN}\eta_L$ due to heat "loss" has already been demonstrated in the earlier analysis, and the term $\nu_S\eta_H(1-\eta_L)$ illustrates the fall in efficiency due to supplementary heating between the plants. This general result thus embraces the cases considered in Sections 3.2.5 and 3.4.1.

An alternative presentation and interpretation is given by Haywood [4] who defines combined cycle efficiency in terms of different ratios of heat quantities,

$$\nu_H \equiv \frac{Q_H}{Q_B} = \frac{Q_H}{(Q_H+Q_S)}, \quad \nu_L \equiv \frac{Q_L}{Q_B} = \frac{Q_L}{(Q_H+Q_S)},$$

so that

$$\eta_{CP} = \frac{W_H + W_L}{Q_H + Q_S}$$
$$= \nu_H\eta_H + \nu_L\eta_L. \tag{3.45}$$

Haywood suggests that this is a more fundamental statement than that of Buxmann. He argues that for series cyclic plants, if the performance parameters of the higher plant are fixed and η_H thus specified, the only way that η_H can then influence η_{CP} is through the ratio of mass flows in the two cycles which then changes the conditions in the lower cycle.

3.5 Combined Plants Consisting of an Open Circuit (Higher) Plant and a Closed Cycle (Lower) Plant

So far we have considered binary (double closed cycle) combined plants. We now study the thermodynamics of an open circuit/closed cycle plant.

The most successful combined power plant has proved to be that using exhaust heating, with an open circuit gas turbine rejecting heat from the turbine exhaust to a closed circuit steam turbine plant in a Heat Recovery Steam Generator (HRSG) (Section 2.4.1.1). Fuel may also be burnt in the exhaust, where there is excess oxygen available for further combustion (Sections 2.4.1.2 and 2.4.1.3). We first consider how the simple analysis of Section 3.2 for the combined doubly cyclic series plant is modified for the open circuit/closed cycle plant. Subsequently we study the cases of supplementary firing of the exhaust, and of pressurised combustion.

3.5.1 Open Circuit/Closed Cycle Combined Plant (No Supplementary Firing)

Figure 3.10a shows the simplest plant with an unfired HRSG. The work output from the gas-turbine plant is

$$W_H = (\eta_O)_H F = (\eta_O)_H[(H_R)_0 - (H_P)_0], \tag{3.46}$$

where $(\eta_O)_H$ is the arbitrary overall efficiency and F is the energy supplied in the fuel, $F = M_F(CV)_0 = M_F[(H_R)_0 - (H_P)_0]$. $(CV)_0$ is the enthalpy of combustion of the fuel, of mass flow M_F. We shall later take unit mass flow of air of enthalpy $(h_A)_0$ to the compressor so that M_F is the fuel-air ratio (f), and $(H_R)_0 = (h_A)_0 + f(CV)_0$, $(H_P)_0 = (1+f)(h_p)_0 = M_G(h_P)_0$, where $M_G = (1+f)$. The work output of the gas turbine W_H is the specific work output (per unit *air* flow), $W_H = w_H$.

The work output from the steam cycle is

$$W_L = \eta_L Q_L, \tag{3.47}$$

where η_L is the *thermal* efficiency of the lower (steam) cycle and Q_L is the heat transferred from the gas-turbine exhaust.

Thus the arbitrary *overall* efficiency of the whole plant is

$$(\eta_O)_{CP} = \frac{(W_H + W_L)}{F} = (\eta_O)_H + \frac{\eta_L Q_L}{F}. \tag{3.48}$$

But if combustion is adiabatic, then the steady flow energy equation for the open-circuit gas turbine (with exhaust of enthalpy $(H_P)_S$ leaving the HRSG and entering the exhaust stack, where the temperature T_S is greater than that of the atmosphere, T_0) is

$$(H_R)_0 = (H_P)_S + W_H + Q_L, \tag{3.49}$$

so that

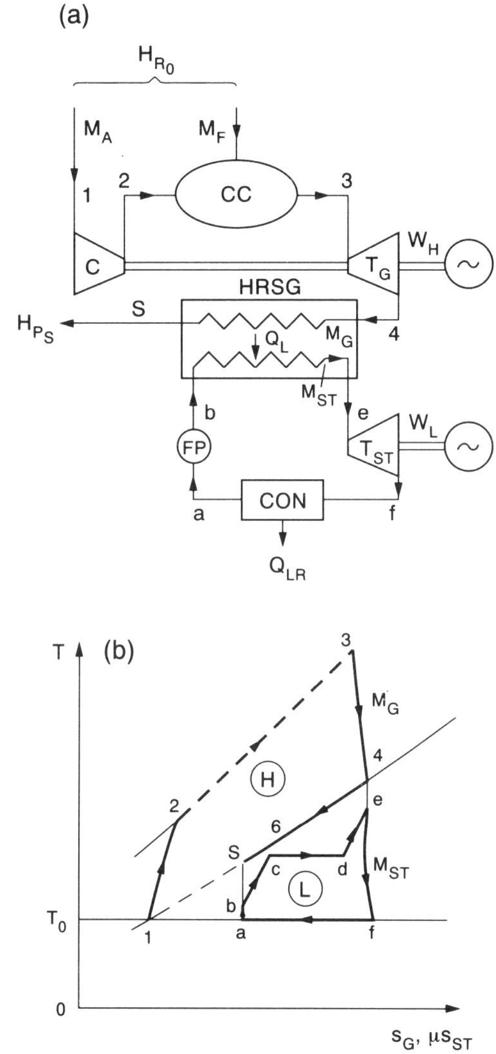

FIGURE 3.10 Open circuit/closed cycle combined plant (no supplementary firing)

$$
\begin{aligned}
Q_L &= (H_R)_0 - (H_P)_S - W_H \\
&= [(H_R)_0 - (H_P)_0] - [(H_P)_S - (H_P)_0] - W_H \\
&= F - [(H_P)_S - (H_P)_0] - W_H, \qquad (3.50)
\end{aligned}
$$

where $(H_P)_0$ is the enthalpy of products leaving the calorimeter in a "calorific value" experiment, after combustion of fuel at temperature T_0.

The arbitrary overall efficiency of the combined plant is therefore

$$(\eta_O)_{CP}=(\eta_O)_H+\eta_L\left[1-\frac{W_H}{F}-\frac{[(H_P)_S-(H_P)_O]}{F}\right]$$

$$=(\eta_O)_H+\eta_L-(\eta_O)_H\eta_L-\frac{\eta_L[(H_P)_S-(H_P)_O]}{F}, \tag{3.51a}$$

$$=(\eta_O)_H+(\eta_B)\eta_L-(\eta_B)(\eta_O)_H\eta_L, \tag{3.51b}$$

$$=(\eta_O)_H+(\eta_O)_L-(\eta_O)_H(\eta_O)_L, \tag{3.51c}$$

where $(\eta_B)=1-[(H_P)_S-(H_P)_O]/F[1-(\eta_O)_H]$, $(\eta_O)_L=(\eta)_B\eta_L$.

The expression for overall efficiency (3.51a) is similar to that for the combined doubly cyclic plant; the term $\eta_L[(H_P)_S-(H_P)_O]/F$ corresponds to the heat "unused" term of Section 3.2.5. The extent of this reduction in overall efficiency depends on how much the exhaust gases can be cooled and could theoretically be zero if they emerged from the HRSG at the (ambient) temperature of the reactants. In practice this is not possible, as the stack temperature cannot be allowed to drop below a level at which corrosion takes place on the tubes of the HRSG (when the feed water temperature is equal to the dew point temperature of the exhaust gases).

An interesting empirical approach to the analysis of the open gas turbine/steam turbine combined plant (exhaust heated, unfired HRSG) has been given by Timmermans [5] who places great emphasis on the importance of achieving a high specific work from the gas turbine set ($W_H=w_H$, for unit air mass flow). He draws particular attention to the role of the pinch point (see Section 2.4.1) in determination of the work output from the lower (steam) plant.

Timmermans introduces an *empirical* expression for the work output from the steam plant

$$W_L=K[(H_P)_4-(H_P)_6]$$

$$=KM_G(\bar{c}_p)_G(T_4-T_6), \tag{3.52}$$

where T_6 is the temperature at the pinch point in the exhaust heat exchanger, K is a constant and $(\bar{c}_p)_G$ is a mean specific heat. (The pinch point is shown on the temperature–entropy chart of Fig. 3.10b.) This may be compared with the work output expressed in terms of the more conventional thermal efficiency of the steam turbine plant used earlier in this section,

$$W_L=\eta_L[(H_P)_4-(H_P)_S]$$

$$=\eta_L M_G(\bar{c}_p)_G(T_4-T_S), \tag{3.47}$$

where $(H_P)_S$ is the enthalpy of the gas leaving the HRSG.

Dividing equation (3.52) by equation (3.47) shows that the ratio of K to η_L is simply

$$\frac{K}{\eta_L} \approx \frac{(T_4 - T_S)}{(T_4 - T_6)}. \tag{3.53}$$

Use of equation (3.47), involving the thermal efficiency, led to equation (3.51) which was readily compared to equations (3.33), for the doubly cyclic "series" plant. However, Timmermans argues that the steam plant efficiency (η_L) cannot be specified independently and that it depends critically on the temperature leaving the gas turbine (T_4) and the temperature of the pinch point (T_6) as indicated in equation (3.52), with K an empirical constant for a limited range of designs.

It then follows that

$$\begin{aligned}(\eta_O)_{CP} &= \frac{W_H + W_L}{F} \\ &= (\eta_O)_H + \frac{K M_G (\bar{c}_p)_G (T_4 - T_6)}{F} \\ &= (\eta_O)_H + \frac{K M_G (\bar{c}_p)_G [(T_4 - T_0) - (T_6 - T_0)](\eta_O)_H}{W_H} \\ &= (\eta_O)_H + K[1 - (\eta_O)_H] - \frac{K (\bar{c}_p)_G (T_6 - T_0)(\eta_O)_H (1 + f)}{w_H}\end{aligned} \tag{3.54}$$

[The same result could be obtained from substituting K for η_L (from equation (3.53)) in equation (3.51).]

This equation emphasises that with K assumed constant and $(\eta_O)_H$ defined, the combined cycle efficiency is strongly dependent upon the pinch point temperature T_6 and the gas turbine specific work w_H, rather than the steam turbine plant efficiency.

3.5.2 Open Circuit/Closed Cycle Combined Plant (With Supplementary Firing)

We consider next a combined plant consisting of an open circuit (gas turbine)/closed cycle (steam) plant with supplementary firing of the gas turbine exhaust (either in a fired HRSG or a fully fired steam boiler (see Fig. 3.11a). For a mass flow of air (M_A) to the compressor of the gas turbine plant, a mass flow M_F of fuel [of specific enthalpy $(h_F)_0$] is supplied to the *two* combustion chambers. This combined open circuit/closed cycle combined plant may be compared to the doubly cyclic plant of Section 3.42, with joint heating (primary plus secondary) intermediate heat transfer and heat "unused".

3.5.2.1 HAYWOOD'S ANALYSIS

A simple analysis of this plant has been given by Haywood [2] who defines a quantity x as the ratio of the work output of the gas turbine W_H

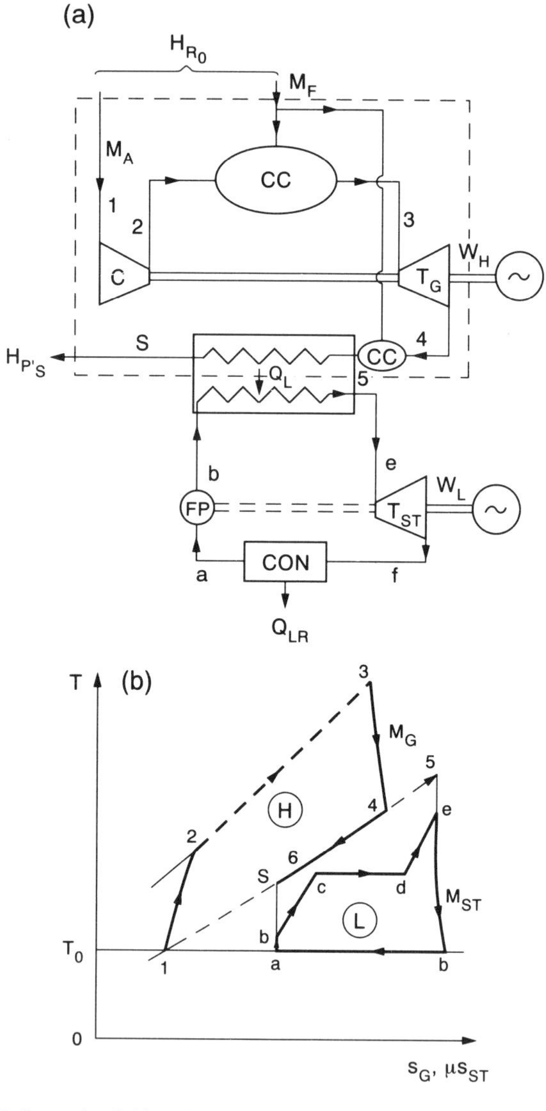

FIGURE 3.11 Open circuit/closed cycle combined plant (with supplementary firing)

to the heating value of the fuel $F=M_F(CV)_0=(H_R)_0-(H_P)_0$. Then applying the steady flow energy equation to the dotted control surface of Fig. 3.11a it follows that

$$(H_R)_0=(H_{P'})_S+W_H+Q_L \tag{3.55}$$

where

$$(H_R)_0=M_A(h_A)_0+M_F(h_F)_0,$$
$$(H_{P'})_S=M_G(h_{P'})_S=M_A(1+f)(h_{P'})_S,$$

and Q_L is the total heat supplied to the closed steam cycle of thermal efficiency η_L, $f=M_F/M_A$ and the prime (′) refers to products after supplementary firing.

If, after Haywood, we define an effectiveness of energy conversion for the plant (*not* a boiler efficiency)

$$\varepsilon_B\equiv\frac{(H_R)_0-(H_{P'})_S}{(H_R)_0-(H_P)_0}=\frac{(H_R)_0-(H_{P'})_S}{M_F(CV)_0}, \tag{3.56}$$

then

$$\begin{aligned}\varepsilon_B M_F(CV)_0&=(H_R)_0-(H_{P'})_S\\&=W_H+Q_L\\&=xM_F(CV)_0+Q_L.\end{aligned} \tag{3.57}$$

The work output from the lower steam cycle is $\eta_L Q_L$ so that

$$\varepsilon_B M_F(CV)_0=xM_F(CV)_0+\frac{W_L}{\eta_L}, \tag{3.58a}$$

and

$$W_L=\eta_L M_F(CV)_0(\varepsilon_B-x). \tag{3.58b}$$

The power ratio is

$$\frac{W_H}{W_L}=\frac{x}{\eta_L(\varepsilon_B-x)}. \tag{3.59}$$

The overall efficiency of the combined plant with supplementary heating is then

$$(\eta_O)_{CP} = \frac{W_H + W_L}{M_F(CV)_0}$$
$$= x + \eta_L(\varepsilon_B - x)$$
$$= x + \eta_L \varepsilon_B - \eta_L x. \tag{3.60}$$

This equation may be compared directly with the alternative form of Buxmann's expression for the efficiency of the doubly cyclic plant, equation (3.43c), noting that $(1 - \nu_{UN})$ used there is analogous to the effectiveness of energy conversion (ε_B) used here. It is also tempting to compare this expression with that for the "series" doubly cyclic plant, $\eta_{CPS} = \eta_H + \eta_L - \eta_H \eta_L$, but x is not an efficiency comparable to η_H, since $x = W_H/M_F(CV)_0$, where M_F is fuel supplied to the whole plant, not just the higher plant. However, if we hypothesise that qM_F goes to the gas turbine combustion chamber and define $(\eta_O)_H = W_H/[qM_F(CV)_0]$, then $x = (\eta_O)_H q$ and equation (3.60) becomes

$$(\eta_O)_{CP} = \varepsilon_B \eta_L + q(\eta_O)_H - q(\eta_O)_H \eta_L, \tag{3.61}$$

which may be compared with equation (3.38a) for the doubly cyclic series/parallel plant,

$$\eta_{CPSP} = \eta_L + q\eta_H - q\eta_H \eta_L, \tag{3.38a}$$

where q was that fraction of the total heat supplied which goes to the higher plant and there was no heat loss ($\nu_{UN} = 0$ and $\varepsilon_B = 1 - \nu_{UN} = 1$).

A weakness in this analysis remains in the definition of ε_B, which is an "effectiveness" of energy extraction for the plant, rather than a conventional boiler or combustion efficiency.

As we indicated in Chapter 2, the combined plant with a fully fired exhaust is often a conversion of a conventional steam plant, with the gas turbine as a "topping" plant in place of the original boiler. If we define $(\eta_O)_L = (\eta_B)\eta_L$ as the overall efficiency of the original steam plant (the thermal efficiency being assumed not to change in the conversion), then the difference in overall efficiency between the converted (combined) plant and the original is

$$(\eta_O)_{CP} - (\eta_O)_L = \eta_L[\varepsilon_B - (\eta_B)] + x(1 - \eta_L). \tag{3.62}$$

The efficiency is increased over the original steam plant if

$$x(1 - \eta_L) > \eta_L[(\eta_B) - \varepsilon_B]. \tag{3.63}$$

The power output is usually substantially increased, more fuel being supplied to the combined plant than the original steam plant.

3.5.2.2 THE ANALYSIS OF MANGAN AND PETTIT

In an early paper Mangan and Pettit [6] presented a more complex analysis for the overall efficiency of the combined plant with supplementary firing of the gas turbine exhaust before entry to the HRSG. They took account of the fuel flows to the main and supplementary combustion chambers $[(M_F)_H$ and $(M_F)_S]$.

The overall efficiency of the combined plant is

$$(\eta_O)_{CP} = \frac{W_H + W_L}{M_F(CV)_0} = \frac{W_H + W_L}{[(M_F)_H + (M_F)_S](CV)_0}$$

$$= \frac{(\eta_O)_H}{\left[1 + \dfrac{(M_F)_S}{(M_F)_H}\right]} + \frac{\eta_L Q_L}{(M_F)_H(CV)_0\left[1 + \dfrac{(M_F)_S}{(M_F)_H}\right]}. \tag{3.64}$$

If the temperature of air and fuel at entry to the gas turbine is equal to the ambient temperature ($T_1 = T_0$), then for the gas turbine alone

$$(M_F)_H(h_F)_0 + M_A(h_A)_0 = W_H + [M_A + (M_F)_H](h_P)_4$$
$$= W_H + (H_P)_4,$$

and for the calorific value of fuel

$$(M_F)_H(h_F)_0 + M_A(h_A)_0 = (M_F)_H(CV)_0 + [(M_F)_H](h_A)_0$$
$$= (M_F)_H(CV)_0 + (H_P)_0,$$

so that

$$(M_F)_H(CV)_0 - W_H = (H_P)_4 - (H_P)_0,$$
$$(M_F)_H(CV)_0[1 - (\eta_O)_H] = (H_P)_4 - (H_P)_0. \tag{3.65}$$

For the supplementary burner also receiving fuel at ambient temperature T_0

$$[M_A + (M_F)_H](h_P)_4 + (M_F)_S(h_F)_0 = [M_A + (M_F)_H + (M_F)_S](h_{P'})_5$$

or

$$(H_P)_4 + (M_F)_S(h_F)_0 = (H_{P'})_5,$$

which together with an equation for calorific value of fuel

$$[M_A + (M_F)_H](h_P)_0 + (M_F)_S(h_F)_0 = [M_A + (M_F)_H + (M_F)_S](h_{P'})_0 + (M_F)_S(CV)_0,$$

or

$$(H_P)_0 + (M_F)_S(h_F)_0 = (H_{P'})_0 + (M_F)_S(CV)_0,$$

yields

$$(H_{P'})_5-(H_{P'})_0=(H_P)_4-(H_P)_0+(M_F)_S(CV)_0, \tag{3.66}$$

when P indicates products after primary combustion and P′ indicates products after supplementary combustion.

From the energy balance for the HRSG

$$\begin{aligned} Q_L &= (H_{P'})_5-(H_{P'})_S \\ &= [(H_{P'})_5-(H_{P'})_0]-[(H_{P'})_S-(H_{P'})_0]; \end{aligned} \tag{3.67}$$

from equation (3.66), eliminating $[(H_{P'})_5-(H_{P'})_0]$,

$$Q_L=[(H_P)_4-(H_P)_0]+(M_F)_S(CV)_0-[(H_{P'})_S-(H_{P'})_0], \tag{3.68}$$

and from equation (3.65)

$$Q_L=(M_F)_H[1-(\eta_O)_H](CV)_0+(M_F)_S(CV)_0-[(H_{P'})_S-(H_{P'})_0]. \tag{3.69}$$

Substitution of this expression for Q_L into equation (3.64) yields

$$\begin{aligned} (\eta_O)_{CP} &= \frac{(\eta_O)_H}{\left[1+\dfrac{(M_F)_S}{(M_F)_H}\right]}+\frac{\eta_L}{\left[1+\dfrac{(M_F)_S}{(M_F)_H}\right]} \\ &\quad\times\left\{\frac{(M_F)_S}{(M_F)_H}+[1-(\eta_O)_H]-\frac{[(H_{P'})_S-(H_{P'})_0](M_F)_S}{(M_F)_H(CV)_0}\right\}, \end{aligned} \tag{3.70}$$

where $H_{P'}=[M_A+(M_F)_H+(M_F)_S]h_{P'}$.

If $(M_F)_S=0$, then this equation reduces to that for no supplementary heating, equation (3.51a).

Equation (3.70) may be written in terms of "heating" quantities

$$Q_H=(M_F)_H(CV)_0 \quad \text{and} \quad Q_S=(M_F)_S(CV)_0,$$

and a "heat loss"

$$Q_{UN}=[M_A+(M_F)_H+(M_F)_S][(h_{P'})_S-(h_{P'})_0].$$

Then with $v_S/(Q_S+Q_H)$, and $v_{UN}\equiv Q_{UN}/(Q_S+Q_H)$, it follows that

$$(M_F)_S/(M_F)_H=v_S/(1-v_S), \tag{3.71}$$

and

$$Q_{UN}/[(M_F)_H(CV)_0]=v_{UN}/(1-v_S), \tag{3.72}$$

so that equation (3.70) becomes

$$(\eta_O)_{CP}=(\eta_O)_H+\eta_L-(\eta_O)_H\eta_L$$
$$-\eta_L v_{UN}-(\eta_O)_H(1-\eta_L)v_S, \tag{3.73}$$

which is directly comparable with equation (3.41).

3.5.2.3 TIMMERMANN'S ANALYSIS

An alternative approach by Timmermans again emphasises the importance of the gas turbine specific work, for the plant with supplementary heating. Referring to Fig. 3.11 and equation (3.56) we may note that

$$1-\varepsilon_B=\frac{M_G\bar{c}_p(T_S-T_0)}{M_F(CV)_0}, \tag{3.74}$$

and that

$$\eta_L=\frac{W_L}{M_G\bar{c}_p(T_5-T_S)}. \tag{3.75}$$

Timmermans suggests η_L is more or less a constant ($\eta_L=K'$). Hence, with $x=W_H/M_F(CV)_0$ we may rewrite equation (3.60) as

$$(\eta_O)_{CP}=x-x\eta_L+\eta_L\varepsilon_B$$
$$=\frac{W_H}{M_F(CV)_0}-K'\left\{\frac{W_H}{M_F(CV)_0}-\left[1-\frac{M_G\bar{c}_p(T_S-T_0)}{M_F(CV)_0}\right]\right\} \tag{3.76}$$

But from Fig. (3.11a)

$$M_F(CV)_0=(H_R)_0-(H_P)_0=W_H+M_G\bar{c}_p(T_5-T_0), \tag{3.77}$$

so that

$$(\eta_O)_{CP}=\frac{W_H+M_G\bar{c}_pK'(T_5-T_S)}{W_H+M_G\bar{c}_p(T_5-T_0)}, \tag{3.78a}$$

where $\bar{c}_p$ has been assumed to be constant.

With $M_G=1+M_F\approx 1$, $W_H=w_H$,

$$(\eta_O)_{CP}=\frac{w_H+\bar{c}_pK'(T_5-T_S)}{w_H+\bar{c}_p(T_5-T_0)} \tag{3.78b}$$

Timmermans argues that K', T_5 and T_S are more or less constant, so that w_H is the main variable, $(\eta_O)_{CP}$ increasing markedly with w_H.

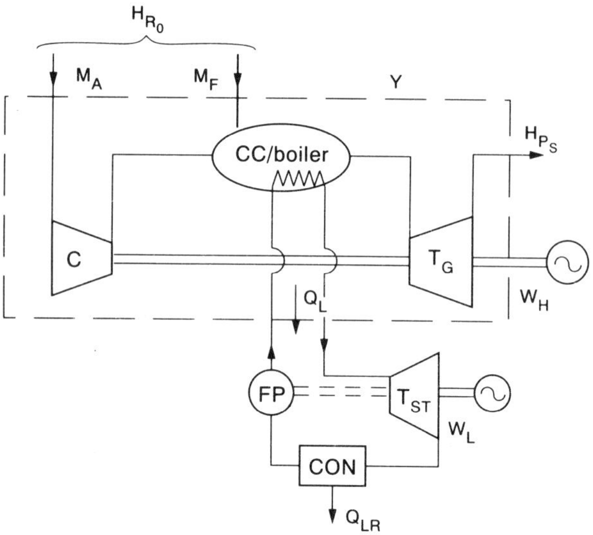

FIGURE 3.12 Open circuit/closed cycle combined plant with pressure charged combustion chamber/boiler

3.5.3 Open Circuit/Closed Cycle Combined Plant (Pressurised Combustion or Pressure Charged Boiler)

The approach developed in Section 3.5.2 may be applied to a combined gas turbine/steam turbine plant with pressurised combustion (Fig. 3.12). Now the fuel supply M_F (per unit air mass flow) is supplied to a single pressure charged boiler rather than to the gas turbine combustion chamber and the HRSG. The analysis developed by Haywood [2] may be applied with an effectiveness ε_B again defined for the gas turbine and pressurised boiler as

$$\varepsilon_B = \frac{(H_R)_0 - (H_P)_S}{(H_R)_0 - (H_P)_0} = \frac{(H_R)_0 - (H_P)_S}{M_F(CV)_0}. \tag{3.79}$$

For the dotted control surface Y we may write

$$Q_L + W_H = (H_R)_0 - (H_P)_S = \varepsilon_B[M_F(CV)_0]$$

and so

$$W_L = \eta_L Q_L = \eta_L[\varepsilon_B M_F(CV)_0 - W_H].$$

If $x = W_H / M_F(CV)_0$, it follows that the overall efficiency of the combined plant is

$$(\eta_O)_{CP} = \frac{W_H + W_L}{M_F(CV)_0}$$
$$= x + \eta_L(\varepsilon_B - x)$$
$$= x + \varepsilon_B \eta_L - x\eta_L. \qquad (3.80)$$

Although the expressions for the supplementary fired and pressure charged plants are notionally the same, the quantities x and η_L are very different for the two plants, as Haywood points out. He suggests, for example, that for the plant with a supplementary fired HRSG, a high steam plant thermal efficiency ($\eta_L = 0.44$ say) can be obtained with a correspondingly low value of x, say 0.05, giving $(\eta_O)_{CP} = 0.424$ for $\varepsilon_B = 0.9$. For a pressure charged plant Haywood suggests a lower value of η_L is more representative (say 0.26) together with a higher value of x, say 0.28. With the same value of ε_B assumed ($\varepsilon_B = 0.9$) this would give a comparable combined plant efficiency of $(\eta_O)_{CP} = 0.441$, but it cannot be assumed that ε_B would be the same in the two very different systems. The pressure charged combined plant has not yet proved very popular, but this may change if pressurised fluidised bed combustion of coal is further developed, as we shall discuss in later chapters.

3.6 The Seippel–Bereuter Analysis of Combined Gas Turbine/Steam Turbine Plants

An early but comprehensive analysis of combined gas turbine/steam turbine plants was given by Seippel and Bereuter [1]. They described a whole range of combined plants, in comparison with a basic steam plant. The analysis is of historical importance, since it gave considerable impetus to the development of combined plant; it is partially reproduced here, although its results are not now often used.

The various combinations proposed by Seippel and Bereuter are illustrated in Fig. 3.13, which is their original figure. However, they concentrated on the performance of

(i) the basic steam plant (A, or ST in what follows here), of overall efficiency $(\eta_O)_{ST}$;
(ii) a combined plant (CP1) in which the unfired exhaust from the gas turbine supplies heat to the steam turbine, of overall efficiency $(\eta_O)_{CP1}$;
(iii) a combined plant (G or CP2) in which the gas turbine exhaust is

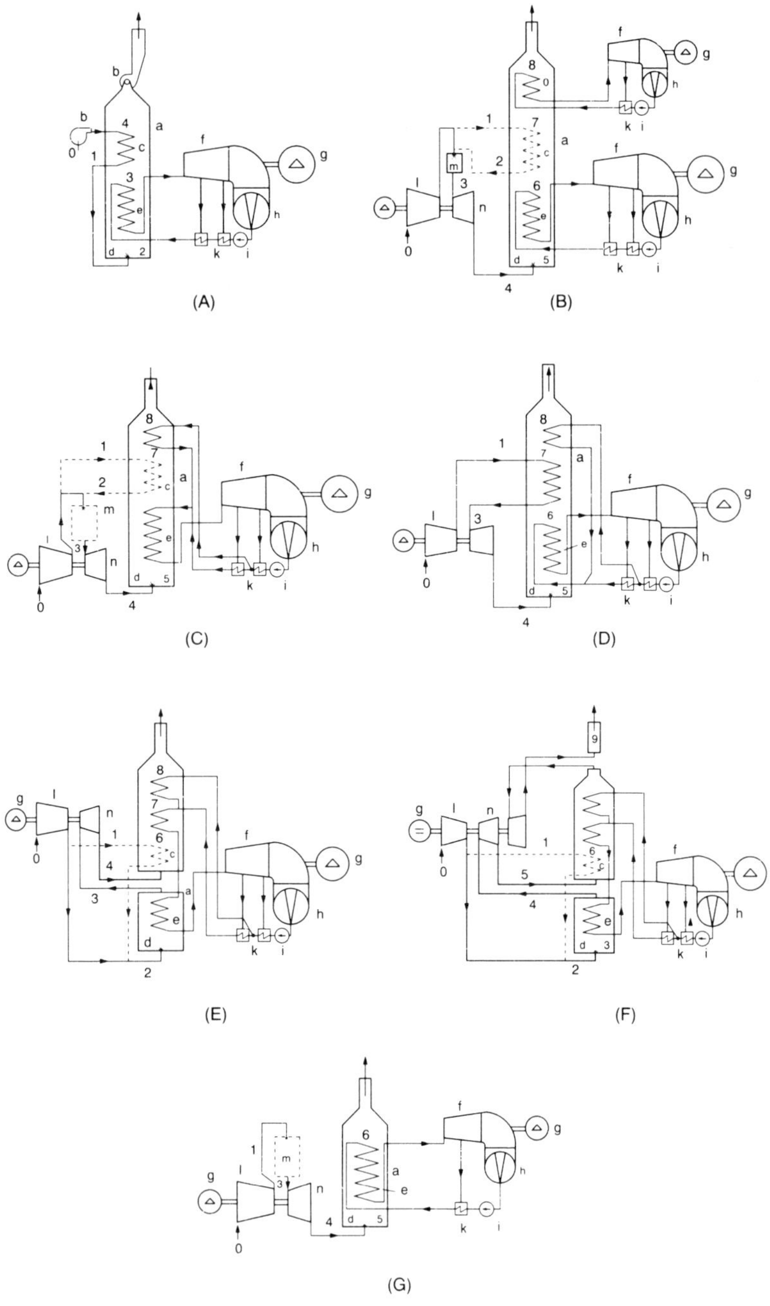
(A)
(B)
(C)
(D)
(E)
(F)
(G)

fired before supplying heat to the steam turbine, with an overall efficiency $(\eta_O)_{CP2}$.

Their purpose was to compare the efficiency of these three plants in order to determine when one plant should be chosen rather than the others, mainly on the basis of superior overall efficiency. In doing so they made some major assumptions, in particular that the *thermal* efficiency of the steam turbine η_{ST} used in each plant was the same. (This was a somewhat doubtful assumption since the maximum temperature of the gas supply will be lower in the case CP1.)

In their analysis, Seippel and Bereuter use an interesting "unfolded" temperature-enthalpy plot. Figure 3.14 shows such a plot for the air supply and combustion gases in the boiler of the basic steam plant (ST). The enthalpy of the air increases across the boiler fans due to the work input (W_{AUX}) and the air is preheated by an amount Q_{PH}. The dotted line shows the change from air to products in the combustion process as an equivalent heating process, which we discuss further below. For the combustion gases, the plot is unfolded, as the temperature and enthalpy drop. Heat $(Q_L)_{ST}$ is transferred to the lower steam plant of thermal efficiency η_{ST}, heat Q_{PH} is transferred to the inlet air flow and there is a stack "loss" $(Q_{UN})_{ST}$ in the discharged gases.

FIGURE 3.13 (*opposite*). *Circuit diagrams of combined steam and gas turbine installations* (*Seippel and Bereuter* [1])

a = Boiler
b = Boiler fans
c = Air heater
d = Combustion chamber with burner
e = Steam part of boiler
f = Steam turbine
g = Generator
h = Condenser
i = Feedwater pump
k = Feedwater heater and de-aerator
l = Compressor
m = Combustion and swirl chamber
n = Gas turbine
o = Autonomous heat-recovery steam circuit

A: *Plain steam installation* with air heater and bled-steam feedwater heating
B: *Combined installation* with combustion turbine preceding boiler. Heater recovery by autonomous recuperation steam circuit
C: *Combined installation* with preceding combustion turbine. Heat recovery by transfer to part of feedwater
D: *Combined installation* as C, but with air indirectly heated before expansion
E: *Combined installation* with pressure-charged boiler firing (Velox principle)
F: *Combined installation* with pressure-fired boiler, expansion being divided between a hot and a cold stage to reduce the temperature of the exhaust gases
G: *Gas turbine* without preceding steam cycle, but followed by a heat-recovery steam system

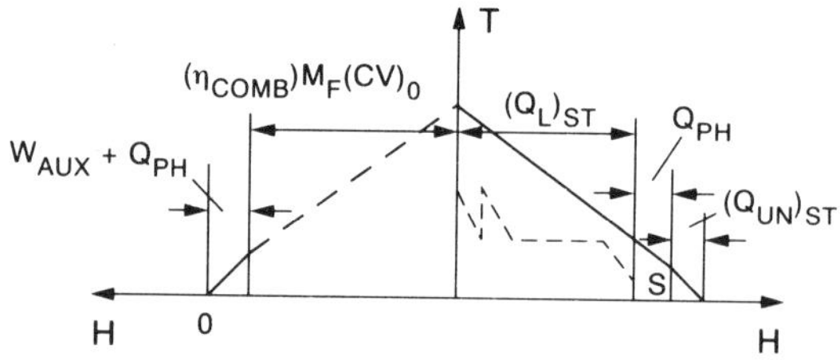

FIGURE 3.14 Unfolded temperature–enthalpy diagram for basic steam plant (ST)

Seippel and Bereuter write the energy balance treating the combustion as a heating process. More rigorously, we may write the steady flow energy equation between air entering (at temperature T_0) and products leaving (at temperature T_S) as follows

$$(H_A)_0 + W_{AUX} + Q_{PH} + M_F(h_F)_0 = Q_R + (Q_L)_{ST} + Q_{PH} + (H_P)_S, \tag{3.81a}$$

where M_F is the mass of fuel supplied of specific enthalpy $(h_F)_0$ and Q_R is the heat lost (mainly by radiation) in combustion. We subtract the enthalpy of products at temperature T_0 from each side to give

$$\begin{aligned} W_{AUX} + (H_A)_0 - (H_P)_0 + M_F(h_F)_0 - Q_R &= [(H_P)_S - (H_P)_0] + (Q_L)_{ST} \\ &= (Q_{UN})_{ST} + (Q_L)_{ST}, \end{aligned} \tag{3.81b}$$

where $(Q_{UN})_{ST}$ is the excess enthalpy or "heat loss" in the exhaust stack.

But in a calorific value experiment at temperature T_0,

$$(H_A)_0 + M_F(h_F)_0 - (H_P)_0 = M_F(CV)_0, \tag{3.82}$$

where $(CV)_0$ is the calorific value. If we define an efficiency of combustion as

$$(\eta_{COMB}) \equiv \frac{M_F(CV)_0 - Q_R}{M_F(CV)_0}, \tag{3.83}$$

then we may rewrite equation (3.81b) as

$$W_{AUX} + (\eta_{COMB})M_F(CV)_0 = (Q_L)_{ST} + (Q_{UN})_{ST}, \tag{3.84}$$

which is essentially the form of equation given by Seippel and Bereuter. [In subsequent discussion we shall adopt their term "combustion heating" for $(\eta_{COMB})M_F(CV)_0$.]

The work output is

$$W_{ST} = \eta_{ST}(Q_L)_{ST}$$
$$= \eta_{ST}[(\eta_{COMB})M_F(CV)_0 - (Q_{UN})_{ST} - W_{AUX}] \tag{3.85}$$

Neglecting the work input to the boiler fans, W_{AUX}, the overall efficiency of the plant is then as described in Chapter 1,

$$(\eta_O)_{ST} = \frac{W_{ST}}{M_F(CV)_0} = \left[\eta_{COMB} - \frac{(Q_{UN})_{ST}}{M_F(CV)_0}\right]\eta_{ST}$$
$$= (\eta_B)_{ST}\eta_{ST}, \tag{3.86}$$

where $(\eta_B)_{ST}$ is a boiler efficiency,

$$(\eta_B)_{ST} \equiv \left[(\eta_{COMB}) - \frac{(Q_{UN})_{ST}}{M_F(CV)_0}\right].$$

The unfolded temperature-enthalpy plot for plant CP2 is shown in Fig. 3.15. Now air is compressed through an enthalpy rise W_C before preheating Q_{PH} and combustion "heating" $(\eta_{COMB})(M_F)_H(CV)_0$. The gas enthalpy drops by the turbine work W_T before further "heating" $(\eta_{COMB})(M_F)_S(CV)_0$ in the supplementary combustion stages. Heat $(Q_L)_{CP2}$ is transferred to the steam turbine *before* heat Q_{PH} is transferred to preheat the air, in Seippel and Bereuter's example. They also considered a further "recovery" stage when the gas enthalpy drops by R_2, supplying heat to a low pressure stream turbine (Fig. 3.13B), but we shall not consider use of such a second steam turbine here. We shall simply assume a stack "loss" $(Q_{UN})_{CP2} = R_2 + (Q_{UN})_{ST}$, where R_2 is the "excess" stack loss over and above that in the steam turbine plant $(Q_{UN})_{ST}$.

The steady flow energy equation now yields

$$W_C + Q_{PH} + (\eta_{COMB})(M_F)_H(CV)_0 - W_T + \eta_{COMB}(M_F)_S(CV)_0$$
$$= (Q_L)_{CP2} + Q_{PH} + (Q_{UN})_{CP2}, \tag{3.87a}$$

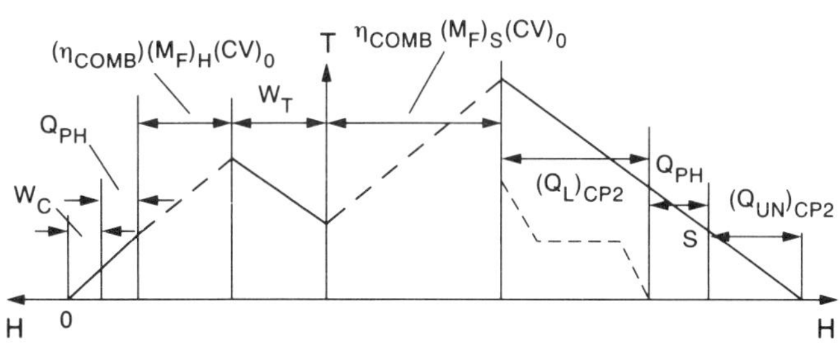

FIGURE 3.15 Unfolded temperature-entropy diagram for combined plant with supplementary firing (CP2)

so that

$$(Q_L)_{CP2} = (\eta_{COMB})[(M_F)_H + (M_F)_S](CV)_0 - (W_T - W_C) - (Q_{UN})_{CP2}, \quad (3.87b)$$

and the work output from the plant is

$$\begin{aligned} W_{CP2} &= \eta_{ST}(Q_L)_{CP2} + (W_T - W_C) \\ &= \eta_{ST}(\eta_{COMB})[(M_F)_H + (M_F)_S](CV)_0 \\ &\quad + (1 - \eta_{ST})(W_T - W_C) - \eta_{ST}(Q_{UN})_{CP2} \end{aligned} \quad (3.88)$$

where it has been assumed that the combustion efficiency in the combined plant is the same as in the steam plant.

A comparison may now be made between the two plants of their overall efficiency in using a total fuel energy of

$$M_F(CV)_0 = [(M_F)_H + (M_F)_S](CV)_0.$$

The efficiency of the combined plant CP2 is

$$(\eta_O)_{CP2} = \frac{W_{CP2}}{M_F(CV)_0} = \eta_{ST}\left[(\eta_{COMB}) - \frac{(Q_{UN})_{CP2}}{M_F(CV)_0}\right] + \frac{(W_T - W_C)(1 - \eta_{ST})}{M_F(CV)_0}. \quad (3.89)$$

If after Haywood [2] we write $x \equiv \frac{(W_T - W_C)}{M_F(CV)_0}$ and define an energy effectiveness for the *plant* as

$$(\varepsilon_B)_{CP2} \equiv (\eta_{COMB}) - \frac{(Q_{UN})_{CP2}}{M_F(CV)_0}, \quad (3.90a)$$

then it follows that

$$(\eta_O)_{CP2} = \eta_{ST}(\varepsilon_B)_{CP2} + x(1 - \eta_{ST}), \quad (3.91)$$

which is the expression derived by Haywood. Note that Haywood takes (η_{COMB}) to be unity, referring to $(\varepsilon_B)_{CP2}$ as a form of energy conversion efficiency,

$$(\varepsilon_B)_{CP2} = \frac{(Q_L)_{CP2} + (W_T - W_C)}{M_F(CV)_0} = 1 - \frac{(Q_{UN})_{CP2}}{M_F(CV)_0} \quad (3.90b)$$

The overall efficiency of plant CP2 can be compared with that of the basic steam plant (ST):

$$(\eta_O)_{ST} = \eta_{ST}(\eta_B)_{ST}; \tag{3.86}$$

$$(\eta_O)_{CP2} = \eta_{ST}(\varepsilon_B)_{CP2} + x(1-\eta_{ST}); \tag{3.91}$$

so that

$$(\eta_O)_{CP2} - (\eta_O)_{ST} = x(1-\eta_{ST}) - \eta_{ST}[(\eta_B)_{ST} - (\varepsilon_B)_{CP2}]. \tag{3.92}$$

Efficiency is thus increased if

$$x(1-\eta_{ST}) > \eta_{ST}[(\eta_B)_{ST} - (\varepsilon_B)_{CP2}]. \tag{3.93}$$

Seippel and Bereuter then make a further comparison between $(\eta_O)_{CP2}$ and $(\eta_O)_{ST}$, introducing the efficiency η^*_{GT} of a gas turbine plant with an ideal heat exchanger, i.e. with no temperature difference between air at exit from the exchanger and gas at entry to the exchanger (exit from the turbine). The efficiency is given by

$$\eta^*_{GT} \equiv \frac{W_T - W_C}{W_T}, \tag{3.94}$$

since the heat supply is equal to the turbine work W_T for this particular plant.

From equation (3.89)

$$(\eta_O)_{CP2} = \eta_{ST}\left[(\eta_{COMB}) - \frac{(Q_{UN})_{ST}}{M_F(CV)_0}\right] + \frac{(W_T - W_C)(1-\eta_{ST})}{M_F(CV)_0} - \frac{\eta_{ST}[(Q_{UN})_{CP2} - (Q_{UN})_{ST}]}{M_F(CV)_0}, \tag{3.95a}$$

i.e.

$$(\eta_O)_{CP2} - (\eta_O)_{ST} = \frac{(W_T - W_C)(1-\eta_{ST})}{M_F(CV)_0} - \frac{\eta_{ST} R_2}{M_F(CV)_0}. \tag{3.95b}$$

Seippel and Bereuter point out that if the temperature difference between the air at entry to the air heater and the gas exit temperature is maintained the same as plant ST and in the combined plant CP2, then the quantity $R_2 = [(Q_{UN})_{CP2} - (Q_{UN})_{ST}]$ is equal to the compressor work W_C. With this assumption, and with

$$(W_T - W_C) = W_T \eta^*_{GT}, \tag{3.96}$$

$$R_2 = W_C = W_T(1-\eta^*_{GT}), \tag{3.97}$$

both substituted into equation (3.95b), it follows that

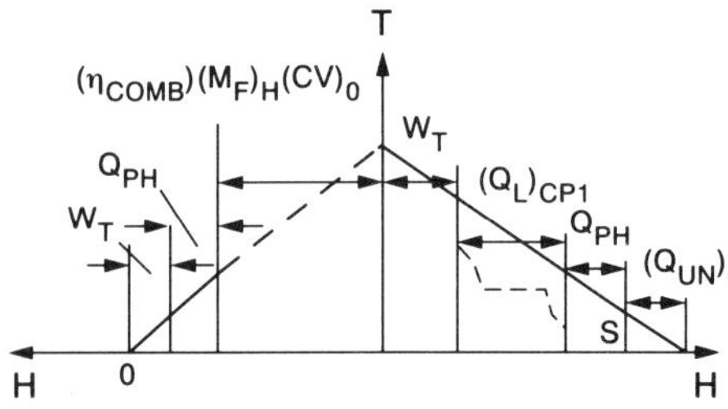

FIGURE 3.16 Unfolded temperature-enthalpy diagram for combined plant with no supplementary firing (CP)

$$(\eta_O)_{CP2}-(\eta_O)_{ST}=\eta^*_{GT}(1-\eta_{ST})\frac{W_T}{M_F(CV)_0}$$

$$-\eta_{ST}(1-\eta^*_{GT})\frac{W_T}{M_F(CV)_0}$$

$$=(\eta^*_{GT}-\eta_{ST})W_T/M_F(CV)_0. \tag{3.98}$$

The "unfolded" temperature-enthalpy chart for the combined plant CP1 (with no supplementary heating) is shown in Fig. 3.16. From equation (3.87a) with $(M_F)_S=0$ the energy balance is now

$$W_C+Q_{PH}+(\eta_{COMB})(M_F)_H(CV)_0-W_T=(Q_L)_{CP1}+Q_{PH}+(Q_{UN})_{CP1} \tag{3.99}$$

where $(Q_{UN})_{CP1}=(Q_{UN})_{ST}+R_1$ and R_1 is now the excess stack loss, over and above the stack loss of the steam turbine plant $(Q_{UN})_{ST}$. The efficiency of plant CP1 is now

$$(\eta_O)_{CP1}=\frac{W_{CP1}}{(M_F)_H(CV)_0}$$

$$=\frac{(W_T-W_C)+\eta_{ST}(Q_L)_{CP1}}{(M_F)_H(CV)_0}$$

$$=\frac{(W_T-W_C)+\eta_{ST}[(\eta_{COMB})M_F(CV)_0-(W_T-W_C)-(Q_{UN})_{CP1}]}{(M_F)_H(CV)_0}$$

$$=\frac{(W_T-W_T)(1-\eta_{ST})}{(M_F)_H(CV)_0}+\eta_{ST}\left[(\eta_{COMB})-\frac{(Q_{UN})_{CP1}}{M_F(CV)_0}\right] \tag{3.100}$$

If $(Q_{UN})_{CP1}=(Q_{UN})_{CP2}$, i.e. $R_1=R_2$, we can compare the efficiency of plant CP2 (with supplementary firing), equation (3.89), with that of plant CP1 (with no supplementary firing), equation (3.100), and derive the equation

$$M_F(CV)_0[(\eta_O)_{CP2}-\eta_{ST}(\eta_{COMB})]=(M_F)_H(CV)_0[(\eta_O)_{CP1}-\eta_{ST}(\eta_{COMB})]$$
$$=(W_T-W_C)(1-\eta_{ST})$$
$$-\eta_{ST}[(Q_{UN})_{ST}+R_1],$$

so that

$$M_F(CV)_0[(\eta_O)_{CP2}-\eta_{ST}(\eta_{COMB})]$$
$$=[M_F-(M_F)_S](CV)_0[(\eta_O)_{CP1}-\eta_{ST}(\eta_{COMB})]$$

and

$$(\eta_O)_{CP1}-(\eta_O)_{CP2}=\frac{(M_F)_S}{M_F}[\eta_{ST}(\eta_{COMB})-(\eta_O)_{CP1}]. \quad (3.101)$$

However, as we shall see later, $(Q_{UN})_{CP1}$ for a plant with no supplementary firing $[(M_F)_S=0]$ is usually less than $(Q_{UN})_{CP2}$, so that the validity of this equation is now doubtful.

Seippel and Bereuter introduce a further "efficiency" $\eta_{GT}^{**} \equiv(\eta_O)_{CP1}/(\eta_{COMB})$ so that

$$(\eta_O)_{CP2}-(\eta_O)_{CP1}=\frac{(M_F)_S}{M_F}(\eta_{ST}-\eta_{GT}^{**})(\eta_{COMB}). \quad (3.102)$$

We now summarise all the results obtained from this presentation of the analyses by Haywood and of Seippel and Bereuter:

$$(\eta_O)_{CP2}-(\eta_O)_{ST}=x(1-\eta_{ST})-[(\eta_B)_{ST}-(\varepsilon_B)_{CP2}]\eta_{ST} \quad (3.92)$$

$$(\eta_O)_{CP2}-(\eta_O)_{ST}=(\eta_{GT}^{*}-\eta_{ST})\frac{W_T}{M_F(CV)_0}; \quad (3.98)$$

$$(\eta_O)_{CP2}-(\eta_O)_{CP1}=\frac{(M_F)_S}{M_F}[\eta_{ST}(\eta_{COMB})-(\eta_O)_{CP1}]$$
$$=\frac{(M_F)_S}{M_F}(\eta_{COMB})(\eta_{ST}-\eta_{GT}^{**}). \quad (3.102)$$

The first of these equations shows that the overall efficiency of the combined plant CP2 [using a steam turbine of the *same* thermal efficiency (η_{ST}) as that of the basic steam turbine plant (ST)] exceeds the efficiency of the basic plant only if the quantity $x(1-\eta_{ST})$ exceeds the loss in "boiler" efficiency $[(\eta_B)_{ST}-(\varepsilon_B)_{CP2}]$ multiplied by η_{ST}.

Seippel and Bereuter argue from the second and third of these equations as follows.

(i) If $\eta_{ST} > \eta^*_{GT}$ (i.e. the thermal efficiency of the "common" steam turbine system exceeds that of gas turbine plant with an ideal heat exchanger) then the basic steam plant should be chosen and should not be replaced.

(ii) If $\eta^{**}_{GT} > \eta_{ST}$ (i.e. the hypothetical efficiency $\eta^{**}_{GT} = (\eta_O)_{CP1}/(\eta_{COMB})$ exceeds the thermal efficiency of the steam turbine plant) then the basic steam plant should be replaced by the combined plant (CP1), i.e. without supplementary firing.

(iii) Only if $\eta^*_{GT} > \eta_{ST} > \eta^{**}_{GT}$ should the basic steam plant (ST) be replace by the combined plant with supplementary firing (CP2).

However, we should note that the various assumptions made in deriving these equations (including η_{ST} unchanged for all three plants, and $(Q_{UN})_{CP2} = (Q_{UN})_{CP1}$) now make these conclusions of doubtful validity for today's combined plant.

3.7 Summary of Efficiency Relationships

We summarise below the major efficiency relationships derived or referred to in this chapter.

Doubly cyclic plants (thermal efficiency)

"Series" plant with no temperature drop between plants and no intermediate irreversibility in plant:

$$\tilde{\eta}_{CPS} = \eta_H + \eta_L - \eta_H \tilde{\eta}_L, \tag{3.103}$$

where

$$\tilde{\eta}_L = E^Q_{HR}/Q_{HL}.$$

"Series" plant with temperature drop between plants and intermediate irreversibility in plant:

$$\eta_{CPS} = \eta_H + \eta_L - \eta_H \eta_L, \tag{3.104}$$

where

$$\eta_L = (\tilde{\eta}_L)(\eta_{ht})(\eta_{td}),$$

$$(\eta_{ht}) = E^Q_L/E^Q_{HR} = 1 - \frac{I^Q_{HL}}{E^Q_{HR}},$$

$$(\eta_{td}) = 1 - \frac{I^{CR}_L}{E^Q_L}.$$

"Series" plant with intermediate heat loss:

$$\eta_{CPS}=\eta_H+\eta_L-\eta_H\eta_L-\nu_{UN}\eta_L, \tag{3.105a}$$

where

$$\nu_{UN}\equiv Q_{UN}/Q_B;$$

or

$$\eta_{CPS}=\eta_H-(\eta_B)\eta_L-(\eta_B)\eta_H\eta_L, \tag{3.105b}$$

where

$$(\eta_B)=Q_L/Q_{HR}=1-(Q_{UN}/Q_{HR});$$

or

$$\eta_{CPS}=\eta_H+(\eta_O)_L-\eta_H(\eta_O)_L, \tag{3.105c}$$

where

$$(\eta_O)_L=\eta_B\eta_L.$$

"Parallel" plant (M, N) with joint heating:

$$\eta_{CPP}=\eta_M-\nu_N(\eta_M-\eta_N), \tag{3.106}$$

where

$$\nu_N\equiv Q_N/(Q_M+Q_N)=Q_N/Q_B.$$

"Series/parallel" plants (i.e. series plus supplementary heating but no heat loss ($Q_{UN}=0$)):

$$\eta_{CPSP}=\eta_H+\eta_L-\eta_H\eta_L-\nu_S\eta_H(1-\eta_L) \tag{3.107a}$$

where

$$\nu_S\equiv Q_S/Q_B=Q_S/(Q_H+Q_S);$$

or

$$\eta_{CPSP}=\eta_L+(1-\nu_S)\eta_H-(1-\nu_S)\eta_H\eta_L, \tag{3.107b}$$

where

$$(1-\nu_S)=Q_H/Q_B=Q_H/(Q_H+Q_S).$$

"Series/parallel" plants (general case of supplementary heating and heat loss):

$$\eta_{CP} = \eta_H + \eta_L - \eta_H \eta_L - v_{UN}\eta_L - v_S \eta_H (1 - \eta_L); \tag{3.108a}$$

or

$$\eta_{CP} = x + (1 - v_{UN})\eta_L - x\eta_L, \tag{3.108b}$$

where

$$x \equiv W_H/Q_B,\ (1 - v_{UN}) = (Q_B - Q_{UN})/Q_B,$$

or

$$\eta_{CP} = v_H \eta_H + v_L \eta_L, \tag{3.108c}$$

where

$$v_H \equiv Q_H/Q_B = Q_H/(Q_H + Q_S),$$
$$v_L \equiv Q_L/Q_B = Q_L/(Q_H + Q_S).$$

Open circuit/closed cycle plants (overall efficiency)

"Series" plant:

$$(\eta_O)_{CP} = (\eta_O)_H + \eta_L - (\eta_O)_H \eta_L - \frac{\eta_L[(H_P)_S - (H_P)_0]}{M_F(CV)_0}; \tag{3.109a}$$

or

$$(\eta_O)_{CP} = (\eta_O)_H + K[1 - (\eta_O)_H] - \frac{K[(H_P)_6 - (H_P)_0](\eta_O)_H(1+f)}{W_H} \tag{3.109b}$$

where K = constant, subscript 6 refers to the pinch point and W_H is the gas turbine specific work.

"Series" plant with supplementary firing:

$$(\eta_O)_{CP} = x + \eta_L(\varepsilon_B - x), \tag{3.110a}$$

where

$$x \equiv \frac{W_H}{M_F(CV)_0},\ \varepsilon_B \equiv \frac{(H_R)_0 - (H_{P'})_S}{(H_R)_0 - (H_P)_0} = \frac{W_H + Q_L}{(H_R)_0 - (H_P)_0};$$

or

$$(\eta_O)_{CP} = \varepsilon_B \eta_L + q(\eta_O)_H - q(\eta_O)_H \eta_L, \tag{3.110b}$$

where q is the fraction of $M_F(CV)_0$ "supplied" to the gas turbine; or

$$(\eta_O)_{CP} = \frac{(\eta_O)_H}{\left[1+\dfrac{(M_F)_S}{M_F}\right]} + \frac{\eta_L}{\left[1+\dfrac{(M_F)_S}{M_F}\right]}$$
$$\times\left[\frac{(M_F)_S}{(M_F)_H} + [1-(\eta_O)_H] - \frac{(M_A+M_F)}{(M_F)_H(CV)_0}[(h_{P'})_S-(h_P)_0](M_F)_S\right] \tag{3.110c}$$

or

$$(\eta_O)_{CP} = \frac{W_H + K'[(H_{P'})_5 - (H_{P'})_S]}{W_H + (H_{P'})_5 - (H_P)_0} \tag{3.110d}$$

where $K' = \text{constant}$.

Seippel and Bereuter relationships:

$$(\eta_O)_{CP2} - (\eta_O)_{ST} = x(1-\eta_{ST}) - [(\eta_B)_{ST} - (\varepsilon_B)_{CP2}]\eta_{ST}; \tag{3.111}$$

$$(\eta_O)_{CP2} - (\eta_O)_{ST} = (\eta^*_{GT} - \eta_{ST})\frac{W_T}{M_F(CV)_0}; \tag{3.112}$$

$$(\eta_O)_{CP2} - (\eta_O)_{CP1} = \frac{(M_F)_S(\eta_{COMB})}{M_F}(\eta_{ST} - \eta^{**}_{GT}), \tag{3.113}$$

where $\eta^*_{GT}, \eta^{**}_{G}$ are defined in Section 3.6.

References

[1] Seippel, C. and Bereuter, R. The Theory of Combined Steam and Gas Turbine Installations. *Brown Boveri Review*, **47**, 783–799, 1960.

[2] Haywood, R. W. *Analysis of Engineering Cycles*, 4th edition. Pergamon Press, Oxford, 1991.

[3] Buxmann, J. *Combined Cycles for Power Generation—Introduction*. Von Karman Institute for Fluid Dynamics, Lecture Series 6, Vol. 1, 1978.

[4] Haywood, R. W. Private communication, 1989 [see also reference 2 above].

[5] Timmermans, A. R. J. *Combined Cycles and Their Possibilities*. Von Karman Institute for Fluid Dynamics, Lecture Series 6, Vol. 1, 1978.

[6] Mangan, J. L. and Pettit, J. L. *A Highly Efficient Steam Turbine Gas Turbine Cycle*. ASME Paper 63-GT-51, 1963.

CHAPTER 4

Parametric Studies of Combined Power Plants

4.1 Introduction

In the last chapter we developed simple relationships for combined plant efficiency [thermal, η_{CP} or overall, $(\eta_O)_{CP}$], for plants "in series", plants "in parallel" (joint heating) and series/parallel combinations. We derived expressions for the way η_{CP} (for doubly cyclic plants) and $(\eta_O)_{CP}$ (for open circuit/closed cycle plants) varied with the efficiencies of the component plants, with the degree of "heat loss" between the two plants and with the division of the total heat supply between the two plants.

However, the component plant efficiencies are not necessarily unrelated, and in general may not be specified independently. For example, a higher plant of low efficiency is likely to reject heat at relatively high temperature, which may allow the lower cycle to have a relatively high efficiency. It is therefore more realistic to vary the important thermodynamic parameters in the whole combined cycle, and this is the approach we adopt later in this chapter (following calculations by several authors). Clearly with the large number of possible combined plants, a comprehensive parametric study of all combinations would be an enormous task. Here we therefore limit our presentation to the following:

in Section 4.2

simple calculations using the analyses of Chapter 3, to show how the efficiency of a combined plant varies with component plant efficiencies (mainly for doubly cyclic plant);

in Section 4.3

parametric studies of binary plants ("Joule"/"Rankine" plants in series and series/parallel),

more complex studies of the open circuit (gas turbine)/closed cycle (steam turbine) combined plant (often called the combined "cycle" gas turbine, or CCGT, plant), and some limited parametric studies of the gas turbine plant with steam injection (STIG);

in Section 4.4

a discussion of the results of the various parametric studies we have described, together with others found in the literature.

4.2 Parametric Studies of Combined Plants Based on Component Efficiencies

4.2.1 The Doubly Cyclic Series Plant

The simplest of all the relationships we have derived is that which holds for the ideal "series" plant, a doubly cyclic binary plant in which all the heat from the upper (closed) cycle is rejected to the lower (closed) cycle:

$$\eta_{CPS} = \eta_H + \eta_L - \eta_H \eta_L . \tag{4.1}$$

The increased efficiency of this series combined plant may be compared to

(i) the efficiency of a "higher" plant, to which a bottoming plant is added;
(ii) the efficiency of a "lower" plant, to which a topping plant is added.

For the first comparison, we can show the combined efficiency η_{CPS} plotted against η_H for various values of the added lower cycle η_L (Fig. 4.1). The gain in efficiency (in comparison with η_H) increases linearly with η_L when the bottoming cycle is added, but the increase is greater for a poor higher cycle, of low efficiency η_H.

Alternatively, for the second comparison, we can show the combined plant efficiency η_{CPS} plotted against η_L, with η_H as a parameter (Fig. 4.2). [The relationship is of course exactly the same as those shown in Fig. 4.1, but with η_H and η_L interchanged since equation (4.1) is symmetrical.] The gain in efficiency now increases linearly with η_H, if the "topping" cycle is added to the lower plant, η_L.

However, such simple plots conceal the point made in the introduction to this chapter. It is not always feasible for η_L to be held constant as η_H is increased; η_H and η_L themselves change in a coupled way, as basic thermodynamic parameters are varied.

4.2.2 Joint Heating of Two Closed Cycles

For parallel cyclic plant, with joint heating and no series coupling, we obtained the simple relationship, equation (3.35), for the combined plant thermal efficiency, which may be written

$$\eta_{CPP} = \frac{\eta_M Q_M + \eta_N Q_N}{Q_M + Q_N},$$
$$= \eta_M(1 - \nu_N) + \eta_N \nu_N,$$
$$= \eta_M - \nu_N(\eta_M - \eta_N), \qquad (4.2a)$$
$$= \eta_N - (1 - \nu_N)(\eta_N - \eta_M), \qquad (4.2b)$$

where

$$\nu_N = \frac{Q_N}{Q_M + Q_N} = \frac{Q_N}{Q_B}.$$

The expressions show that, as to be expected, there is always a loss in efficiency in joint heating. If η_M is greater than η_N, then equation (4.2a) shows that $\eta_{CPP} < \eta_M$. If η_N is greater than η_M, then equation (4.2b) shows that $\eta_{CPP} < \eta_N$.

Since there is thus no thermodynamic advantage, we do not explore the variation of η_{CPP} with η_M and η_N as parameters for joint heating of parallel cyclic plant.

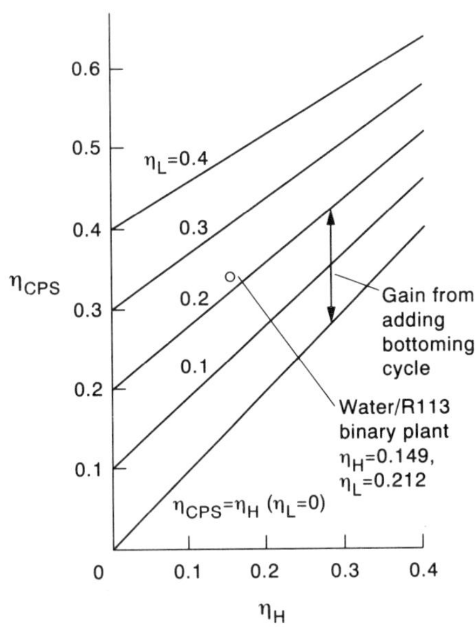

FIGURE 4.1 "Series" efficiency η_{CPS}—addition of bottoming cycle η_L

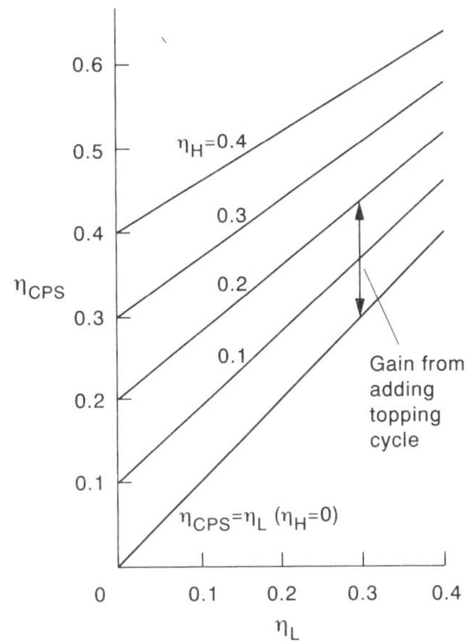

FIGURE 4.2 "Series" efficiency η_{CPS}—addition of topping cycle η_H

4.2.3 Heat Unused Between Plants in Series

If not all the heat rejected from an upper cycle is supplied to a lower cycle in series, there being heat unused ($Q_{UN} = \nu_{UN} Q_B$) between the plant (a heat "loss"), then from Section 3.2.5 the efficiency of the combined plant is

$$\eta_{CP} = \eta_H + \eta_L - \eta_H \eta_L - \nu_{UN} \eta_L, \tag{4.3a}$$

$$= \eta_{CPS} - \Delta\eta_{UN}, \tag{4.3b}$$

where $\Delta\eta_{UN}$ is the loss in efficiency due to the heat unused.

We can illustrate this relation simply by taking $\nu_{UN} = 0.25$ so that

$$\eta_{CP} = \eta_H + \eta_L(0.75 - \eta_H). \tag{4.3c}$$

For this level of heat "loss", Fig. 4.3 shows a plot of η_{CP} against η_H with η_L as a parameter ($\eta_L = 0.2$, 0.4). The thermal efficiency of the ideal series combined plant is also shown, for $\eta_L = 0.2$, 0.4. The loss in efficiency $\Delta\eta_{UN}$ is marked, obviously more so for higher efficiency of the lower plant η_L.

This analysis has been presented for the thermal efficiency of doubly cyclic binary plants. In Section 3.5.2.2 we referred to the work of Mangan and Pettit [1] on an open circuit/closed cycle plant leading to the relation

$$(\eta_O)_{CP} = (\eta_O)_H + \eta_L - (\eta_O)_H \eta_L - \frac{\eta_L[(H_P)_S - (H_P)_O]}{M_F(CV)_0} \tag{4.4}$$

or

$$(\eta_O)_{CP} = (\eta_O)_H + \eta_L[1 - (\eta_O)_H](\eta_B), \tag{4.5}$$

for a plant with no supplementary heating. Mangan and Pettit's plot of $(\eta_O)_{CP}$ against "boiler efficiency" (η_B) for various $(\eta_O)_H$ and η_L is shown in Fig. 4.4; this figure, for an open circuit/closed cycle plant, may be compared with Fig. 4.3 for a doubly cyclic plant.

4.2.4 Supplementary Heat Supply Between Plants in Series

If supplementary heat is supplied to a lower cycle, between two cyclic plants, and that heat supply is a fraction ν_S of the total heat supply $(\nu_S = Q_S/Q_B)$, then if there is no heat "loss" $(\nu_{UN} = 0)$ the combined cycle efficiency is, from equation (3.37), Section 3.4.1,

$$\eta_{CPSP} = \eta_H + \eta_L - \eta_H\eta_L - \nu_S\eta_H(1 - \eta_L), \tag{4.6a}$$

$$= \eta_{CPS} - \Delta\eta_S, \tag{4.6b}$$

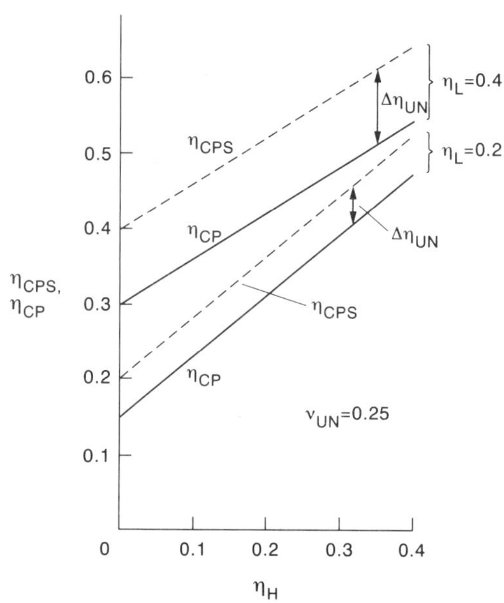

FIGURE 4.3 Effect of intermediate heat "loss" $(\Delta\eta_{UN} = \eta_{CPS} - \eta_{CP} = \mu_{UN}\eta_L)$

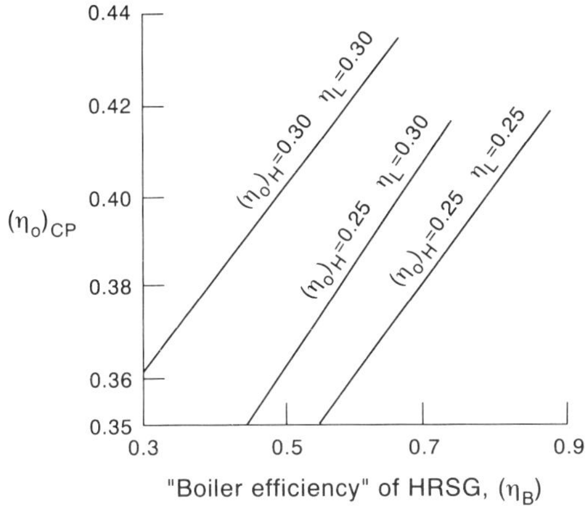

FIGURE 4.4 Open circuit/closed cycle plant—overall efficiency $(\eta_O)_{CP}$ (after Mangan and Pettit [1]). $(\eta_O)_{CP}=(\eta_O)_H+\eta_L-(\eta_O)_H\eta_L-(1-\eta_B)[1-(\eta_O)_H]\eta_L$

where $\Delta\eta_S=\nu_S\eta_H(1-\eta_L)$. It should be emphasised that this efficiency is not the same as the efficiency of the "parallel" combined plant (with joint heating) since the heat supply to the lower cycle now comes from two sources—heat rejection from the upper cycle and a supplementary heat supply—a series/parallel scheme.

We illustrate this case simply by taking $\nu_S=0.5$ (equal division of heat supply between the top cycle and the (supplementary) heat supply to the lower cycle), so that

$$\begin{aligned}\eta_{CPSP}&=\eta_H+\eta_L-\eta_H\eta_L-0.5\eta_H(1-\eta_L)\\&=0.5\eta_H+\eta_L-0.5\eta_H\eta_L\\&=0.5\eta_H+\eta_L(1-0.5\eta_H).\end{aligned}\tag{4.6c}$$

This relationship is again shown as a plot of η_{CPSP} against η_H, with η_L as a parameter ($\eta_L=0.2, 0.4$) in Fig. 4.5. We also show η_{CPS} (for $\eta_L=0.2, 0.4$) and the penalty on overall efficiency ($\Delta\eta_S$) due to supplementary heat supply, which increases with η_L.

This analysis is for the thermal efficiency of a doubly cyclic plant, with two sources of heat supply. In Section 3.5.2.2 we referred to the analysis by Mangan and Pettit for an open circuit/closed cycle combined plant, leading to the relationship for overall plant efficiency

$$(\eta_O)_{CPSP}=\frac{(\eta_O)_H}{\left[1+\dfrac{(M_F)_S}{(M_F)_H}\right]}+\frac{\left[\dfrac{(M_F)_S}{(M_F)_H}+1-(\eta_O)_H\right]\eta_L}{\left[1+\dfrac{(M_F)_S}{(M_F)_H}\right]}\tag{4.7}$$

if $(H_{P'})_S = (H_{P'0})$, i.e. if $\nu_{UN} = 0$. Here the fuel ratio $(M_F)_S/(M_F)_H$ compares with the heat ratio $Q_S/Q_H = \nu_S/(1-\nu_S)$ for the doubly cyclic binary system.

4.2.5 Heat Loss and Supplementary Heating Between Plants in Series

If both heat "loss" and supplementary heating take place between two cycles (upper and lower) then, from equation (3.41), the combined plant efficiency is

$$\eta_{CP} = \eta_H + \eta_L - \eta_H\eta_L - \nu_{UN}\eta_L - \eta_H(1-\eta_L)\nu_S, \tag{4.8a}$$

$$= \eta_{CPS} - \Delta\eta_{UN,S}, \tag{4.8b}$$

where

$$\Delta\eta_{UN,S} = \Delta\eta_{UN} + \Delta\eta_S.$$

With

$$\nu_{UN} = 0.25,\ \nu_S = 0.5,$$

$$\eta_{CP} = 0.5\eta_H + 0.75\eta_L - 0.5\eta_H\eta_L. \tag{4.8c}$$

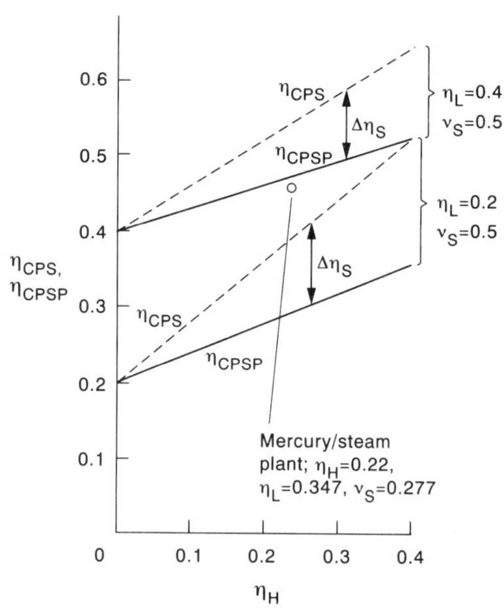

FIGURE 4.5 Effect of supplementary heat supply, $\Delta\eta_S = \eta_{CPS} - \eta_{CPSP} = \mu_S\eta_H(1-\eta_L)$

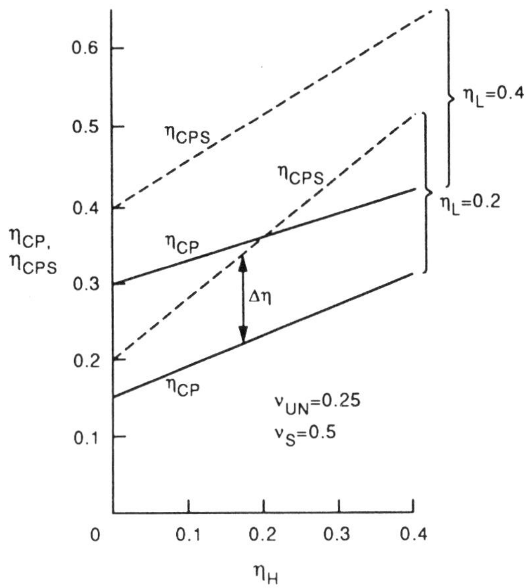

FIGURE 4.6 Effect of intermediate heat loss and supplementary heat supply $\Delta\eta = \eta_{CPS} - \eta_{CP} = \nu_{UN}\eta_L - \nu_S\eta_H(1-\eta_L)$

Figure 4.6 shows a plot of this expression for η_{CP} against η_H with η_L as a parameter ($\eta_L = 0.2$, 0.4). Also shown is η_{CPS}, together with the loss of efficiency $\Delta\eta = \Delta\eta_{UN,S}$ due to the two effects, which is now substantial.

The full Mangan and Pettit equation for an open circuit/closed cycle plant, corresponding to equation (4.8a) is

$$(\eta_O)_{CP} = \frac{(\eta_O)_H}{\left[1 + \dfrac{(M_F)_S}{(M_F)_H}\right]} + \frac{\eta_L}{\left[1 + \dfrac{(M_F)_S}{(M_F)_H}\right]} \times \left[\frac{(M_F)_S}{(M_F)_H} + [1 - (\eta_O)_H] - \left(\frac{(H_{P'})_S - (H_{P'})_0}{(M_F)_H(CV)_0}\right)\right] \tag{4.9}$$

where $(M_F)_S/(M_F)_H$ corresponds to $\nu_S/(1-\nu_S)$ and $[(H_{P'})_S - (H_P)_0]/M_F(CV)_0$ corresponds to ν_{UN}.

4.2.6 Discussion

These simple parametric analyses illustrate the interplay between component cycle efficiency. They show

(i) the importance of minimising the heat loss between cycles "in series";

(ii) the use of joint heating for cycles "in parallel" always lowers combined plant efficiency (in comparison with the efficiency of the better cycle); and

(iii) the effect of supplementary heating in "series" plants introduces an element of joint heating and similarly lowers efficiency (but it usually substantially increases plant output and the drop in efficiency may be acceptable).

In this preliminary discussion we concentrated on the thermal efficiency of combined plants comprising two closed cycles, with external heat supplies using known component cycle efficiencies as main parameters. But the conclusions have relevance to open circuit/closed cycle plant. For example, we noted in Chapter 3 the similarity between the expression for a combined cyclic plant [equation (4.3a), for thermal efficiency with heat "loss" between cycles] and that for an open circuit/closed cycle plant [and equation (4.4), for overall efficiency with exhaust stack "loss", or "excess" enthalpy]. Mangan and Pettit have provided the parallel analyses for open circuit/closed cycle combined plants, with fuel burnt in the main gas turbine combustion chamber and in a supplementary burner before the HRSG, and with a stack "loss" [equations (4.8a) and (4.9)].

4.3 Parametric Studies of Combined Power Plants Based on Variations of Main Thermodynamic Parameters

The analysis of Section 4.2 was over-simplified because the component plant efficiencies were taken as known quantities. However, because these efficiencies are coupled rather than independent we shall now study how combined plant efficiency varies with fluid thermodynamic properties, with selection of state points (pressure and temperature levels) and major parameters (e.g. pressure ratio) in the component plants.

We first continue to examine binary plants (doubly closed cyclic plants), in Section 4.3.1; for such plant we can determine *thermal* efficiency as a function of thermodynamic properties and states in the cycles. Initially, in Section 4.3.1.1, we comment on "Rankine"/"Rankine" combinations and in particular how they are affected by choice of the two fluids. We then concentrate on "Joule"/"Rankine" combined binary plants, firstly for the plants in series (Section 4.3.1.2) and subsequently for series/parallel operation, i.e. plants with an element of joint heating (Section 4.3.1.3). Since the higher (gas) cycle is closed, the pressure levels within it are to some extent arbitrary (not related to atmospheric pressure as they are in an open circuit gas turbine plant) but subject to a maximum permissible level. The doubly cyclic closed plant therefore has the possibility of increased mass flow and power output.

In Section 4.3.2 we review some of the many parametric studies of the open circuit (gas turbine)/closed cycle (steam turbine) plant reported in the literature. [This type of plant is often referred to as a combined cycle gas turbine (CCGT) although the gas turbine is open circuit.] We refer in Section 4.3.2.1 to three comprehensive analyses, by Cerri [2], Rufli [3] and Kehlhofer [4], which lead to estimates of *overall* efficiency of such combined plants. We then discuss in Section 4.3.2.2 four particular aspects of CCGT plant

(i) the importance of gas turbine specific work;
(ii) the role of regenerative feed heating in the steam cycle;
(iii) the possible introduction of reheat into the open circuit gas turbine plant; and
(iv) off-design or part load performance.

In Section 4.3.3 we review some limited parametric studies of the doubly open plant, in which steam is injected into an open circuit gas turbine plant—the STIG (steam injection gas turbine plant).

We should note that several published papers give analyses of performance based on the use of exergy. It has been explained in Chapter 1 that the rational efficiency (η_R) (based on maximum possible work output) is frequently little different from the overall efficiency (η_O) (based on the calorific value of the fuel). The main advantage of analyses based on exergy flows is that the lost work due to local irreversibilities (of various components) can be estimated. We defer reference to such approaches until Chapter 5, where such component irreversibilities are identified for three particular combined plants.

4.3.1 Binary Plants (Doubly Cyclic)

There was substantial development of the binary plant (doubly cyclic, "Rankine"/"Rankine") in the 1940s and 1950s, mainly through work by the General Electric Company in the United States, on the mercury/steam plant. However, the main stream of subsequent development has been on other forms of combined plant.

We do not therefore undertake detailed numerical parametric studies here of the "Rankine"/"Rankine" binary plant, but we do discuss in Section 4.3.1.1 the choice of possible fluids, the essential parameters for such a combined plant, and quote two examples with efficiency levels. Subsequently we study the "Joule"/"Rankine" (gas/vapour) binary plant, both in series and series/parallel operation (in Sections 4.3.1.2 and 4.3.1.3 respectively), referring to full parametric calculations.

4.3.1.1 "RANKINE"/"RANKINE" BINARY PLANT

The critical parameter for the "Rankine"/"Rankine" combined plant is the choice of the fluid for one of the cycles, the other invariably being water. However, this choice is strongly influenced by the intended application of the combined plant, as Wood [5] pointed out in a comprehensive survey paper in 1970. Wood searched for an alternative fluid to water for power generation:

(a) in a binary superposition ("topping") cycle above steam;
(b) in a binary subposition ("bottoming") cycle below steam;
(c) in low temperature recuperation applied to process heat or gas turbine exhaust heat.

With regard to application (c), use of the alternative fluid in a closed cycle receiving waste heat from a process will usually involve a single "Rankine" cycle, but a binary plant may be involved as we shall see below. Use of the alternative fluid in a closed cycle utilising exhaust heat from a gas turbine immediately implies a bottoming cycle for a combined power plant.

Choice of Fluid for the Higher (Topping) Cycle

The purposes of application (a) is to improve power plant efficiency, essentially by "Carnotization", making the combined plant approach Carnot efficiency. Water in the lower cycle achieves the desired objective of rejecting all heat at the lowest (condenser) temperature. Mercury is the prime candidate for the higher cycle, as it takes in substantially all its heat at the top temperature. A temperature-entropy diagram proposed by Wood [5] is shown in Fig. 4.7 and illustrates the steep slope of the

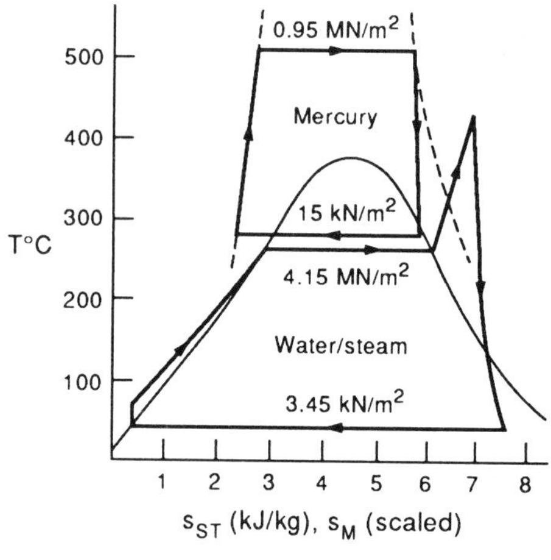

FIGURE 4.7 Mercury/steam binary cycle—temperature entropy diagram (after Wood [5])

saturated fluid line for mercury (almost isentropic). Further the critical temperature of mercury (about 1500° C) is much higher than that of water and it can be evaporated near the maximum allowable cycle temperature. The saturation pressure at this temperature, say 510° C, is also relatively low at 0.95 MN/m^2. Mercury's very low pressure at a normal condensing temperature is avoided by use of the binary mercury/water cycle rather than a single-vapour mercury cycle, condensation taking place at about 280° C in a condenser-boiler (at about 15 kN/m^2).

Potassium is virtually the only other candidate for use in a topping cycle above steam, but its inconveniently low saturation pressure at top temperature (10.5 kN/m^2 at 560°C) and its other properties (incompatibility with water, the requirement for heated pipes to prevent solidification) have virtually ruled it out.

Thus for the "topping" cycle mercury proved to be the first choice. The pressure and temperature levels are virtually chosen by the properties of the two fluids, so we do not here undertake a detailed numerical study involving their variation as parameters. We simply quote the details of a plant representative of developments in the United States in the forties and fifties.

There was an element of joint heating in these plants and the relationship

$$\eta_{CP} = \eta_H + \eta_L - \eta_H \eta_L - \nu_S \eta_H (1 - \eta_L) \tag{4.6a}$$

holds. Of the total heat supplied from the boiler, part ($\nu_S Q_B$) went to heating the feed water of the lower (water) cycle and superheating the steam and part $[(1-\nu_S)Q_B]$ to evaporating the mercury in the higher cycle.

Haywood [6] quotes figures similar to those of the Schiller station, New Hampshire, for which

$\nu_S = 0.277$

$\eta_H = 0.220$ (for boiler and condenser pressures of 900 kN/m^2 and 17 kN/m^2)

$\eta_L = 0.347$ (for boiler and condenser pressures of 4 MN/m^2 and 4 kN/m^2).

Thus

$$\begin{aligned}\eta_{CP} &= 0.220 + 0.347 - 0.220 \cdot 0.347 - 0.277 \cdot 0.220 \cdot 0.653\\ &= 0.567 - 0.0762 - 0.0398\\ &= 0.451\end{aligned}$$

and this point is shown in Fig. 4.5.

With a boiler efficiency of 0.85, the overall efficiency is 0.383.

Choice of Fluid for the Lower (Bottoming) Cycle

The purpose of choosing an alternative fluid to water for a bottoming cycle is to reduce the size of the L.P. turbine (in comparison with the large steam turbine of today's big power plants). Horn and Norris [7] argue that since low specific volume (v) at (ambient) condensing temperature (T_s) is required, consideration of the perfect gas law,

$$v \approx \frac{\bar{R}T}{Mp_s}, \tag{4.10}$$

shows that the product of saturation pressure (p_s) and molecular weight (M), at condensing temperature, should be as high as possible. Wood argues that in addition the interchange with the higher water cycle should be at about 250° F (120° C) or above. In spite of ammonia's low critical temperature of 270° F (132° C) he considers its possible use in detail, but rejects it on grounds of low Rankine efficiency. Wood examines a number of other refrigerants, as do Horn and Norris. The latter consider not only thermodynamic properties and how they would affect turbine design but also other general factors (toxicity, corrosion, flammability, cost and availability and thermal decomposition). They conclude that dichlorofluoromethane is the most practicable choice.

Although the possibility of an alternative fluid in the bottoming cycle was seriously considered in the late sixties, major development has not subsequently taken place, although a pilot project using ammonia in a low pressure vapour turbine has been undertaken in France.

Choice of Fluids in a Binary Plant Utilising Process Heat

A comprehensive study of small "Rankine" power plants (single fluid or binary cycles) utilising heat from a waste heat source has been carried out by Colosimo, Rose and Decker [8]. They consider five single vapour cycles and two binary cycles utilising a hot gas source (at 700° F, 371° C) and with an available energy flow of 143,500 Btu/hour/° F temperature drop (75.7 kW/K). The two binary cycles are steam/R11 (or R113) and a two level steam cycle (strictly a doubly cyclic single fluid plant rather than a binary plant, using the previous definitions of Chapter 2). The analysis of Colosimo *et al.* is thorough and complex, involving not only thermodynamic properties but also practical design features and plant economics.

Again we do not describe detailed parametric studies here, but note the performance of the particular binary plant they select for construction and test. Colosimo, Rose and Decker calculate the efficiency of an ideal steam/R113 plant with (i) steam evaporating at 2.4 MN/m^2, being superheated to 343° C, and condensing at 0.47 MN/m^2 (149° C); and (ii) R113 (trichloro-trifluoroethane) evaporating at 1.2 MN/m^2 (149° C) and condensing at 73 kN/m^2 (34° C). They assume that all the heat rejected from the steam cycle is used to evaporate the R113, so the simple ideal relationship

$$\eta_{CPS} = \eta_H + \eta_L - \eta_H \eta_L \tag{4.1}$$

applies directly. For the conditions they proposed

$$\eta_H = 0.149$$

$$\eta_L = 0.212$$

so that $\eta_{CPS} = 0.149 + 0.212 - 0.149 \cdot 0.212 = 0.329$. This point is shown on Fig. 4.1. (We should note that the choice of operating conditions for this plant was not based on thermodynamic criteria alone, but economic factors were considered, including the capital cost per unit power output.)

The thermodynamic case for the Kalina cycle (described in Section 2.4.3.1) is strong for those applications using process heat, but again we do not discuss parametric studies here; applications to particular conditions have been described by Kalina and his colleagues [9, 10].

4.3.1.2 "JOULE"/"RANKINE" (GAS/VAPOUR) SERIES PLANT

Parametric analyses of open circuit/closed cycle combined plants are complex, but some of the main features may be determined by the simple study of a binary plant (doubly cyclic). In this section, the higher cycle is taken to be a closed gas turbine cycle, with no supplementary heating of the turbine exhaust, and the lower cycle a closed steam cycle. But even with this binary system (usually of series type, with no heat unused in that rejected from the higher cycle) some parameters must be fixed while others are varied, in order to simplify the calculations (Buxmann and Bormann [11]).

To illustrate the method of performance calculation we work through a combined "series" cycle from an assumed top temperature T_3 in the higher gas turbine cycle (Fig. 4.8). For a given turbine efficiency and a selected pressure ratio r the temperature at turbine exit is determined. For constant specific heats (c_p, c_v) it is given by

$$T_4 = T_3 - (\eta_T)\left(1 - \frac{1}{\rho}\right), \tag{4.11}$$

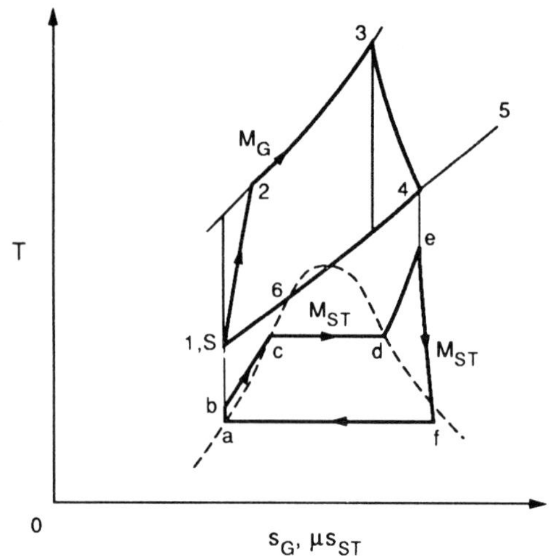

FIGURE 4.8 Binary "Joule"/"Rankine" plant—temperature-entropy diagram

where $\rho = r^{(\kappa - 1/\kappa)}$. Alternatively, gas tables (e.g. Keenan and Kaye [12]) may be used to find T_4.

We next refer to the heat exchange process between the two cycles to see the limitations that are imposed there. If the upper temperature difference $\Delta T_{4e} = T_4 - T_e$ is fixed, with the gas temperature at exit from the turbine T_4 now determined, the top temperature in the steam cycle is obtained, T_e. If the evaporation pressure (p_c) is pre-selected, then both the enthalpy (h_e) and the saturation temperature (T_c) are known, and hence, from an assumed pinch point temperature difference (ΔT_{6c}) the gas temperature T_6 is determined. If there is no heat loss, the heat balance in the HRSG is

$$M_G(h_4 - h_6) = M_{ST}(h_e - h_c), \tag{4.12}$$

where M_G and M_{ST} are the gas and steam mass flows. All the values of enthalpy are known, so the mass flow ratio (M_G/M_{ST}) is obtained. Further, if the condensing pressure is fixed, the enthalpy h_a is known, together with the enthalpy h_b at exit from the feed pump (for given pump efficiency). A further application of the heat balance in the whole exchanger,

$$M_G(h_4 - h_S) = M_{ST}(h_e - h_b), \tag{4.13}$$

yields the enthalpy h_S and temperature T_S. In Buxmann and Bormann's calculations this temperature is taken as the temperature at entry to the compressor ($T_1 = T_S$), there being no "unused" heat rejection from the closed gas cycle. Temperature and pressure levels are independent of atmospheric conditions.

Determination of the location of the remaining state points in the two cycles is simply completed. For the vapour cycle, an assumed turbine isentropic efficiency $(\eta_T)_{ST}$ yields the enthalpy (and the dryness fraction) at the turbine exit (state f). For the gas cycle, from the initially assumed pressure ratio (r) the temperature at exit from the compressor is obtained if the compressor isentropic efficiency (η_C) is specified. With constant specific heats

$$T_2 = T_1 + \frac{(\rho - 1)T_1}{(\eta_C)}, \tag{4.14}$$

or the gas tables can be used again to find h_2 and T_2. Finally, the required heat input can be obtained,

$$Q_B = M_G(h_3 - h_2), \tag{4.15}$$

the total work output is given by

$$\begin{aligned} W &= W_H + W_L \\ &= M_G[(h_3 - h_4) - (h_2 - h_1)] \\ &\quad + M_{ST}[(h_e - h_f) - (h_b - h_a)], \end{aligned} \tag{4.16}$$

and the combined plant thermal efficiency is

$$\eta_{CPS} = W/Q_B. \tag{4.17}$$

Pressure losses can also be allowed for in such a calculation if required.

For most parametric calculations the following are assumed to be held constant:

$$(\eta_C), (\eta_T)_G, (\eta_T)_{ST}, \Delta T_{4e}, \Delta T_{6c}.$$

p_c and T_3 can then also be specified and the gas turbine pressure ratio r varied. With the same p_c, another T_3 can be prescribed and r again varied. Thermal efficiency can thus be determined for the specified p_c and two variable parameters, r and T_3. Such calculations are illustrated in Fig. 4.9.

Alternatively, for a given T_3, thermal efficiency can be determined for two variable parameters, p_c and r. The results of such calculations are shown in Fig. 4.10.

These results were obtained by Buxmann and Bormann [11] who made the following assumptions for the "constant" parameters:

$$(\eta_T)_G = 0.88,$$
$$(\eta_C) = 0.86,$$
$$p_a = 0.05 \text{ bar},$$
$$(\eta_T)_{ST} = 0.84,$$
$$\Delta T_{4e} = 50^\circ \text{ C},$$
$$\Delta T_{6c} = 20^\circ \text{ C}.$$

In addition they introduced pressure losses of 7% of the upper pressure in the gas cycle and 12% of the upper pressure in the steam cycle; assumed $(h_b - h_a)$ to be 2% of $(h_e - h_f)$; and that the lower pressure level p_1 in the gas cycle was constant at 8 bar.

Figure 4.9 shows that the optimum pressure ratio increases with maximum turbine inlet temperature in the gas cycle. At pressure ratios higher than the optimum, the inlet temperature to the compressor increases and the work output of the gas cycle decreases. At pressure ratios lower than the optimum, the work output of the gas turbine also decreases, and is obviously zero at a pressure ratio of unity, when all the work is produced by the steam turbine. The mass flow ratio varies sharply with pressure ratio as is indicated on Fig. 4.9 for $T_3 = 700^\circ$ C.

The effect of changing the steam pressure (p_c) is illustrated in Fig. 4.10. Increase in p_c gives some increase in efficiency at lower pressure ratio and also lowers the optimum pressure ratio; specifying higher levels than 98 bar for p_c does not increase efficiency further.

Some comments on other possible effects may be made.

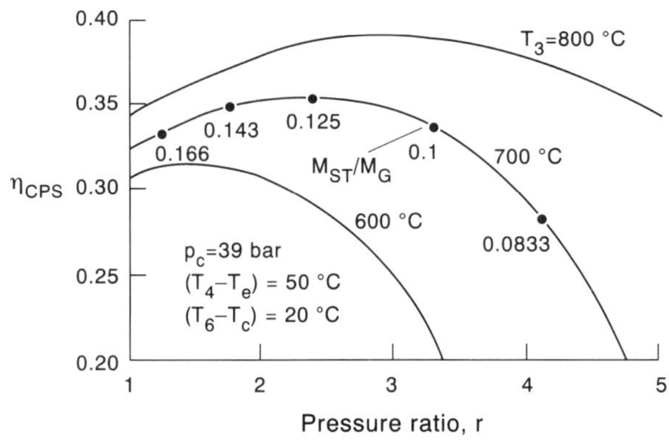

FIGURE 4.9 Thermal efficiency η_{CPS} of binary "Joule"/"Rankine" plant—constant live steam pressure (p_c) (after Buxmann and Bormann [11])

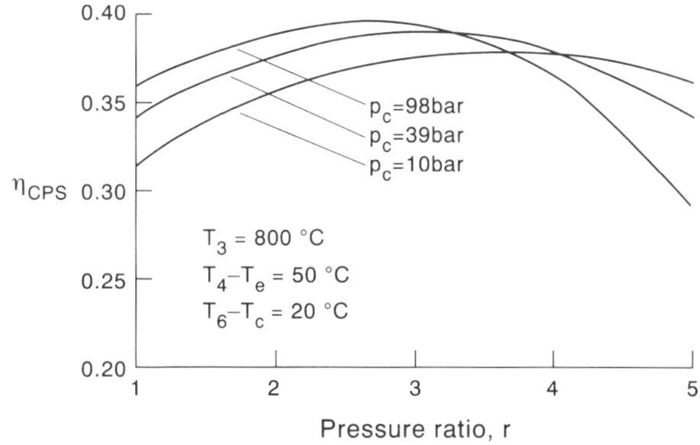

FIGURE 4.10 Thermal efficiency of binary "Joule"/"Rankine" plant—constant maximum temperature T_3 (after Buxmann and Bormann [11])

(i) Cooling the gas before entry to the compressor would increase the power output from the gas turbine cycles, but would lower the overall efficiency.
(ii) Use of alternative gases would have a major effect; for example the use of helium would lower the optimum pressure ratio by 20% to 30%, compared with air.
(ii) Although regenerative feed heating would improve the thermal efficiency of the steam cycle, it would mean a lower evaporation pressure, reducing combined plant efficiency because of the increased temperature differences between gas and steam (see Fig.

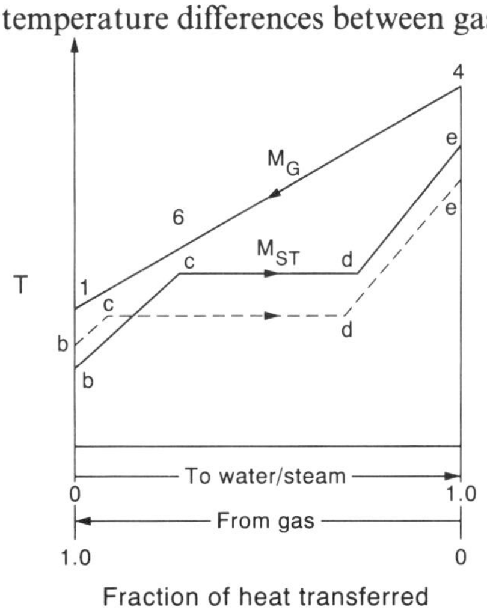

FIGURE 4.11 Temperature in HRSG—effect of feed heating in steam cycle with gas turbine cycle fixed (T_4, T_1, constant, T_b unrestricted) (after Buxmann and Bormann [11])

4.11). We shall return to a discussion of the effects of feed heating later in Section 4.3.2.2.2.

(iv) The distribution of work output between the gas cycle and the steam cycle is much affected by whether a heat exchanger is used in the former. The expressions for work output are

$$W_H = \eta_H Q_B, \tag{4.18}$$

$$W_L = \eta_L(1-\eta_H)Q_B, \tag{4.19}$$

if, as assumed initially, all the heat rejected by the gas cycle is transferred to the steam cycle. Thus the ratio

$$\frac{W_H}{W_L} = \frac{\eta_H}{\eta_L(1-\eta_H)}, \tag{4.20}$$

and we may expect that if η_H is improved by the introduction of a heat exchanger, (W_H/W_L) will increase substantially. A further contribution to this increase is made by the resulting drop in the lower cycle efficiency due to the decrease in steam turbine inlet temperature.

For a binary "Joule/Rankine" series plant, calculations can be carried out to illustrate this effect, comparing the Joule cycle with and without a heat exchanger (CBT and CBTX cycles). Dibelius and Ziemann [13] made assumptions for turbine and compressor efficiencies comparable to those of Buxmann and Bormann, but held the *upper* gas cycle pressure constant at 60 bar, and the condenser temperature at 37° C. The optimum pressure ratio (between 2 and 2.8) was found to depend little on maximum turbine inlet temperature, which was however the main influence on both combined cycle efficiency and the power ratio, as Fig. 4.12 shows (presumably for optimum pressure ratio). η_{CPS} is substantially higher for the cycle with the heat exchanger, as is the power ratio. This is essentially because of the higher gas cycle efficiency, as explained above.

For the non-heat exchanger case Dibelius and Ziemann introduced some feed heating into the steam cycle which will have increased its thermal efficiency, but the effect on combined plant efficiency is likely to be adverse—see Section 4.3.2.2.2. In discussing these results, Buxmann and Bormann point out that higher efficiency can be obtained with the CBT gas turbine plant by the addition of reheat in the steam cycle.

4.3.1.3 "JOULE"/"RANKINE" SERIES/PARALLEL PLANT

For two plants joined in series and in parallel (through supplementary heating), we have already considered in a general way the effect on the combined plant efficiency of variations in efficiency of the component

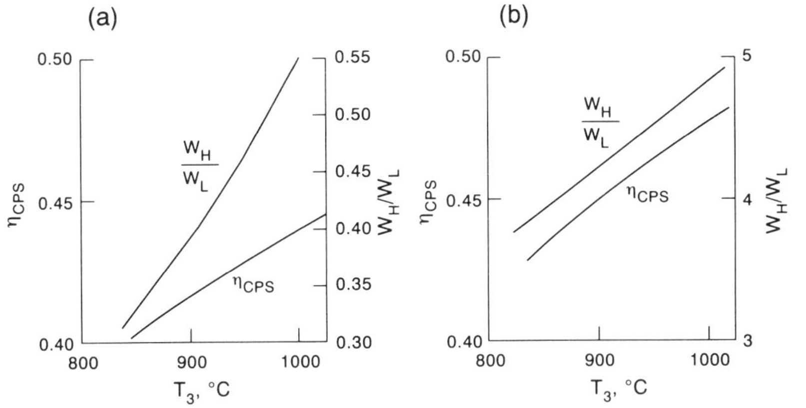

FIGURE 4.12 Binary "Joule"/"Rankine" series plant—thermal efficiency (η_{CPS}) and work ratio (W_G/W_{ST})
(a) CBT higher cycle
(b) CBTX higher cycle
(after Buxmann and Bormann [11])

plants and the fraction of total heat supply that is absorbed by the second cycle ($\nu_S = Q_S/Q_B$) (see Section 4.2.4). The resulting expression for efficiency is

$$\eta_{CPSP} = \eta_H + \eta_L - \eta_H\eta_L - \nu_S\eta_H(1-\eta_L), \tag{4.6a}$$

or

$$\eta_{CPSP} = \eta_L + q\eta_H - q\eta_H\eta_L \tag{4.6d}$$

where

$$q = (1-\nu_S) = Q_H/Q_B = Q_H/(Q_H + Q_S),$$

when all the heat rejected from the higher plant is supplied to the lower plant ($Q_{UN} = 0$).

If, however, there is also heat unused (or "lost") between the two plants, then there is a further drop in efficiency of the combined plant so that

$$\eta_{CP} = \eta_{CPS} - \nu_S\eta_H(1-\eta_L) - \nu_{UN}\eta_L, \tag{4.8a}$$

where

$$\eta_{CPS} = \eta_H + \eta_L - \eta_H\eta_L,$$

and

$$\nu_{UN} \equiv \frac{Q_{UN}}{Q_B} = \frac{Q_{UN}}{(Q_H + Q_S)}.$$

There are many ways in which the two plants may be linked, in series and in parallel, with both supplementary heat supply and heat "loss" involved. Some practical open circuit/closed cycle combinations were considered in considerable detail by Seippel and Bereuter [14]; Buxmann and Bormann [11] have studied binary (doubly closed cyclic) plants, undertaking some parametric studies which mainly involve the variation of parameters in the higher (gas turbine) cycle, and we summarise their results here.

Four series-parallel binary plants are proposed by Buxmann and Bormann. In the first, parallel heating of both the gas turbine and the steam turbine plant (for evaporation) is employed; but an element of series heating is achieved by the gases exhausting from the gas turbine heating both the feed water and the evaporated steam (to a superheated state) in the steam turbine cycle. An air cooler then cools the air before entry to the compressor (i.e. some heat is "lost" between the two plants).

In the second (Fig. 4.13) the introduction of a heat exchanger in the upper cycle may obviate the need for the air cooler; the condenser pressure then controls the temperature at entry to the compressor. The only series coupling between the plants is now in the heat transferred from the gases to the feed water. A third, more complicated, plant involves reheating of the steam, retention of the heat exchanger and the reintroduction of the air cooler. Again the only series coupling between the plants is in the feed water heater.

Finally, in the most complex plant proposed by Buxmann and Bormann intercooling between two compressors is introduced, the heat exchanger in the gas turbine plant retained, and the steam is reheated. Steam is also bled between the H.P. and L.P. turbines to heat the feed water. Again the only series coupling between the plant is in the water heater used before the feed heater. The condenser pressure again controls the temperature at entry to the low pressure compressor.

Buxmann and Bormann's parametric studies of joint heating (series/parallel operation) concentrated on the configuration of Fig. 4.13, but with and without an air cooler, and with and without steam reheating. In order to restrict the very large range of possible variable parameters, they specified a number of fixed quantities as shown in Table 4.1.

The results of Buxmann and Bormann are as follows.

Combined Plant Without Steam Reheat

The thermal efficiency of the steam cycle, without feed water preheating, was calculated to be $\eta_L = 0.319$ for the steam parameters chosen. The ratio of combined plant efficiency η_{CPSP} to η_L is plotted in Fig. 4.14, against the two major gas turbine plant parameters, compressor pressure ratio (r) and gas turbine top temperature (T_3). The corresponding variation of gas turbine plant efficiency is shown at the bottom of Fig. 4.14. Pressure losses and real gas effects (i.e. specific heat variation with

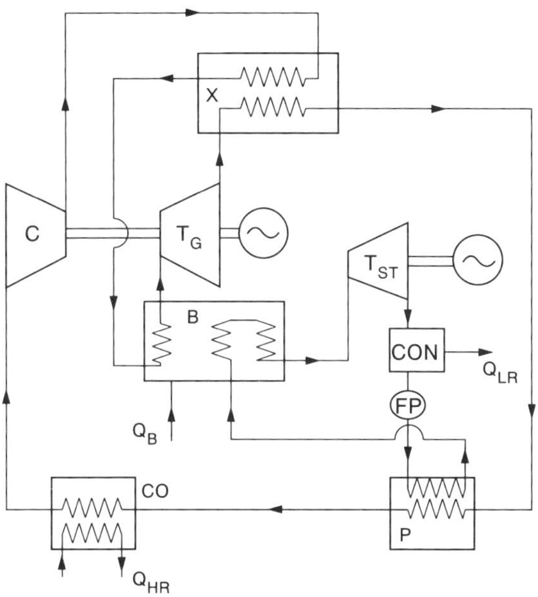

C	Compressor
T_G	Gas turbine
X	Heat exchanger
B	Heater (joint heating of gas) evaporation and superheating of water/steam
T_{ST}	Steam turbine
CON	Condenser
FP	Feed pump
P	Preheater (economiser)
CO	Gas cooler

FIGURE 4.13 Joint heating of "Joule"/"Rankine" combined plant (after Buxmann and Bormann [11])

temperature) were included in the calculations, and the pinch-point temperature difference (at the cold end of the heat exchanger) was taken as 30° C. The ratio of external heat supplied to the two cycles $(Q_H/Q_S) = (1/\nu_S - 1)$ is shown in Fig. 4.15 together with the corresponding work ratio (W_H/W_L).

At low pressure ratios (less than about 5), for a given top temperature T_3, the combined cycle efficiency increases with pressure ratio, in spite of the decreasing gas turbine plant efficiency. Since the upper pressure in the gas turbine plant is held constant, the compressor inlet pressure drops (but inlet temperature is held constant) as pressure ratio is increased, and the air temperature at compressor outlet increases. Less heat is transferred to this air in the heat exchanger and more heating of the feed water becomes possible; the fraction of external heating of the steam cycle in the joint heater (ν_S) is therefore decreased. From equation (4.6a), the decrease in the term $\nu_S(1-\eta_L)\eta_H$ compensates for the drop in η_H, while η_L remains constant. This effect continues up to the point when the (specified) saturation temperature of the boiler water is reached; combined plant efficiency then falls with further increase of pressure ratio because of the developing temperature difference in the water/steam heater.

It is important to realise that the general equation for the combined plant efficiency of this series/parallel plant (with no intermediate heat loss from the gas turbine plant),

$$\eta_{CPSP} = \eta_H + \eta_L - \eta_H\eta_L - \nu_S\eta_H(1-\eta_L), \tag{4.6a}$$

is valid for all these calculations, even though η_H and μ_S are changing as the gas turbine plant parameters, pressure ratio and top temperature, are varied. For example, for $T_3 = 950°$ C and $r = 3.8$, from Fig. 4.15, $(Q_H/Q_S) = 0.55$ and $\nu_S = 1/1.55 = 0.645$. η_H for the gas turbine plant is 0.349, and $\eta_L = 0.319$ so that the equation for overall efficiency is

TABLE 4.1 *Assumptions of Buxmann and Bormann* [11]

Gas Turbine Plant
(i) pressure at turbine inlet, 40 bar
(ii) cooling water temperature, 15° C
(iii) compressor efficiency, 0.88
(iv) turbine efficiency, 0.90
Steam Turbine Plant
(i) turbine entry conditions (no reheat) 50 bar, 520° C
(ii) turbine entry conditions (with reheat) 190 bar, 530° C; reheat to 40 bar, 530° C
(iii) turbine efficiency 0.86
(iv) feed pump efficiency 0.75
(v) condenser pressure 0.05 bar
Combined Heater Efficiency, 0.92

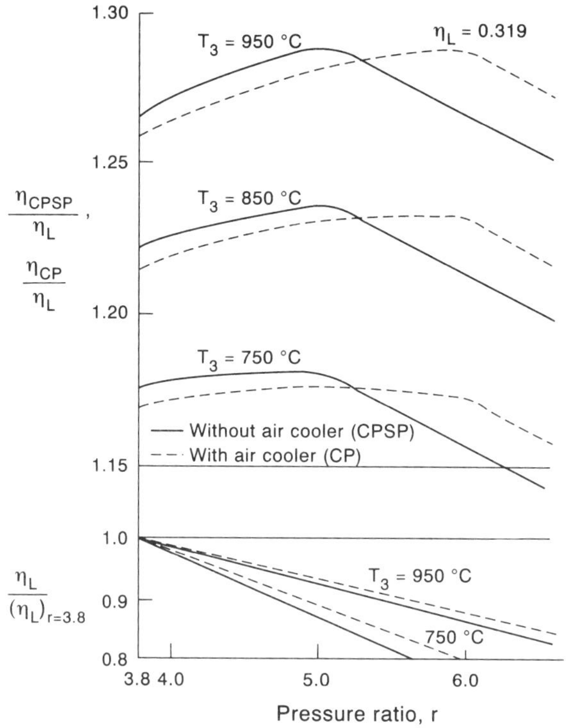

FIGURE 4.14 Thermal efficiency of "Joule"/"Rankine" plant with joint heating (after Buxmann and Bormann [11])

$$\eta_{CPSP} = 0.349 + 0.319 - 0.349 \cdot 0.319 - 0.645 \cdot 0.349 \cdot 0.681$$
$$= 0.404.$$

From Fig. 4.14, for $T_3 = 950°$ C and $r = 3.8$, $(\eta_{CPSP}/\eta_L) = 1.265$, so that $\eta_{CPSP} = 1.265 \cdot 0.319 = 0.404$ as we have just calculated.

For a higher pressure ratio of 5.0 (with T_3 still 950° C),

$$\eta_H = 0.349 \cdot 0.925 = 0.323,$$
$$\nu_S = 1/1.715 = 0.583,$$

so that

$$\eta_{CPSP} = 0.323 + 0.319 - 0.323 \cdot 0.319 - 0.583 \cdot 0.323 \cdot 0.681$$
$$= 0.411,$$

which again agrees with Buxmann's calculations, $\eta_{CPSP} = 1.287 \cdot 0.319$

$=0.411$. This illustrates how η_{CPSP} increases although η_H falls, because of the decrease in the last term, $\nu_S\eta_H(1-\eta_L)$.

When an additional air cooler is added, the change in combined plant efficiency is as illustrated in Fig. 4.14. The combined plant efficiency drops, for a given (low) pressure ratio, in spite of the increase in the gas turbine cycle efficiency. However, the pressure ratio at which maximum combined plant efficiency is reached (when boiler saturation temperature is attained at exit from the feed water heater) is now higher. The ratio of gas turbine to steam turbine work is as illustrated in Fig. 4.15, together with the ratio of heat supplied.

Now the overall equation relating efficiencies is

$$\eta_{CP}=\eta_{CPS}-\nu_S\dot{\eta}_H(1-\eta_L)-\nu_{UN}\eta_L \tag{4.8a}$$

where

$$\nu_{UN}=\text{(heat "loss" between plants)/total heat supplied}$$
$$=\text{(heat rejected to air cooler)}/(Q_H+Q_S)$$

and

$$\nu_S=Q_S/(Q_H+Q_S).$$

As the gas turbine plant parameters (r and T_3) are varied, so η_H, ν_S, ν_{UN}

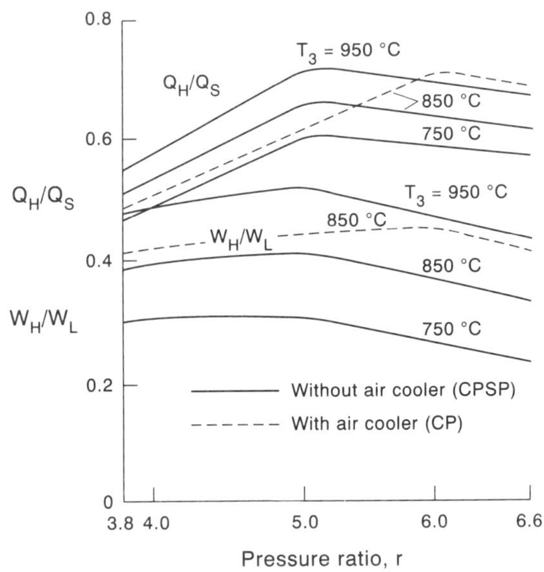

FIGURE 4.15 "Joule"/"Rankine" plant with joint heating. Ratios of heat supply and work output (after Buxmann and Bormann [11])

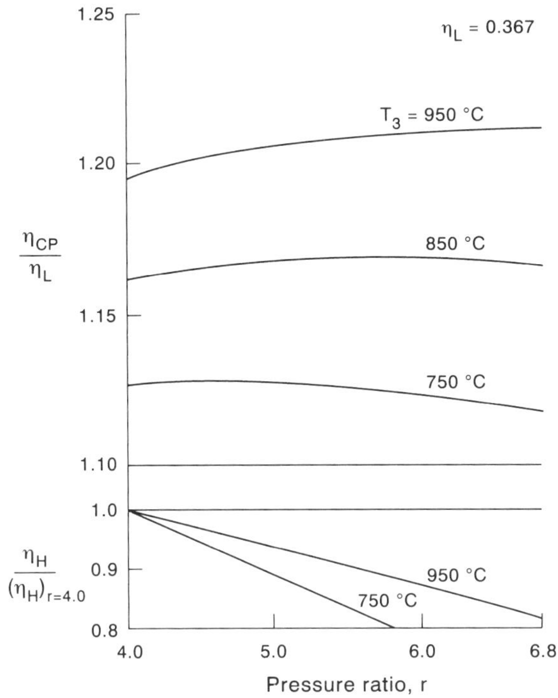

FIGURE 4.16 "Joule"/"Rankine" plant with joint heating and steam reheating—thermal efficiency (after Buxmann and Bormann [11])

and now η_L are changed, but the results shown in Fig. 4.14 all obey the overall equation (4.8a).

Combined Plant With Steam Reheat

Similar calculations for the plant with steam reheat, with the air cooler included, are shown in Fig. 4.16. The efficiency of the basic steam cycle is increased to 0.367, but it now absorbs a bigger fraction of the total heat supply (at lower pressure ratios), ν_S increases and the overall efficiency is reduced (for a given pressure ratio and top temperature).

The equation for combined plant efficiency,

$$\eta_{CP} = \eta_{CPS} - \nu_S \eta_H (1 - \eta_L) - \nu_{UN} \eta_L, \tag{4.8a}$$

is again valid for these calculations, with η_L now equal to 0.367 because of the introduction of reheat. Other fixed quantities remained unchanged.

TABLE 4.2 *Parametric Studies of Open Circuit (Gas Turbine)/Closed Circuit (Steam Turbine) Combined Plant*

	Note
Cerri [2]	Detailed parametric calculation of thermal efficiency (no feed heating)
Cerri and Colagé [15]	Effect of regenerative feed heating
Rufli [3]	Detailed parametric calculation of thermal efficiency (feed heating included)
Kehlhofer [4]	Comprehensive handbook based on a number of types of gas turbine/steam turbine combined plant (CCGT)
Pfenninger [16]	General statement of parametric results
Wunsch [17]	General statement of parametric results
Negri di Montenegro *et al.* [18]	Limited optimisation study
Korpela [19]	Limited parametric study
Timmermans [20]	Limited study emphasising effect of gas turbine specific work
Rice [21]	Study emphasising advantages of reheat gas turbine combined plant
Mangan and Pettit [1]	Early limited study of advantages of combined plant
Hemsley and Roberts [22]	Emphasis on selection of steam turbines for combined plant
Mayers *et al.* [23]	Early limited study of advantages of combined plant
El-Masri [24]	Exergy (2nd Law) analysis (see Chapter 5)
Manfrida *et al.* [25]	Exergy (2nd Law) analysis (see Chapter 5)

4.3.2 Open Circuit/Closed Cycle (CCGT) Plants

There have been many studies of the performance of combined open circuit (gas turbine)/closed cycle (steam turbine) plants, which we shall refer to as CCGT plants (although they are not strictly combined *cycles*). Table 4.2 gives a list of references to this work.

In Section 4.3.2.1 we report first on two of the more comprehensive parametric analyses, those by Cerri [2] and Rufli [3]. In each of these papers an open circuit gas turbine supplies an HRSG (both with or without supplementary firing of the exhaust in Cerri's work, but without in Rufli's calculations); the steam cycle is a simple one without reheat. The variation of the overall efficiency of the combined plant with some important thermodynamic properties and parameters is calculated. In Cerri's analysis it is assumed that there is no regenerative feed heating in the steam cycle, but Rufli assumes it is present. The role of feed heating (with or without supplementary firing of the exhaust, but not fully or maximally fired) is discussed later.

Also in Section 4.3.2.1 we summarise a comprehensive review of this type of plant by Kehlhofer [4], based on his wide experience with Asea Brown Boveri. In his practical handbook Kehlhofer presents separate results for

(a) a simple plant with a single pressure steam system;

(b) a single pressure system, with a pre-heater loop;
(c) a two pressure system for fuels containing sulphur;
(d) a two pressure system for fuels containing no sulphur;
(e) a plant with supplementary firing;
(f) a maximally fired plant.

Subsequently in Section 4.3.2.2 we discuss four other aspects of the open circuit/closed cycle combined plant.

In Section 4.3.2.2.1 we emphasise the importance of gas turbine specific work on overall plant efficiency, using arguments put forward by Timmermans.

In Section 4.3.2.2.2 we present a simple analysis by Horlock [26] of the effects of feed heating.

In Section 4.3.2.2.3 we discuss a strong case presented by Rice [21] for reheating in the gas turbine rather than supplementary heating of the gas turbine exhaust.

In Section 4.3.2.2.4 we consider the "off-design" performance of open circuit (gas turbine)/closed cycle (steam turbine) combined plants.

4.3.2.1 ESTIMATES OF CCGT PLANT PERFORMANCE

4.3.2.1.1 Cerri's Calculations

Cerri's approach is similar to that followed by Buxmann and Bormann for the binary (doubly cyclic) plant but a greater degree of optimisation is carried out. Provision is made for small corrections (e.g. pressure losses, mechanical efficiencies). Referring to Fig. 4.17a which gives the notation used, Cerri starts from the statement that overall efficiency depends on the following parameters

$$(\eta_O)_{CP} = f(T_3, r, \lambda, p_c, T_e, \delta), \tag{4.21}$$

where r is the pressure ratio, λ is the ratio of steam mass flow to air mass flow $\left(\frac{M_{ST}}{M_A}\right)$ and δ is the ratio of the "heat" supplied in the afterburner to that supplied in the gas turbine combustion chamber, $\delta = \frac{Q_S}{Q_H} = \frac{\nu_S}{(1-\nu_S)}$. This statement implies that a number of other parameters are pre-selected and held constant, viz. compressor and turbine efficiencies (polytropic), mechanical efficiency, inlet air temperature (T_1), condenser pressure and temperature (p_a, T_a), combustion chamber efficiency (η_{CC}), pressure losses, heating value of the fuel and stoichiometric air–fuel ratio. (A list of the values assumed by Cerri for these quantities is given in Table 4.3.)

In addition a number of limits are imposed, on

(i) upper steam temperature ($T_e = 542°$ C), or a minimum value of the temperature difference ($T_4 - T_e = 50$ K);
(ii) maximum wetness at steam turbine exhaust (0.30);
(iii) a minimum pinch-point temperature difference of 30 K;
(iv) a maximum allowable value of δ, corresponding to a

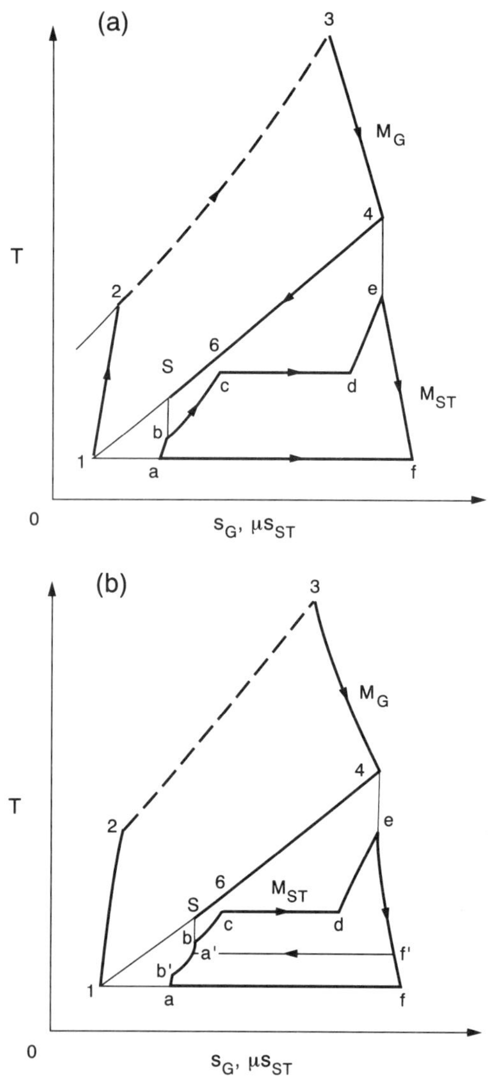

FIGURE 4.17 Open circuit/closed cycle combined plant—temperature-entropy diagrams: (a) no regenerative feed heating, (b) regenerative feed heating

stoichiometric air–fuel ratio of 14, and allowing for a minimum excess of air after supplementary combustion of 6%;

(v) a minimum value of the exhaust stack temperature, $T_S = 160°C$ (Cerri argues that this minimum value is sufficient to prevent corrosive condensation in the HRSG for his conditions, with no feed heating, but we return to this point later when referring to Kehlhofer's work.)

It should be emphasised that these quantities are not necessarily imposed (with one exception) but are limits on the validity of the calculations. The one exception is the upper steam temperature, which is prescribed when supplementary firing is used, and then effectively determines the degree of afterburning and the parameter δ.

With these boundary conditions set, Cerri's method of analysis is as follows. An initial decision is made on the value of δ—either zero (no supplementary firing) or a value which is sufficient to raise the steam temperature to its maximum (prescribed) level.

For no supplementary firing, equation (4.21) becomes

$$(\eta_O)_{CP} = f(T_3, r, \lambda, p_c, T_e). \tag{4.22}$$

Cerri now advances the calculation by prescribing T_3, selecting two of the remaining parameters (r and λ) and optimising the others (p_c, T_e) to give "maximal" efficiency, $(\eta_O)^* = f(r, \lambda)$, within the prescribed limits set by conditions (i) to (v) above. Figure 4.18 shows the result of a set of such calculations of $(\eta_O)^*$ (for $T_3 = 1000°$ C, $\delta = 0$) against pressure ratio r, for a range of values of λ. The specific work (work output per unit mass of air flow) can also be calculated.

For supplementary firing, δ has to be sufficient to raise the steam temperature to the prescribed upper limit of T_e. Then

$$(\eta_O)_{CP} = f(T_3, r, \lambda, p_c). \tag{4.23}$$

T_3 may be set, values of r and λ selected and p_c again varied to give maximal efficiency, $(\eta_O)^*$. Figure 4.19 shows the result of a set of such

TABLE 4.3 *Cerri's Assumptions*

$p_1 = 1$ bar	$T_1 = 15°$ C
$p_a = 0.05$ bar	$T_c = 32.5°$ C
$(\eta_p)_C = 0.89$	$(\eta_p)_{GT} = 0.88$
$(\eta_p)_{ST} = 0.86$	$(\eta_{mechanical}) = 0.98$
$(\eta_{COMB}) = 0.98$	$(\eta_B) = 0.96$
$(p_2 - p_3)/p_2 = 0.03$	$(p_4 - p_1) = 0.05$ bar

$(CV)_0 = 42{,}700$ kJ/kg (it is assumed that the combustion process approximates to heating of air at T_2 to products at T_3)

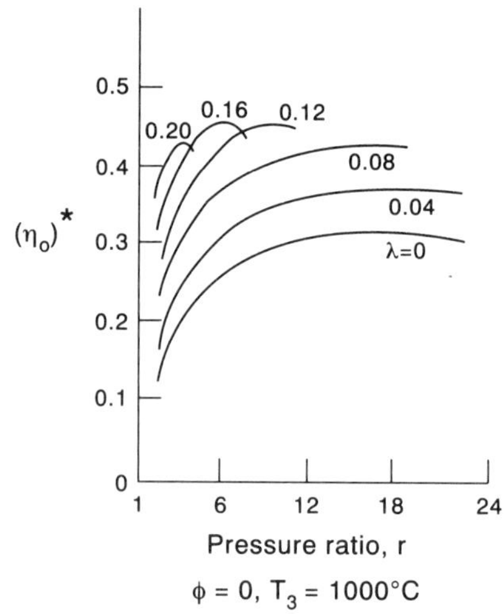

FIGURE 4.18 "Maximal" efficiency, $(\eta_O)^*$—no supplementary heating (after Cerri [2])

calculations of $(\eta_O)^*$ (for $T_3 = 1000°$ C, and δ sufficient to raise a steam temperature T_e of 542° C) against pressure ratio, for various values of λ.

A further step in the optimisation is now possible. For each of the constant λ lines on Fig. 4.18 for $\delta = 0$ (or on Fig. 4.19 for $\delta \neq 0$), there is an optimum pressure ratio to give maximum $(\eta_O)^*$. Thus for each gas

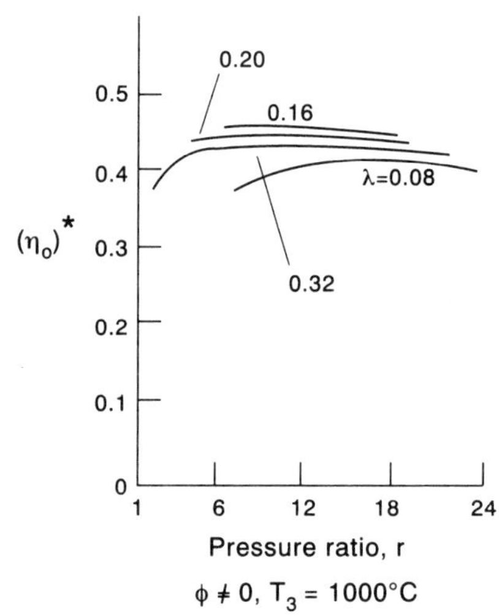

FIGURE 4.19 "Maximal" efficiency, $(\eta_O)^*$ with supplementary heating (after Cerri [2])

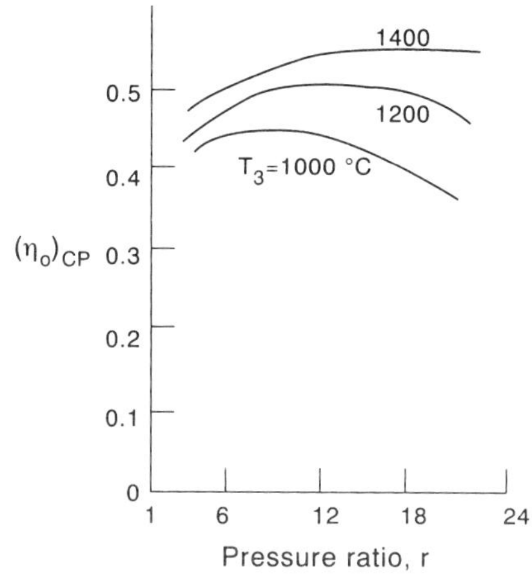

FIGURE 4.20 Best efficiency, $(\eta_O)_{CP}$—no supplementary heating (after Cerri [2])

turbine pressure ratio there is a corresponding λ, and we can draw the envelope of the maxima on Figs. 4.18 and 4.19 to give the best possible combined plant efficiency. Figures 4.20 and 4.21 show these envelopes for different T_3, and these figures summarise Cerri's results for no regenerative feed heating in the steam cycle.

A modification to the method of calculation, if regenerative feed heating is employed, is described by Cerri and Colagé [15]. They define the degree of regeneration R as the ratio of the feed-water enthalpy rise

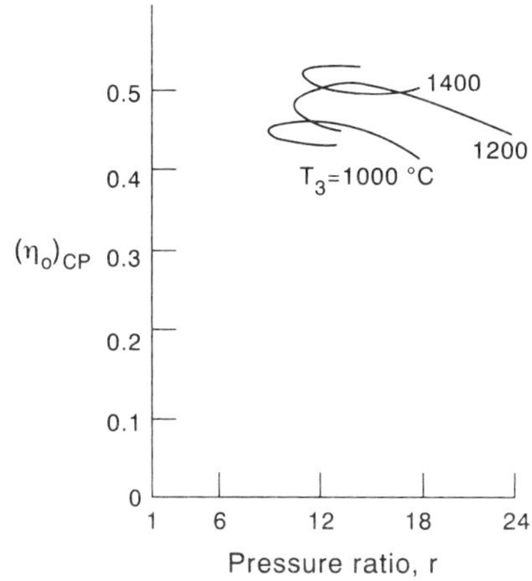

FIGURE 4.21 Best efficiency, $(\eta_O)_{CP}$—with supplementary heating (after Cerri [2])

(due to feed heating) to the maximum possible (the difference between the enthalpy at boiler saturation pressure and that at condenser pressure). Referring to Fig. 4.17b which shows the notation used,

$$R = \frac{(h_{a'} - h_a)}{(h_c - h_a)}. \tag{4.24}$$

In some preliminary calculations, Cerri and Colagé find that increasing the number of feed heaters above $Z=3$ produces little increase in overall plant efficiency, and that equally spacing the intermediate feed-water temperatures produces close to optimum conditions. (Similar conclusions were reached by Haywood [27] in his basic study of feed heating.) Equation (4.21) for combined plant efficiency may now be modified to give

$$(\eta_O)_{CP} = f(T_3, r, \lambda, p_c, T_e, \delta, R, Z), \tag{4.25a}$$

where Z is the number of feed heaters. However, on the basis of their preliminary calculations, Cerri and Colagé select $Z=3$, so the only additional parameter to be considered when feed heating is introduced is the ratio R, and

$$(\eta_O)_{CP} = f(T_3, r, \lambda, p_c, T_e, \delta, R), \tag{4.25b}$$

The parameter δ is determined as in the earlier calculations with no feed heating, i.e. it is either zero, or sufficient to raise the steam temperature to its maximum value of 542° C. The fixed parameters are held at the same values as in Table 4.3, and the various limitations are again imposed, but with an additional limit on the minimum difference between gas outlet temperature at exit from the exhaust heater and the economiser inlet temperature.

The method of calculation is then again one of optimisation. For example, with values of T_3, R and r selected, the remaining parameters (λ, p_c, T_e) are selected to give maximum efficiency. Figure 4.22 shows the results of such a calculation for $\delta=0$. Similar optimisation may be undertaken for δ sufficient to raise the steam temperature (T_e) to its maximum value of 542° C.

Figure 4.23 shows a summary of results of these calculations—combined plant efficiency plotted against gas turbine maximum temperature, for three pressure ratios ($r=4$, 6, 8) and for $R=0$ and 0.2, and $\delta=0$ (no supplementary heating).

Cerri's calculations are indeed so comprehensive that it is difficult to summarise the overall results. However, the important results appear to be as follows.

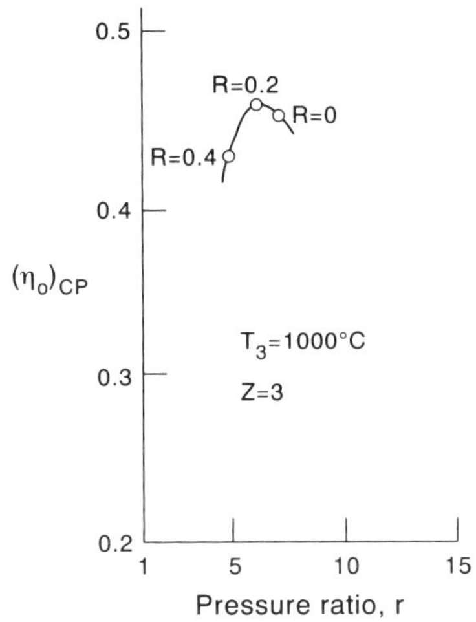

FIGURE 4.22 Maximum efficiency with regenerative feed heating—no supplementary heating (after Cerri and Colagé [15])

For no feed heating

(i) Pressure ratio can be selected over quite a wide range, since given proper choice of other parameters, the combined plant efficiency is largely independent of r.

(ii) Similarly, steam evaporation pressure has but a small effect on combined plant efficiency $(\eta_O)_{CP}$.

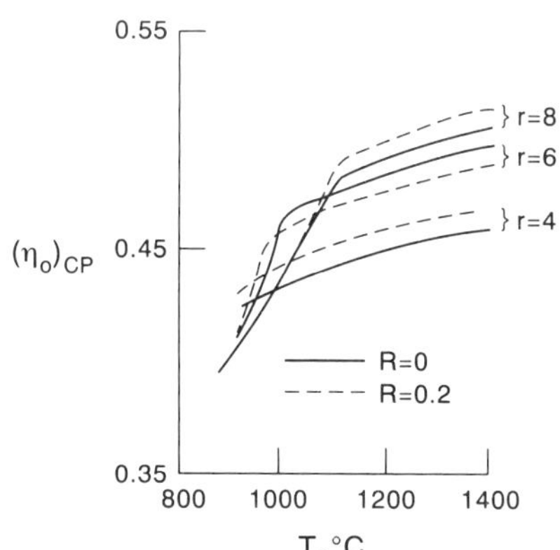

FIGURE 4.23 Maximum efficiency with and without regenerative feed heating (after Cerri and Colagé [15])

(iii) $(\eta_O)_{CP}$ increases with maximum steam temperature T_e.
(iv) With optimum choice of other parameters, supplementary firing increases efficiency.

For feed heating

(i) In general, regenerative feed heating leads to lower $(\eta_O)_{CP}$. However, if the pressure ratio in the gas turbine plant is low, and exhaust gas turbine temperatures correspondingly high, then a small degree of regeneration ($R = 0.2 - 0.3$) increases $(\eta_O)_{CP}$. Feed heating then assists in the attainment of high steam temperatures and increases the steam mass flow and power output.
(ii) For higher pressure ratios and low exhaust temperatures, supplementary firing then again allows a modest degree of regenerative feed heating to increase combined plant efficiency, for the same reason with the same effects (attainment of higher steam temperature, greater steam mass flow and power output).

Cerri's calculations are for a single pressure steam cycle. He limits the final gas temperature (for the case with no feed heating) and the difference between gas and water temperatures at the cold end of the economiser ($T_S - T_b$), but with feed heating appears to allow the water temperature T_b to vary with choice of $R = \dfrac{(h_{a'} - h_a)}{(h_c - h_a)}$.

We shall refer later in Section 4.3.2.1.3 to a practical viewpoint expressed by Kehlhofer, who imposes a limit on T_b, set by the dew point of the exhaust gases at state S, depending on the sulphur content of the fuel. We should note that such a limit on T_b means that a "pre-heating" loop is required before the water enters the economiser, for the case with no regenerative feed heating. However, this does not invalidate calculations such as Cerri's, for Kehlhofer argues that the plant efficiency is independent of T_b in this case; it is simply that a particular way of extracting the heat from the exhaust has to be devised. For the case of regenerative feed heating, the efficiency *is* affected by a prescribed T_b, the amount of steam that has to be bled from the turbine being proportional to $(h_{a'} - h_a)$.

4.3.2.1.2 Rufli's Calculations

The second major parametric calculation that we describe here is that by Rufli [3]. Rufli makes assumptions for pressure losses and efficiencies similar to those made by Cerri, but there are several critical additions or differences in his work:

(i) Rufli allows for single or dual pressure steam cycles, introducing an additional parameter μ, the ratio of the mass flow through the

H.P. evaporator to the total steam flow (for $\mu = 1$, there is only a single evaporator).

(ii) He uses work by Traupel [28] to allow for the effect of cooling air flow on expansion in the gas turbine.

(iii) He considers the effect on HRSG design of various thermodynamic parameters, producing estimates of heat transfer surface area per unit mass flow, an important quantity affecting the cost of the plant.

(iv) He assumes that regenerative feed heating is always used.

Assumption (iv) leads to a different approach to the method of calculation. For any particular calculation point, Rufli specifies both the temperature of the feed water before entry to the HRSG (T_b) *and* the gas temperature at entry to the exhaust stack (T_S) [separate values of the latter for oil fuel (175° C) and natural gas fuel (125 °C)]. These limitations are close to those which Kehlhofer recommends, as we discuss later.

Referring to Fig. 4.17b, if the pressure ratio r and maximum temperature T_3 in the gas turbine plant are specified, then the temperature at exit from the turbine can be found (T_4). For no supplementary heating, the temperature difference ΔT_{4e} is specified, and the maximum temperature in the steam cycle (T_e) is found. In a typical calculation, both the feed-water temperature (T_b, after bled steam feed heating) and the exhaust stack temperature T_S are also specified by Rufli; for single phase heating in the steam cycle, a boiler pressure p_c is then guessed, with a corresponding saturation temperature T_c, so that for a specified pinch-point temperature difference the gas temperature T_6 is obtained. The ratio of gas and steam mass flows then follows,

$$\lambda = \frac{M_{ST}}{M_G} = \frac{(h_4 - h_6)}{(h_e - h_c)}, \tag{4.26}$$

and new values of "stack" enthalpy and temperature are obtained from an energy balance in the economiser.

The stack temperature thus determined is then compared with the specified stack temperature. The process is repeated with different guessed values of boiler pressure p_c until the specified stack temperature is obtained. The steam cycle is then completely specified and the combined plant efficiency can be determined.

This calculation can be repeated for a different value of T_b; the bled steam pressure is set by each choice of $T_b (\approx T_{a'})$ and the amount of bled steam determined by the enthalpy rise $(h_{a'} - h_a)$.

Similar calculations can be performed for a dual pressure steam cycle, the boiler pressure p_c being determined for a given value of the mass flow ratio μ. A typical set of such calculations is shown in Fig. 4.24.

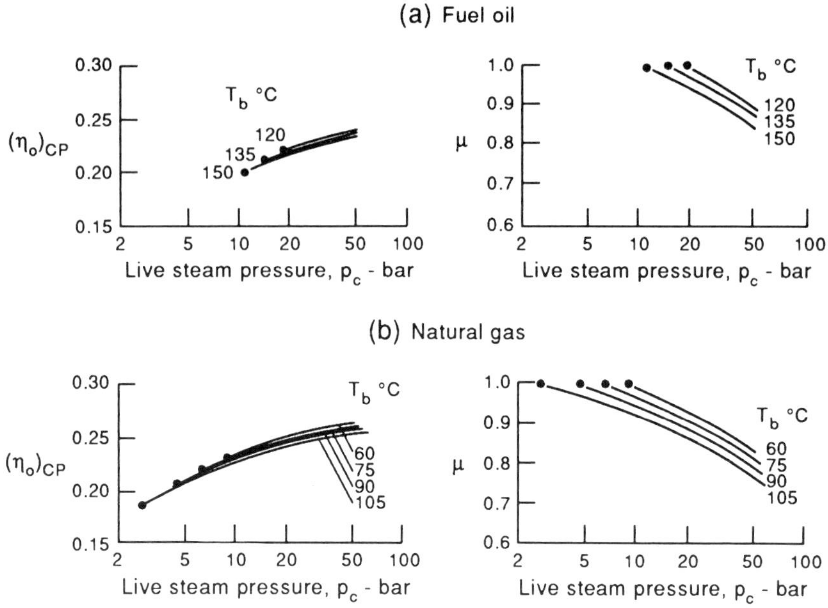

FIGURE 4.24 Overall efficiency $(\eta_O)_{CP}$ and mass flow ratio (μ) as functions of live steam pressure (p_c) and feed-water preheat temperature (T_b) (after Rufli [3])

Using such an approach, Rufli produces a comprehensive set of performance characteristics for an open circuit/closed cycle combined plant; the overall efficiency, heat exchanger surface area/(unit steam power per unit air mass flow site), the unit steam power per unit air mass flow rate and the steam turbine inlet pressure are all determined, for compressor pressure ratio and the maximum gas turbine temperature (T_3). Three particular cases are considered:

(a) that giving maximum overall efficiency (with a dual pressure boiler);
(b) a dual pressure steam cycle giving a 5% reduction in steam power output, and requiring less heat transfer surface area;
(c) a single pressure steam cycle.

Rufli's calculations of overall efficiency $(\eta_O)_{CP}$ for cases (a) and (c) are shown in Fig. 4.25, for natural gas fuel. In reference 3 he also presents results for oil fuel, and shows the sensitivity of the performance characteristics to the pinch-point temperature difference.

It should be emphasised that Rufli's analysis is based on the assumption of regenerative feed heating in the steam cycle to the temperature

$T_b \approx T_{a'}$ corresponding to the dew point of the exhaust gases, which is dependent upon the choice of fuel (distillate oil or natural gas). We discuss the use of such feed-water heating in more detail in Section 4.3.2.2.

4.3.2.1.3 Kehlhofer's Calculations (including those of Wunsch)

Kehlhofer [4] has given a comprehensive account of the CCGT [open current (gas turbine)/closed cycle (steam turbine)] plant, amplifying that of Wunsch [17]. Their work is based on considerable practical experience with Asea Brown Boveri, being more empirical than that of Cerri and Rufli, but is soundly based thermodynamically.

Cerri studied (a) the simple plant with no feed heating and (b) the supplementary fired plant also with no feed heating, for single pressure steam cycles. He then considered regenerative feed heating separately. Rufli assumed regenerative feed heating for all his calculations, allowing for single and dual pressure steam raising, but not for supplementary firing. Kehlhofer's approach is more to regard the gas turbine plant as a "given", and then to concentrate on the optimisation of the steam plant.

Kehlhofer places great emphasis on the feed-water temperature enter-

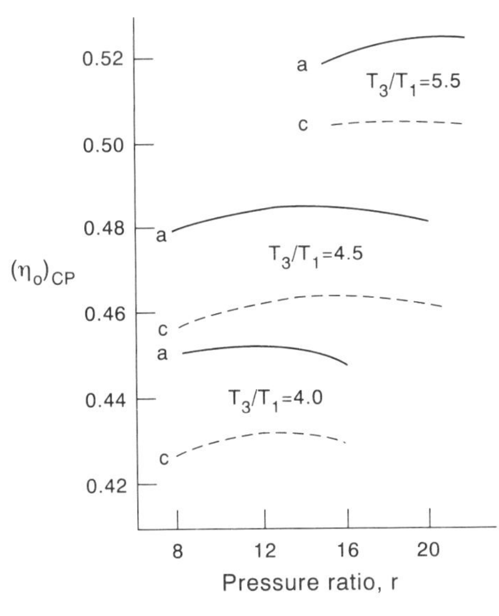

FIGURE 4.25 Combined plant overall efficiency $(\eta_O)_{CP}$ for natural gas
a — dual pressure HRSG.
c -- single pressure HRSG.
(after Rufli [3])

ing the economiser of the HRSG, arguing that it is crucial to prevent corrosion on the metal surfaces. Because of the high heat transfer coefficient on the water side, the metal surface temperature will be little different from the feed-water temperature. Hence the dew point of the gases at exhaust from the HRSG must be less than the feed-water entry temperature; for sulphur-free fuels the water dew-point controls, for fuels with sulphur a "sulphuric acid" dew-point (at a higher temperature) controls. Through these limitations on the exhaust gas temperature, the choice of fuel with or without sulphur content (distillate oil or natural gas respectively) has a critical influence *ab initio* on the choice of the thermodynamic system.

For the simple single pressure system with feed heating (as studied by Rufli, Fig. 4.17b), Kehlhofer first points out that, given the gas turbine conditions, the amount of steam production (M_{ST}) is controlled by the pinch-point condition if the steam pressure (p_c) is selected, as we have indicated earlier [equation (4.12)]. However, with a fuel oil containing sulphur, the feed-water temperature at entry to the HRSG is set quite high (T_b about 130° C), so the heat that can be extracted from the exhaust gases beyond the pinch point [$M_{ST}(h_c - h_b)$] is limited. The condensate can be brought up to T_b by a single stage of bled steam heating, in a direct contact heater, the tapping pressure being set approximately by the temperature T_b, as in Fig. 4.17b. Kehlhofer argues that for low p_c, the steam cycle thermal efficiency (η_L) is low, but the exhaust gas temperature is also low, and the HRSG "boiler efficiency" (η_B) is therefore high; the overall steam cycle efficiency, $(\eta_O)_L = \eta_B \eta_L$, may therefore be quite high. Raising the steam pressure (p_c) increases the cycle thermal efficiency (η_L) but lowers the boiler efficiency (η_B). The optimum $(\eta_O)_L$ [and hence the optimum $(\eta_O)_{CP}$ for a given $(\eta_O)_H$] may occur at quite low live steam pressure (p_c). Wunsch presents a similar argument showing an optimum HRSG exhaust temperature.

Kehlhofer then suggests that more heat can be extracted from the exhaust gases, even if there is a high limiting value of T_b (imposed by use of a fuel oil with a high sulphur content). It is thermodynamically better to do this without regenerative feed heating which leads to less work output from the steam turbine. Wunsch [17] originally showed *a single pressure system with a pre-heating loop*, the extra heat being extracted from the exhaust gases by steam raised in a low pressure evaporator in the loop (as in Fig. 4.26); the evaporation temperature will be set by the "sulphuric acid" dew point (and feed-water entry temperature $T_b \approx 130°$ C). Figure 4.27 illustrates the heat transfer in the HRSG for this system, using the "inverse" temperature scale (T_0/T) discussed in Section 3.2.4.1. The total irreversibility created in raising the feed water to temperature T_b is now split between that arising from the heat transfer from gas to the evaporation (pre-heater) loop, and that in the deaerator/

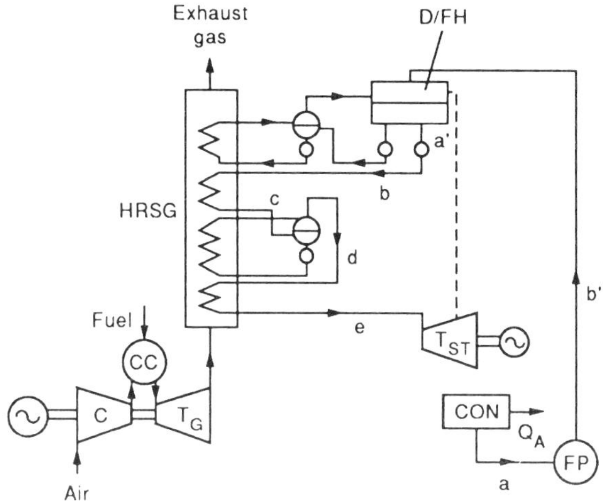

FIGURE 4.26 Single pressure system with L.P. evaporator in a pre-heating loop, as alternative to feed heating by bled steam (shown dotted) (after Wunsch [17])

C	Compressor
T_G	Gas turbine
T_S	Steam turbine
HRSG	Heat recovery steam generator
CON	Condenser
D/FH	Deaerator/feed heater
FP	Feed pump

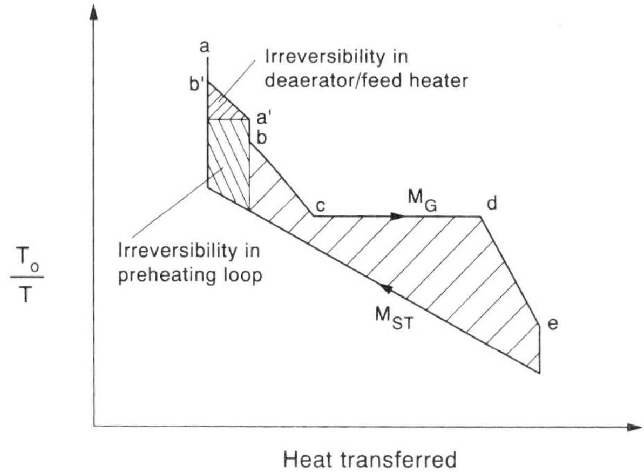

FIGURE 4.27 Pre-heating loop (irreversibility diagram)

feed heater; but that irreversibility is just the same as that which would have occurred if the water had been heated from condenser temperature in the HRSG.

Kehlhofer explains that the pre-heating loop must be designed so that the heat extracted is sufficient to raise the temperature of the fuel feed water flow from condenser temperature T_a to $T_{a'}$. The heat available increases with live steam pressure (p_c), for selected $T_b(\approx T_{a'})$ and given gas turbine conditions, but the heat required to preheat the feed water is set by $(T_{a'} - T_a)$. The live steam pressure is thus determined from the heat balance in the pre-heater if heating of the feed water by bled steam is to be avoided. Kehlhofer's calculations suggest that for this system the optimum (low) live steam pressure may not be achievable because of this requirement on this heat balance in the pre-heater.

Kehlhofer regards the two pressure system as a natural extension of the single pressure cycle with a low pressure evaporator acting as a pre-heater. Under some conditions more steam could be produced in the L.P. evaporator than is required to pre-heat the feed water, and this can be used by admitting it to the turbine at the low pressure. For a fuel with high sulphur content (requiring high feed water temperature (T_b) at entry to the HRSG) *a dual pressure system with no low pressure water economiser* may have two regenerative surface feed heaters and a pre-heating loop. For a sulphur-free fuel (with a lower T_b) *a dual pressure system with a low pressure economiser* may have a single-stage deaerator/direct contact feed heater using bled steam.

Kehlhofer's parametric calculations for these dual pressure systems are carried out essentially by holding most parameters constant and varying one, e.g. by determining

(i) how the overall steam plant efficiency $(\eta_O)_L$ varies with H.P. steam pressure (for given gas turbine conditions and specified L.P. steam pressure and condenser pressure), and
(ii) how $(\eta_O)_L$ varies with L.P. pressure for given gas turbine conditions and specified H.P. steam and condenser pressures.

For the dual pressure system generally, Kehlhofer argues that high H.P. pressure is required in order to reduce temperature differences between gas and water/steam (and hence reduce irreversibility); but that low L.P. pressure is required to give good utilisation of the total "waste heat". These dual pressure systems have also been systematically analysed by Rufli, who for optimum gas turbine pressure ratios gives correspondingly high steam pressure (p_c) (see Section 4.3.2.1.2).

Kehlhofer also reports Wunch's calculations of the overall efficiency of plants with additional firing of the gas turbine exhaust (Fig. 4.28), i.e. with *limited supplementary firing of an HRSG*, or *maximum firing in a conventional boiler*, i.e. with water wall cooling. For low levels of

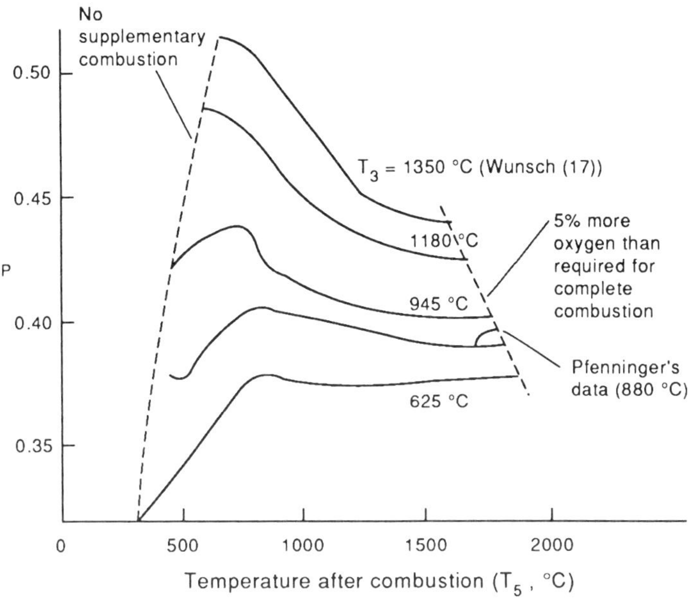

FIGURE 4.28 Variation of overall efficiency with supplementary combustion (after Wunsch [17])

supplementary firing (with the gas temperature T_5 limited to about 750° C), the irreversibility due to temperature differences in the HRSG is small, even with a single pressure steam cycle, but the overall efficiency is relatively low (steam cycle efficiency is low). With more supplementary firing, the HRSG temperature differences increase, but the gas/water temperature paths run close together towards the exit from the HRSG, and the utilisation of "waste heat" is good, the gas exhaust temperature dropping towards the feed-water entry temperature (for the highest T_5, maximum supplementary firing, the minimum temperature difference in the waste heat boiler is at the water end, the pattern of the conventional steam generator). Kehlhofer states that the dual pressure system shows to advantage for supplementary firing (with low irreversibility in the HRSG) but that advantage disappears above $T_5 \approx 750°$ C.

4.3.2.2 DISCUSSION OF SOME IMPORTANT ASPECTS OF CCGT PLANT

We have described three sets of calculations of the gas turbine/steam turbine combined plant. Now we draw out four important points—the effect of gas turbine specific work, the use of feed heating (related to the avoidance of corrosion in the HRSG), the possible replacement of supplementary heating of the gas turbine exhaust by reheating between the H.P. and L.P. turbines, and off design performance.

4.3.2.2.1 The Importance of Gas Turbine Specific Work

In Section 3.5.1 we recorded the emphasis given by Timmermans [20] to the strong influence of the specific work of the higher plant on overall efficiency. To supplement the detailed calculations above, we present some simple estimates of overall efficiency based on Timmermans' approach.

For the series plant [open circuit (gas turbine)/closed cycle (steam turbine)] equation (3.54) gave the relationship between $(\eta_O)_{CP}$ and $(\eta_O)_H$, K (constant) and specific work w_H (per unit air mass flow),

$$(\eta_O)_{CP} \approx (\eta_O)_H + K - K(\eta_O)_H - \frac{K\bar{c}_p(\eta_O)_H(T_6 - T_0)}{w_H} \tag{4.27}$$

the lower plant efficiency (η_L) having been eliminated by the introduction of the equation

$$W_L = KM_G\bar{c}_p(T_4 - T_6), \tag{4.28}$$

in favour of the constant K.

Figure 4.29 shows a plot of $(\eta_O)_{CP}$ against $(\eta_O)_H$ and specific work w_H; K has been taken as 0.375, T_6 as 260° C, T_0 as 20° C and $\bar{c}_p = 1.1$ kJ/kg K.

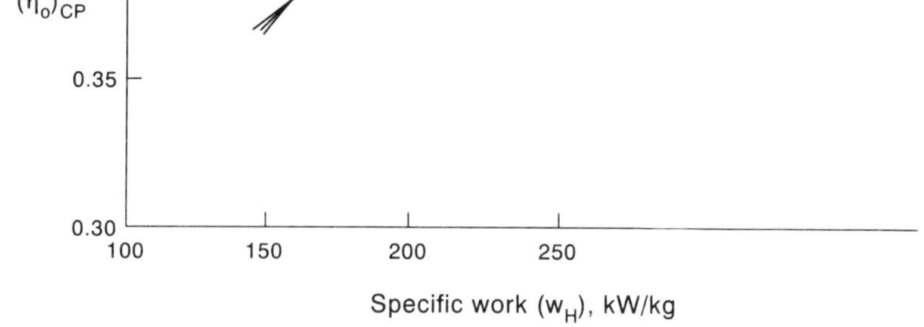

FIGURE 4.29 Variation of overall efficiency with gas turbine specific work (Timmermans' equations)

Timmermans' general point is brought out well—that w_H is the important parameter rather than $(\eta_O)_H$—and this may be confirmed by taking partial differential coefficients of $(\eta_O)_{CP}$. The variation of overall efficiency with $(\eta_O)_H$ (at constant w_H) is

$$\frac{\partial(\eta_O)_{CP}}{\partial(\eta_O)_H}=1-K\left[1+\frac{\bar{c}_p(T_6-T_0)}{w_H}\right], \tag{4.29}$$

which is small and indeed zero when

$$w_H=\frac{K\bar{c}_p(T_6-T_0)}{1-K}=\frac{0.375\times1.1\times240}{0.625}$$
$$=158.4\,\text{kJ/kg}.$$

The rate of change of $(\eta_O)_{CP}$ with (fractional) w_H is

$$w_H\frac{(\partial\eta_O)_{CP}}{\partial w_H}=\frac{K\bar{c}_p(\eta_O)_H(T_6-T_0)}{w_H}. \tag{4.30}$$

This is much larger, of the order of 0.15, for $w_H=200$ kJ/kg, and $(\eta_O)_H=0.30$, say.

For the open circuit/closed cycle combined plant with supplementary firing of the HRSG, Timmermans suggests that the gas turbine specific work is again most important. In Section 3.5.2.3 we reproduced his expression for overall efficiency

$$(\eta_O)_{CP}=\frac{w_H+K'\bar{c}_p(T_5-T_S)}{w_H+\bar{c}_p(T_5-T_0)}, \tag{4.31a}$$

where T_5 is the temperature after supplementary combustion, T_S is the stack temperature, and K' is a constant.

Timmermans suggests that T_5 should be limited to 760° C; with a stack temperature T_S of 160° C, an ambient temperature T_0 of 20° C, a specific heat at constant pressure of 1.1 kJ/kg K and $K'=0.31$, we obtain the relation

$$(\eta_O)_{CP}=\frac{w_H+0.31\times1.1(760-160)}{w_H+1.1(760-20)}$$
$$=\frac{w_H+205}{w_H+814}, \tag{4.31b}$$

which is also shown plotted on Fig. 4.29.

For $w_H=150$ kJ/kg, $(\eta_O)_{CP}=0.369$ and for $w_H=250$ kJ/kg, $(\eta_O)_{CP}=0.428$, so that an increase of 66% in w_H produces a substantial increase of overall efficiency (5.9%). [The corresponding change in the power ratio (W_H/W_L) is from 0.732 to 1.22.]

Kopelka [19], like Timmermans, places similar emphasis on the specific work w_H, and we discuss his calculations in Section 4.4.

4.3.2.2.2 The Effect of Regenerative Feed Heating

In the first of the major parametric analyses of the open circuit [gas turbine/closed cycle (steam turbine plant)] that we have described, Cerri showed that, except for low pressure ratios, the effect of the introduction of regenerative feed heating in the steam cycle was to reduce efficiency. Rufli, on the other hand, specified regenerative feed heating, to a feed-water temperature determined by a specified stack temperature, in each of the calculations. Kehlhofer's approach is more pragmatic, avoiding regenerative feed heating unless absolutely necessary. Its role therefore merits further discussion here.

A simple approach to the general effect of feed heating using bled steam in a doubly cyclic combined plant has been considered by Horlock [26], and we reproduce part of that analysis here. The first point to be made is that for *given* states in the gas turbine cycle, *given* steam pressures and pinch-point temperature differences (at the beginning of evaporation and at superheater outlet), regenerative feed heating would always lower combined plant efficiency, in comparison with the case of no feed heating if there were no lower limit on the feed-water temperature entering or the gas temperature leaving the HRSG. This is illustrated by a simple analysis, assuming that the turbine follows an expansion line such that $(h-h_w)=\beta$ (a constant), where h is the local steam enthalpy and h_w is the saturated water enthalpy at the same pressure. As has been emphasised by Kehlhofer, the problem is whether the increase in the efficiency of the steam cycle by feed heating is offset by an increase in the unused (or "lost") heat in the doubly cyclic plant, because of the increase in the temperature T_S. However, it can be argued that for a plant in which the higher cycle conditions are prescribed (including the heat supply) and there is no minimum feed-water temperature at entry to the HRSG, *any* bleeding of steam from the lower cycle turbine must result in reduction of the overall work output, and hence of combined plant efficiency, if steam turbine inlet pressure and condenser pressure are prescribed.

We illustrate this point by reference to the (T, s) diagram of Fig. 4.30a which shows a combined (doubly cyclic) plant with no supplementary heating and no feed heating. We have explained how the relative mass flows are controlled by the pinch-point condition to give

$$\frac{M_G}{M_{ST}}=\frac{(h_e-h_c)}{(h_4-h_6)}=\frac{\beta}{(h_4-h_6)}, \tag{4.32}$$

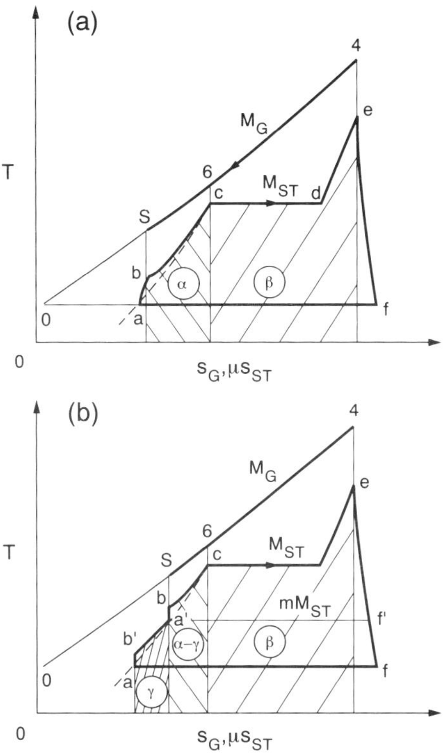

FIGURE 4.30 Heat transferred in HRSG
(a) without regenerative feed heating
(b) with regenerative feed heating

and

$$M_G(h_6 - h_S) = M_{ST}(h_c - h_b), \tag{4.33}$$

so that

$$\frac{(h_6 - h_S)}{(h_4 - h_6)} = \frac{\alpha}{\beta}, \tag{4.34}$$

where

$$\alpha = (h_c - h_b),$$

and where for the present we have assumed that the feed water can enter the HRSG at T_b, just above the condenser temperature.

Equation (4.34) defines h_S (and T_S) if all the other states are fixed.

We can determine the efficiency of the combined plant by first obtaining the total amount of heat rejected,

$$Q_A = M_{ST}(h_f - h_a) + M_G(h_S - h_0)$$
$$= M_G\left[\beta\left(\frac{M_{ST}}{M_G}\right) + (h_S - h_0)\right], \tag{4.35}$$

where h_0 is the enthalpy of the exhaust gas at ambient temperature. Using equation (4.32), it follows that

$$Q_A = M_G[(h_4 - h_0) - (h_6 - h_S)]. \tag{4.36}$$

The overall efficiency of the combined cycle is

$$\eta_{CP} = 1 - \frac{Q_A}{Q_B}, \tag{4.37}$$

where $Q_B = Q_H = M_G(h_3 - h_2)$ is the heat supplied to the gas between states 2 and 3. But the efficiency of the gas turbine plant operating alone would be

$$\eta_H = 1 - \frac{M_G(h_4 - h_0)}{Q_H}, \tag{4.38}$$

so that

$$\eta_{CP} = \eta_H + \frac{(h_6 - h_S)}{(h_3 - h_2)}. \tag{4.39}$$

Thus from (4.34) and (4.39)

$$\eta_{CP} = \eta_H + \left(\frac{\alpha}{\beta}\right)\left(\frac{h_4 - h_6}{h_3 - h_2}\right). \tag{4.40}$$

This analysis can be shown to be identical with the equation

$$\eta_{CP} = \eta_H + \eta_L - \eta_H\eta_L - \nu_{UN}\eta_L. \tag{4.3a}$$

If there is a supplementary heat supply, then the enthalpy increase from the gas turbine exhaust (now state 4) to the state 5 is simply $Q_S = M_G(h_5 - h_4)$ where state 5 now has a different interpretation, and it may be shown that

$$\eta_{CP} = (1 - \nu_S)\left[\eta_H + \left(\frac{h_5 - h_6}{h_3 - h_2}\right)\left(\frac{\alpha}{\beta}\right)\right]. \tag{4.41}$$

If feed heating is introduced into the basic steam cycle (with no supplementary heating) then less heat is extracted from the gas flow and the enthalpy at exit from the HRSG is increased. Thus an increase in the efficiency of the steam cycle by feed heating leads to an increased heat rejection from the gas turbine part of the combined recuperative cycle. At first sight it would appear there might be a balance between these two effects with a possible optimum amount of feed heating. But the introduction of feed heating, with all the other cycle parameters fixed, will inevitably lead to a reduction in steam turbine work per unit mass of steam supplied to the turbine. Since neither the external heat supply nor the gas turbine work output is being changed, M_{ST} is not affected by the introduction of feed heating, since (M_{ST}/M_G) is controlled at the pinch point. The overall cycle efficiency must therefore always fall with the introduction of feed heating, in comparison with that of the basic non-feed heating recuperative combined plant if the main states in the two cycles are specified (including p_c and T_e).

As we discussed in Section 4.3.2.1.3, in practice in an open gas turbine plant there will be a limit on the feed-water temperature at entry to the HRSG, below which corrosive action takes place. However, it is desirable to have a deaerator in the steam cycle, and this is usually also a direct contact feed heater (as in the Brown Boveri Donge/Geertruidenberg plant) (see Wunsch [17]). Horlock [26] shows that if a direct contact feed heater is used to increase the enthalpy of the condensate by $(\alpha/2)$, then the combined plant efficiency is

$$\eta_{CP}=\eta_H+\frac{(h_4-h_6)}{(h_3-h_2)}\left[\frac{\alpha}{\beta}-\frac{\alpha^2}{2\beta(\alpha+2\beta)}\right], \tag{4.42}$$

which in comparison with equation (4.40) shows that the overall efficiency of the combined plant with feed heating is *less* than that without feed heating by an amount

$$\frac{(h_4-h_6)}{(h_3-h_2)}\left[\frac{\alpha^2}{2\beta(\alpha+2\beta)}\right].$$

In practice, as Kehlhofer [4] has emphasised, a lower limit may be placed on the feed-water temperature at entry to the HRSG of the gas turbine plant in open circuit. In a simplified analysis Horlock [26] again studied a doubly cyclic plant, with a reduced enthalpy rise $(\alpha-\gamma)$ achieved in the economiser of the HRSG, and the remaining enthalpy rise (γ) achieved by regenerative feed heating (Fig. 4.30b). For the recuperative steam cycle with no supplementary heating but feed heating in a single heater, Horlock shows the combined plant efficiency is less than that of the basic non-feed-heated plant by

$$\frac{(\alpha\Delta h_{46}-\beta\Delta h_{6S})^2}{\beta(h_3-h_2)[(\alpha+\beta)\Delta h_{46}-\beta\Delta h_{6S}]}$$

where

$$\Delta h_{46} = (h_4 - h_6),\ \Delta h_{6S} = (h_6 - h_S).$$

The efficiency η_{CP} is again decreased by the feed heating.

This analysis adds confirmation to the results obtained by Cerri and Colagé, that for high pressure ratio, there is no advantage in introducing feed heating. However, it should be remembered that Horlock's work [26] is based on fixed levels of steam pressure and temperature in the HRSG. At lower pressure ratios, where feed heating was said to be advantageous by Cerri and Colagé [15], they found that the optimum boiler pressure increased, whereas at higher pressure ratios it remained substantially constant (as Horlock assumed in his simple analysis). Rufli specified regenerative feed heating in the steam cycle, for the reasons advanced by Wunsch and Kehlhofer, that deaeration was required and the feed water must not enter the HRSG below a specified temperature.

Roberts and Hemsley [22] took a somewhat different approach from Horlock. They argued that as condensate temperature after feed heating was increased, for given boiler pressure p_c and specified T_S (the stack temperature in an open circuit gas turbine plant), the effect was to reduce the pinch-point temperature difference. This led to increased combined plant efficiency (because of less irreversibility in the HRSG). However, the results given by Roberts and Hemsley do include a plot of $(\eta_O)_{CP}$ (for a fixed value of $\Delta T_{6c} = 10\,\mathrm{K}$) against increasing feed-water temperature and this does show a reduction of $(\eta_O)_{CP}$ consistent with Horlock's argument.

The complex arguments for different systems in the steam cycle of the CCGT were discussed in Section 4.3.2.1.3. However, the essential point of the analysis described in this section is that the use of regenerative feed heating to raise the feed-water temperature before entry to the HRSG, while increasing the efficiency η_L of the steam cycle, reduces overall combined plant efficiency $(\eta_O)_{CP}$.

However, as explained in Section 4.3.2.1.3, regenerative feed heating can be avoided, even if a minimum T_b at entry to the HRSG is prescribed, by a pre-heating loop (Kehlhofer [4]). The plant then has the same efficiency as a plant in which the feed water is heated all the way from T_a to T_c in the HRSG, the system studied by Cerri.

4.3.2.2.3 The Choice Between Supplementary Heating or Reheating in the Gas Turbine Plant

Cerri's calculations have included the study of the effect of supplementary firing of the gas turbine exhaust, before entry to the HRSG. Broadly, the effect of supplementary heating is to reduce overall efficiency (through the term $[-\nu_S\eta_H(1-\eta_L]$ in equation (4.8a) for combined cycle efficiency for binary plant, or through the term $(M_F)_S/(M_F)_H$ in the Mangan and Pettit equation (4.9) for open circuit/closed cycle plant). However, this decrease can be balanced by an increase in the lower (steam) cycle efficiency. The higher gas temperature after supplementary firing may allow the choice of a higher steam evaporation pressure and a higher steam turbine inlet temperature, both of which, for given condenser pressure and turbine efficiency, will give a higher η_L and hence a higher $(\eta_O)_{CP}$. Such a balance is illustrated in Cerri's calculations (Figs. 4.20 and 4.21) which show comparable efficiency for $\delta=0$ and $\delta\neq 0$; it has also been referred to by Pfenniger [16] and Wunsch [17] of Brown Boveri. A plot of efficiency against the amount of supplementary firing given by these authors is shown in Fig. 4.28, although more recently Davidson and Keeley [29] suggest that any increase in efficiency is now unlikely, because turbine inlet temperatures have gone up. Another benefit of supplementary firing is the increase in work output of the plant.

An interesting additional argument has been introduced by Rice [21]. Basically, Rice argues that if supplementary firing is to be employed it should be used within the higher (gas turbine) plant (reheating between the gas generator and power turbines) rather than as supplementary heating after full expansion in the gas turbine. The thermodynamic argument is simply that "heat" should be added at the highest possible temperature (if the conditions of heat rejection are unchanged).

Rice's point may be illustrated by the following analysis. (We consider a doubly cyclic plant, with heating rather than combustion of fuel.) We refer to Fig. 4.31 which shows a combined plant with supplementary

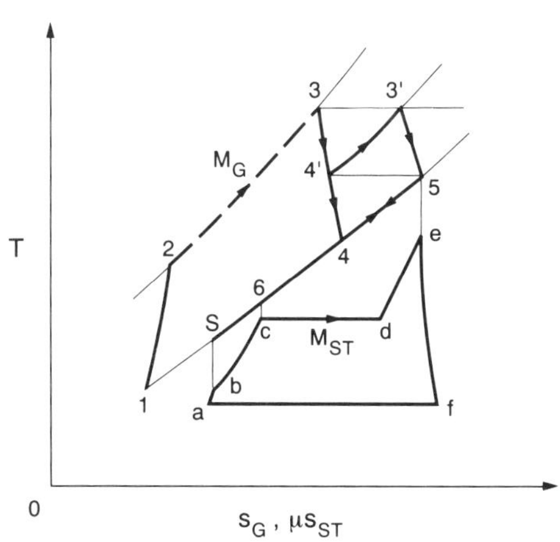

FIGURE 4.31 Reheat in gas turbine cycle of combined plant—temperature-entropy diagram

heating between turbine exhaust (state 4) to state 5 ($T_5 < T_3$, since the steam turbine inlet temperature is generally much lower than the gas turbine inlet temperature).

We next consider an alternative cycle in which the gas is reheated to state 3′ ($T'_3 = T_3$) after partial expansion in the turbine to state 4′. It is then expanded to the same final state 5 as that reached after supplementary heating in the first cycle.

Suppose that the polytropic efficiency in the turbines is η_p, so that all expansion follows the relationship

$$\frac{p}{T^{\left(\frac{\kappa}{\kappa-1}\eta_p\right)}} = \text{constant}, \tag{4.43}$$

where κ is the (constant) ratio of specific heats.

Between states 3 and 4

$$\frac{T_3}{T_4} = \left(\frac{p_3}{p_4}\right)^{\frac{(\kappa-1)}{\kappa}\eta_p} = \rho. \tag{4.44}$$

If the optimum division of pressure ratio is taken so that

$$\frac{p_3}{p_{4'}} = \frac{p_{4'}}{p_4} = \frac{p_{3'}}{p_5} = \rho^{1/2} \tag{4.45}$$

(which Rice shows to be virtually the correct split between gas generator turbine and power turbine), then the overall gas turbine work $(W_T)_G$ in the original and the new cycle are as follows:

Original (supplementary heated)

$$(W_T)_G = M_G c_p T_3 \left(1 - \frac{1}{\rho}\right); \tag{4.46}$$

New (reheated)

$$\begin{aligned}(W_T)'_G &= M_G c_p T_3 \left(1 - \frac{1}{\rho^{1/2}}\right) + M_G c_p T'_3 \left(1 - \frac{1}{\rho^{1/2}}\right) \\ &= 2 M_G c_p T_3 \left(1 - \frac{1}{\rho^{1/2}}\right).\end{aligned} \tag{4.47}$$

Since the final temperature T_5 is the same in the two cycles, we may assume that the choice of pressure and temperature in the steam cycle is unchanged, and that the efficiency η_L and work output W_L will be the same in the lower cycles of the two combined plants.

It is obvious from the T, s diagram (Fig. 4.31) that the difference between the work output of the two plants,

$$(W_T)'_G-(W_T)_G=\Delta W=M_G c_p T_3\left[2\left(1-\frac{1}{\rho^{1/2}}\right)-\left(1-\frac{1}{\rho}\right)\right], \quad (4.48)$$

is the same as the difference in the heat supplied to the two cycles, through the difference between the reheat and the supplementary heat

$$\begin{aligned}\Delta Q&=M_G[(h_{3'}-h_{4'})-(h_5-h_4)]\\&=M_G(h_{3'}-h_5)-(h_{4'}-h_4)\\&=\Delta W. \quad (4.49)\end{aligned}$$

The efficiency of the reheat cycle may therefore be written

$$\eta_{CPRH}=\frac{W_{CP}+\Delta W}{Q_{CP}+\Delta Q}, \quad (4.50)$$

where subscript CP refers to the unreheated plant. Hence

$$\begin{aligned}\Delta\eta&=\eta_{CPRH}-\eta_{CP}\\&=\frac{W_{CP}+\Delta W}{Q_{CP}+\Delta Q}-\frac{W_{CP}}{Q_{CP}}\\&=\frac{Q_{CP}\Delta W-W_{CP}\Delta Q}{(Q_{CP}+\Delta Q)Q_{CP}}\\&=\frac{(1-\eta_{CP})\Delta W}{Q_{CP}+\Delta Q}\\&\approx\frac{M_G c_p T_3\left(1-\frac{1}{\rho}\right)\left[\frac{2\left(1-\frac{1}{\rho^{1/2}}\right)}{\left(1-\frac{1}{\rho}\right)}-1\right](1-\eta_{CP})}{Q_{CP}}\\&=\left[\frac{(W_T)_G}{Q}\right]_{CP}\left[\frac{\left(1-\frac{1}{\rho^{1/2}}\right)^2}{\left(1-\frac{1}{\rho}\right)}\right](1-\eta_{CP}). \quad (4.51)\end{aligned}$$

This is obviously greater than zero. For example, with $\rho=1.69$, the square bracket becomes $\left(\frac{0.3}{1.3}\right)^2\Big/\frac{0.69}{1.69}=0.13$, so that

$$\Delta\eta=0.13\left[\frac{(W_T)_G}{Q}\right]_{CP}(1-\eta_{CP}),$$

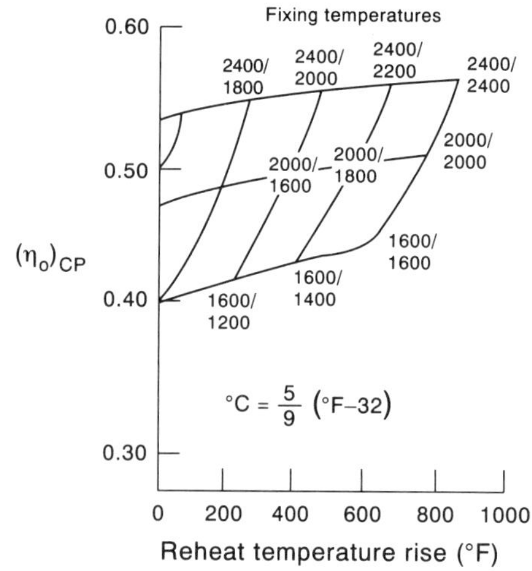

FIGURE 4.32 Effect of gas turbine reheat on overall efficiency of combined plant (after Rice [21])

Steam cycle conditions:

Feed-water temperature	116° C
Steam pressure	5.86 MN/m^2
Steam temperature	482° C/482° C
Turbine efficiency	0.75

which is indeed a worthwhile increase in combined plant performance.

Rice [21] has made a comprehensive study of his "high-temperature turbine technology" (HTTT) reheat gas turbine combined plant. He first analyses the higher (gas turbine) plant with reheat, obtaining $(\eta_O)_H$, turbine exit temperature, and power turbine expansion ratio, all as functions of plant overall pressure ratio and firing temperatures in the main and reheat burners. (The optimum power turbine expansion ratio is little different from the square root of the overall pressure ratio.) Rice then pre-selects the steam cycle conditions rather than undertaking a full optimisation.

Rice also argues that a high temperature at entry to the HRSG (resulting from reheat in the gas turbine plant) leads via the pinch-point restriction to a lower exhaust stack temperature and "heat" loss, in comparison with an HRSG receiving gas at a lower temperature from a simple gas turbine plant. This point is accepted, but there are the additional complications, of higher irreversibility in the HRSG (because of higher temperature differences) and the opportunity of higher steam cycle efficiency by the introduction of regenerative feed heating.

Rice's summary plot of overall plant efficiency $(\eta_O)_{CP}$ against reheat temperature rise is shown in Fig. 4.32, for various gas turbine firing temperatures. Optimum pressure ratios [to give highest $(\eta_O)_H$] are drawn from the earlier calculations and were shown in Fig. 1.11. These pressure ratios are very high in comparison with results from other authors, even for the simple plant. Rice's assumptions for the steam cycle are given in the table below Fig. 4.32.

In subsequent papers Rice has extended his arguments for the reheat gas turbine/steam turbine plant, considering reheat in the steam cycle, and various commercial gas turbine plants which could be used in a combined plant.

4.3.2.2.4 Off-Design Behaviour of CCGT Plants

Much attention has been paid to the off-design performance of the gas turbine/steam turbine combined plant (e.g. Kehlhofer [4], Gyarmathy [30] and Frutschi [31]).

Kehlhofer suggests that the steam turbine of such a plant operates most economically if it is run "uncontrolled", i.e. its performance is controlled only by the exhaust flow and temperature of the gas turbine—in particular the gas mass flow (M_G) and the exhaust temperature from the gas turbine (T_4).

Gyarmathy has followed this approach in analysing a simple single pressure steam system (with preheating). An iterative calculation for the HRSG yields the steam flow (M_{ST}), the live steam pressure and temperature (p_e, T_e) and hence the steam cycle efficiency and power output. (Here the iteration either uses the through flow characteristics of the steam turbine, or if the steam turbine is governed, the steam pressure p_e is specified as an input parameter.)

Frutschi describes possible control methods for the gas turbine (Fig. 4.33). The standard method is to control fuel flow (and hence turbine inlet temperature (Fig. 4.33a)). Alternative methods involve recirculation of compressor air (Fig. 4.33b), intake throttling (Fig. 4.33c), guide vane control (Fig. 4.33d) and compressor intake temperature control by heating the inlet air with steam passed out from the steam turbine (Fig. 4.33e).

For calculations of these various possibilities, the reader is referred to the three original references and to Kehlhofer's early work [4]. A typical calculation for fuel-only (temperature) control by Gyarmathy is given in Fig. 4.34. The air mass flow is held constant, and the fuel input varied, which produces a drop in the gas turbine inlet temperature (T_3). The steam mass flow (M_{ST}) drops as T_3 falls, as does the gas turbine exit temperature (T_4) and the steam turbine inlet temperature (T_e). Feed-water temperature and condensate temperature are kept constant. The drop in the plant overall efficiency and power output of the component plants are also shown, all plotted against combined plant power output.

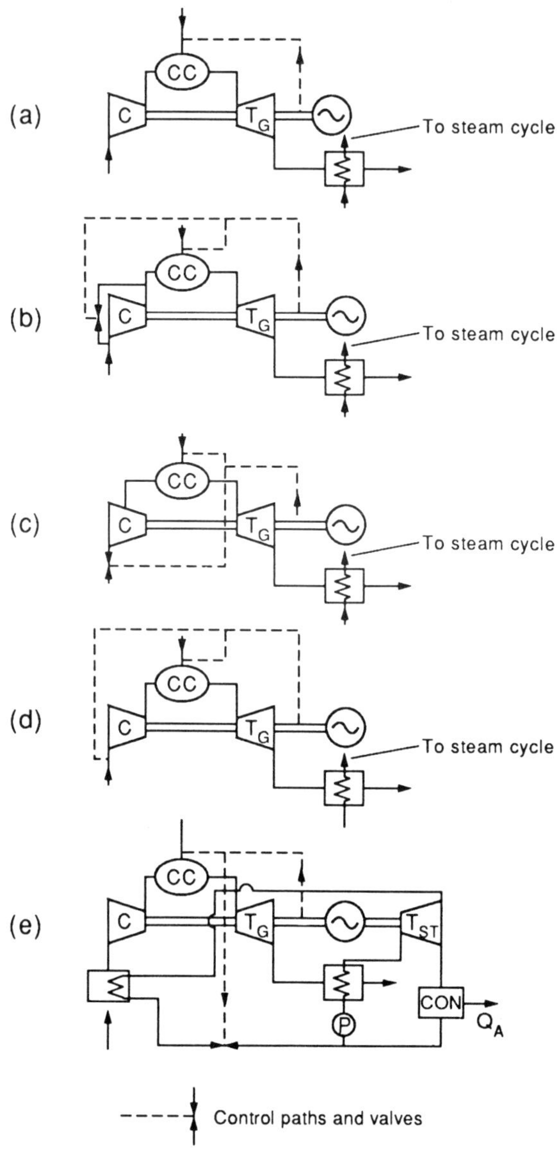

FIGURE 4.33 Control methods for gas turbine with HRSG for combined plant operation (after Frutschi [31]
(a) Temperature control, (b) compressor back flow control, (c) compressor air throttle control, (d) compressor guide vane control, (e) compressor intake temperature control, using pass-out steam

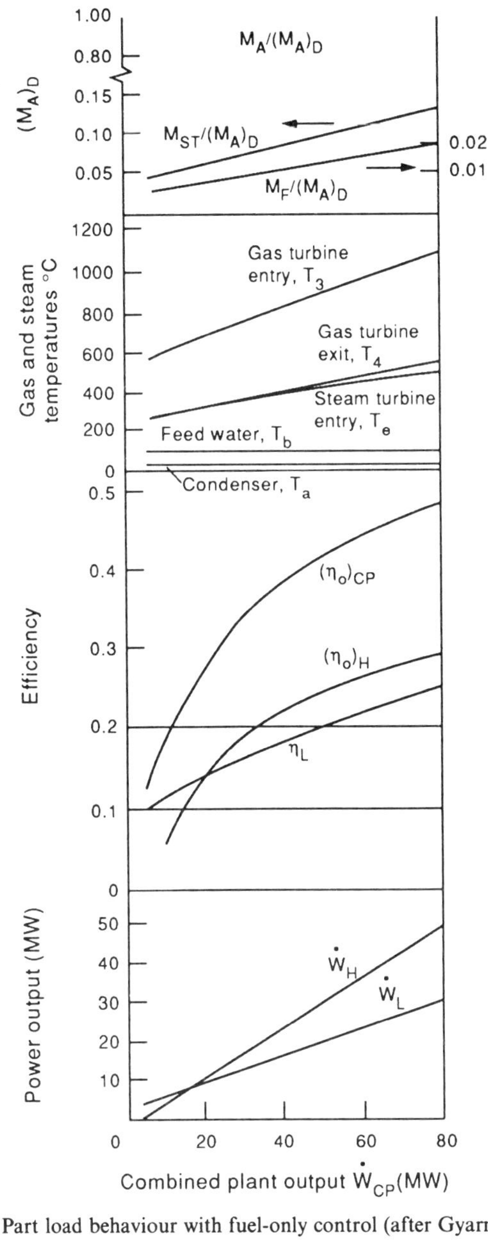

FIGURE 4.34 Part load behaviour with fuel-only control (after Gyarmathy [30])

This kind of plot, of combined plant efficiency against fractional load, is calculated by Frutschi for the various types of control shown in Fig. 4.33. He concludes that compressor throttling is not to be recommended and that guide vane control, possibly coupled with inlet air preheating, is a useful control method for combined power plant.

Of course, where several gas turbines feed a single steam turbine, an effective way of maintaining relatively high efficiency of design is obtained simply by cutting out one or more of the gas turbines. The resulting plot of overall efficiency against fractional power output takes the form shown in Fig. 4.35.

4.3.3 Doubly Open Plants

The basic doubly open combined plant was described in Chapter 2, but there are several variants. Most parametric studies which have been made involve a plant in which steam is produced by the gas turbine exhaust gases in a waste heat boiler, and injected into the air stream at exit from the compressor, before the combustion chamber. This is the basic steam injected gas turbine (STIG) (Fig. 4.36a).

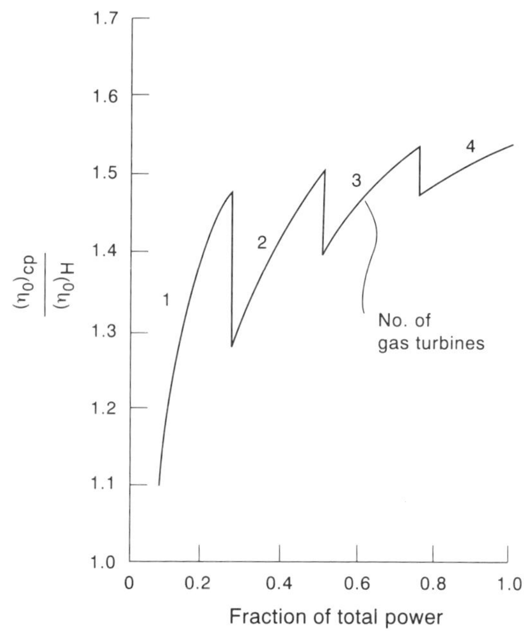

FIGURE 4.35 Part-load efficiency of combined power plant, with 1, 2, 3, 4 gas turbines in operation (after Wunsch [17])

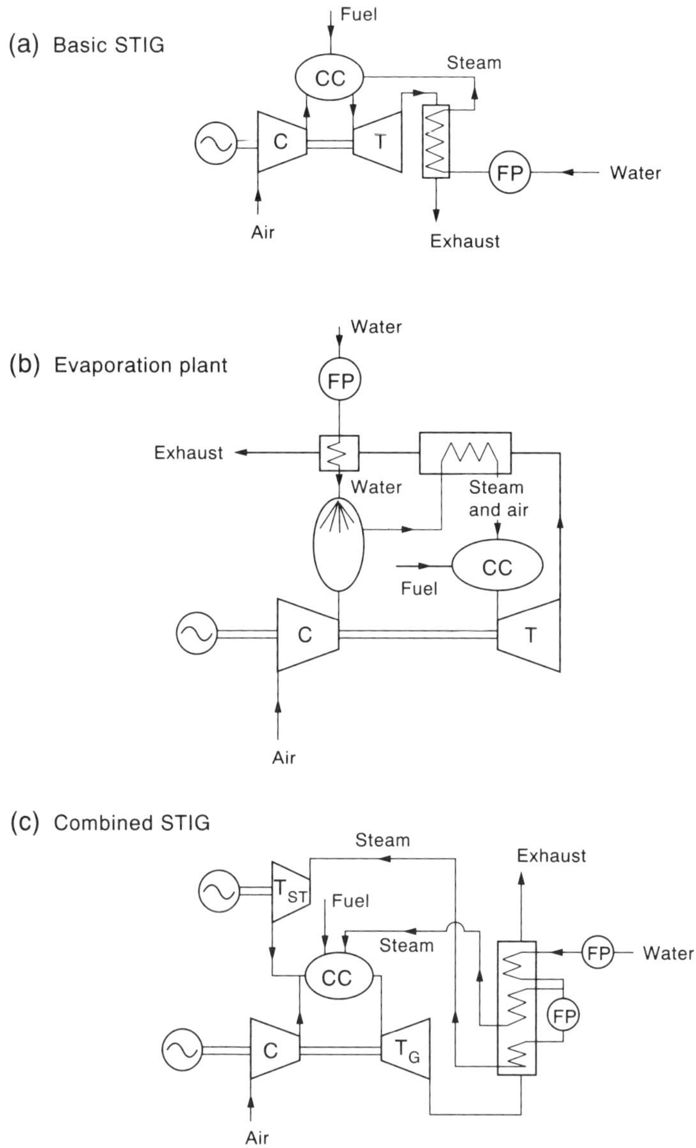

FIGURE 4.36 Steam injection/gas turbine (STIG) plants (after Frutschi and Plancherel [37])

4.3.3.1 THE STEAM INJECTION COMBINED PLANT (STIG)

Two distinct approaches to calculations for the STIG plant have been undertaken. In the first, described by Fraize and Kinney [32], it was assumed that at all points after the steam injection the two fluids are fully mixed and at a common temperature. This analysis is then based upon mean values (for the mixture) of specific heat at constant pressure, and ratio of specific heats (and hence gas constant). The calculation proceeds along the lines of a normal gas turbine type calculation, with a mixture of gas and steam behaving as a new semi-perfect gas in the turbine and in the waste heat boiler. In a second approach, Cerri and Arsuffi [33] treat the two fluids as separate, each expanding through the same ratio of partial pressure in the turbine, the enthalpy drop of the gas being determined from normal pressure–temperature relationships for a perfect gas and the enthalpy drop of the superheated steam being determined from steam property tables, for the partial pressure levels at entry and exit. This second approach implies that temperature differences between the gas and the steam can develop (indeed this would be the case if the steam were treated as a separate perfect gas, with different constants, expanding through the same pressure ratio as the main gas); but if such temperature differences develop, then Cerri and Arsuffi's assumption that the partial pressures remain constant fractions of the overall pressure is no longer entirely valid and the two "separate" fluids would expand through different effective pressure ratios.

We therefore follow here the assumptions of Fraize and Kinney, that there is temperature equilibrium between the gas and superheated steam and that the mixture behaves as a "new" semi-perfect gas. This approach is to some extent justified by earlier work which shows that superheated steam behaves as a perfect gas when expanding with high polytropic efficiency (Horlock [34]).

There is one other critical assumption that has to be made—the effect of the steam injection on the pressure level in the mixing process. Consider the diagrammatic representation of the mixing process in Fig. 4.37. We assume that the air pressure at delivery from the compressor is

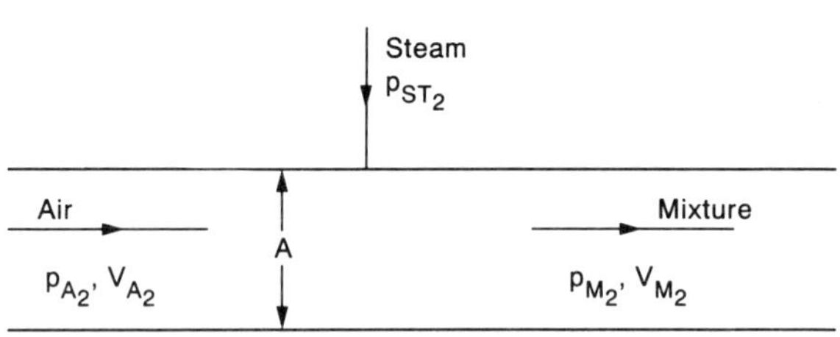

FIGURE 4.37 Injection/mixing model

p_{A_2} and that the steam is supplied at pressure p_{ST_2} which will necessarily be slightly above p_{A_2}. For such a process the momentum equation, for a duct of constant area (A) with the steam entering perpendicular to the main flow, yields

$$p_{A_2}A + \rho_{A_2}AV_{A_2}^2 = p_{M_2}A + \rho_{M_2}AV_{M_2}^2, \tag{4.52a}$$

where ρ and V are density and velocity at the relevant state points A_2 and M_2. (We may note that the steam pressure p_{ST_2} does not appear in the equation, but the leaving mass flow $(\rho_{M_2}AV_{M_2})$ is of course greater than the air mass flow $(\rho_{A_2}AV_{A_2})$ by the entering steam flow.) The equation may also be written

$$p_{A_2}(1 + \kappa_A M_{A_2}^2) = p_{M_2}(1 + \kappa_M M_{M_2}^2), \tag{4.52b}$$

where M_{A_2} and M_{M_2} are the flow Mach numbers. Thus if these are kept low, then there is theoretically little change in pressure in the mixing process, $p_{M_2} \approx p_{A_2}$. It should therefore be sufficient in cycle calculations to assume that the entry pressure to the combustion chamber is equal to the compressor delivery pressure, and simply allow for an empirical overall pressure loss in combustion, as is usual practice.

Fraize and Kinney make this assumption, with several others, in their calculations. The reader is referred to their original paper, particularly for their method of calculating the mean values of $\bar{c}_p$ and the isentropic index $[(\kappa - 1)/\kappa]$ in the turbine expansion process. The quantities in Table 4.4 are held constant, and pressure ratio and mass flow ratio are the parameters varied.

The main results obtained by Fraize and Kinney are shown in Figs. 4.38 and 4.39. The first figure shows how the specific work output of the plant increases substantially with the ratio of steam to air mass flow, principally because of the increased mass flow through the turbine; the

TABLE 4.4 *Assumptions of Fraize and Kinney*

Turbine inlet temperature	1700 K
Compressor inlet temperature	300 K
Entry water temperature	300 K
Temperature difference between gas and steam at high temperature end of heat exchanger	56° C
Pinch-point temperature difference	28° C
Minimum stack temperature	408 K
Turbine efficiency	0.89
Compressor efficiency	0.86
Ratio of turbine gas flow to compressor flow (to allow for cooling air flow)	0.95
Ratio of turbine expansion ratio to compressor pressure ratio	0.96

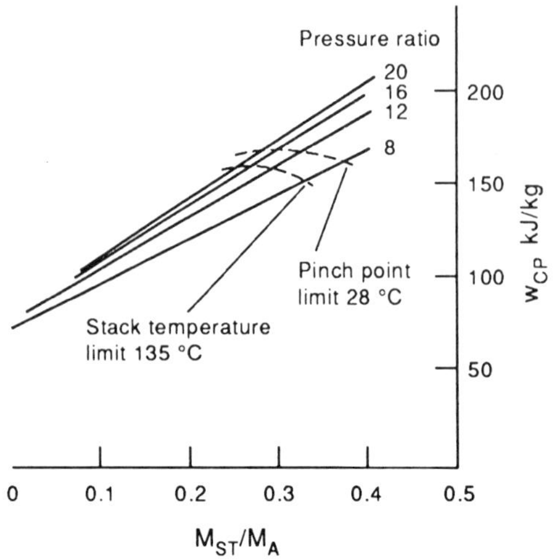

FIGURE 4.38 Variation of specific work output (w_{CP}) with steam/air mass flow ratio (M_{ST}/M_A) for basic STIG plant (after Fraize and Kinney [32])

effect also increases with pressure ratio. The second figure shows that overall efficiency also increases—essentially by the hidden combined plant effect. Although joint "heating" of the gas and steam plants has been introduced, substantial heat is transferred from the gas turbine exhaust to raise the steam and this effect more than compensates for the

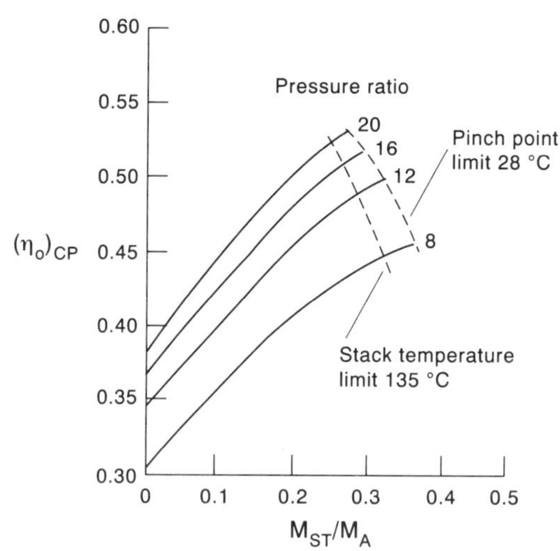

FIGURE 4.39 Variation of overall efficiency $(\eta_O)_{CP}$ with steam/air mass flow ratio (M_{ST}/M_A) for basic STIG plant (after Fraize and Kinney [32])

loss in efficiency resulting from the joint heating. It is effectively a series/parallel combined plant which has been analysed earlier in this chapter.

Broadly similar calculations are presented by Brown and Cohn [35], and the dramatic increases in specific work with steam flow have been confirmed experimentally by Messerlie and Strother [36] and Kolp and Moeller [37]. Cerri and Arsuffi have analysed a STIG system coupled with a water distillation process, and have allowed for variation of the maximum temperature in their work. This shows the expected increase in specific work output with maximum temperature, but the calculations are based on separate expansion of steam and gas as explained earlier.

4.3.3.2 OTHER STIG TYPE COMBINED PLANTS

There are, however, several variations on the combined plant, using gas and steam mixed, and they have been discussed in detail by Frutschi and Plancherel [38]. Two are illustrated in Fig. 4.36.

In the first case, called the evaporation cycle, a heat exchanger is used in the gas turbine cycle (Fig. 4.36b). The water, after heating by the turbine exhaust, is sprayed into the air between the compressor and the heat exchanger. Frutschi and Plancherel argue that the stack temperature of the exhaust gases is lowered and higher overall efficiency than the basic STIG plant results.

In the second case, the combined STIG cycle, steam is raised at two pressure levels in the boiler (Fig. 4.36c). Superheated steam at the higher pressure level expands through a steam turbine before injection into the air stream. Low pressure steam is injected into the combustion chamber directly. Attainable efficiency for this plant may theoretically be of the order of 50%. In a variation of this combined STIG (the Foster-Pegg plant) the steam turbine drives a second high pressure compressor.

4.4 General Discussion

A number of authors have presented results of calculations of the performance of gas turbine/steam turbine combined plant, some of which we have described in this chapter. The simplest approach was that of Buxmann and Bormann [11], based on two closed cycles—a binary plant; it was supplemented by the work of Dibelius and Ziemann [13], a study which included a heat exchanger in the gas turbine plant. Calculations are also reported by Pfenninger [16], Wunsch [17] and Kehlhofer [4], of the Brown Boveri Company, which has extensive experience in the area of open circuit/(gas turbine)/closed cycle (steam turbine plant). Timmermans [20] and Korpela [19] report other results, using somewhat simplified approaches, whereas Cerri [2] and Rufli [3] give details of more complex analyses of open circuit/(gas turbine)/closed cycle (steam tur-

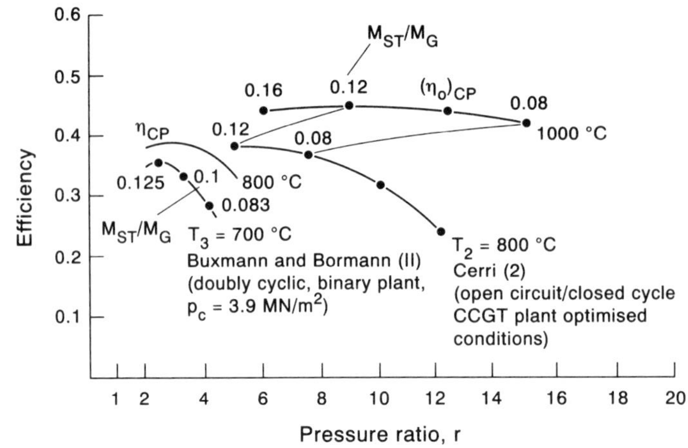

FIGURE 4.40 Efficiency of gas/steam turbine plant (no feed heating, no supplementary heating)

bine) performance. Other work in this general area of combined plant performance is tabulated in Table 4.1.

Because of the different assumptions made by these various authors it is difficult to compare their results and to draw overall conclusions. However, some major points can be made.

For the binary cycle with no supplementary heating (or the unfired open circuit/closed cycle plant)

1. For the doubly (closed) cyclic series plant, the thermal efficiency increases markedly with the maximum temperature in the gas turbine cycle. This is illustrated in Fig. 4.40 which shows results obtained by Buxmann and Bormann based on their assumption that the ratio of steam to gas mass flow is obtained directly from the gas turbine exhaust temperature, the selected steam saturation pressure and a particular pinch-point temperature difference.

   Also shown in Fig. 4.40 are Cerri's calculations of overall efficiency of the open circuit/closed cycle plant (including exhaust heat "losses") based on his rather different approach of optimising the overall efficiency from a range of parameters. The results shown are taken from Cerri's graphs of constant turbine inlet temperature, similar to Fig. 4.20, assuming that at a given gas turbine pressure ratio the steam/air mass flow is chosen to obtain "maximal" efficiency, and that this maximal efficiency is then plotted against varying pressure ratio. This gives efficiency levels little different from those calculated by Buxmann and Bormann, which are found not to be very sensitive to steam saturation pressure, which is prescribed in their calculation but is a variable in Cerri's. The differences between these two sets of calculations are probably due more to difference in assumptions (e.g. the absence of heat "loss" in the Buxmann and Bormann calculations and the restriction they impose on lower gas pressure level for the binary plant) than due to the difference in the approach. However, Buxmann and Bormann suggest that the optimum pressure ratio is much lower than that obtained from Cerri's calculations; but the curve of efficiency with pressure ratio (at a given maximum temperature T_3) is flat near the maximum. The indicated mass flow ratios (steam to air) are comparable, and very small. Essentially, the gas turbine dominates in this combined plant, doubly cyclic or open/closed, with no gas turbine heat exchanger and no feed heating in the steam cycle.

Deaeration is required in the steam cycle, and this may be achieved in a single direct contact feed heater, which is Brown Boveri practice (see Wunsch [17]) and the basis of Rufli's analysis. We compare Rufli's calculations with Cerri's in Fig. 4.41. They appear to give comparable combined plant efficiency, but we must remember the differences in assumptions made by the two authors.

2. The importance of the specific work in the gas turbine plant (itself strongly dependent upon maximum temperature) is emphasised by both Timmermans [20] and Korpela [19]. Their results are indicated in Fig. 4.42 in which combined plant overall efficiency is plotted against gas turbine specific work (the optimum pressure ratio for a given T_3 is selected from Korpela's results). The lower combined plant efficiency

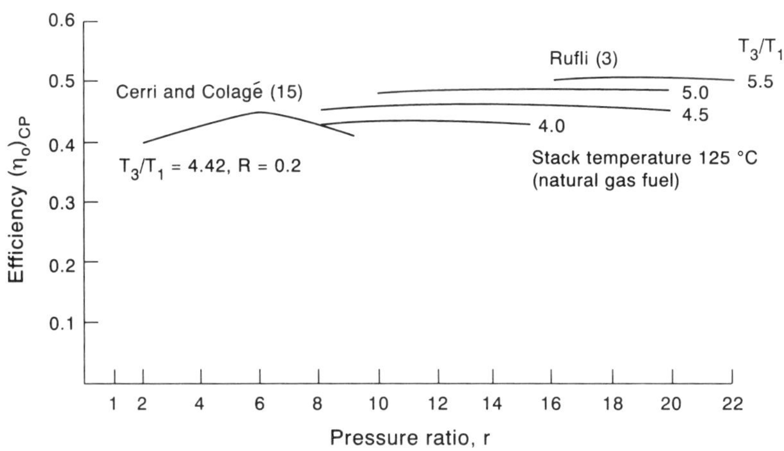

FIGURE 4.41 Overall efficiency of open circuit/closed cycle CCGT plant (with regenerative feed heating)

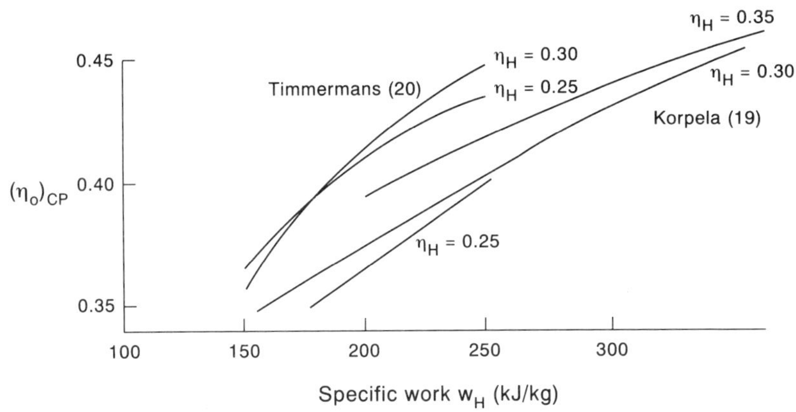

FIGURE 4.42 Overall efficiency as a function of gas turbine specific work

indicated in the latter's results is probably due to the high condenser temperature selected in his calculation (the condensate was assumed to be available for district heating in the temperature range 55°C to 85°C).

3. Based on their contention that specific work is the major criterion for selecting maximum combined cycle efficiency, Timmermans and Korpela would select higher pressure ratios for the open/closed plant (of the order of 10 to 15, for the range of maximum gas turbine

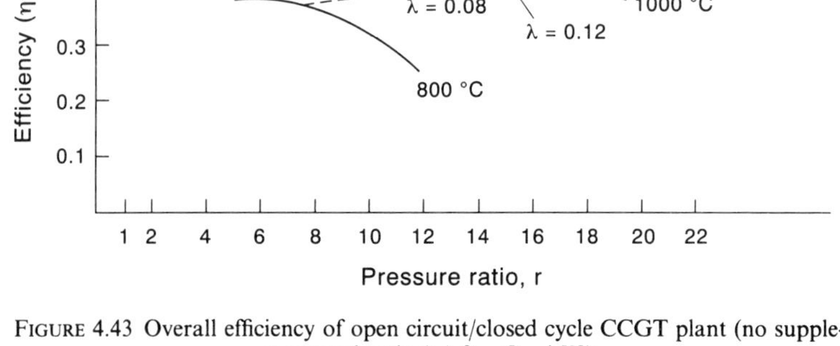

FIGURE 4.43 Overall efficiency of open circuit/closed cycle CCGT plant (no supplementary heating) (after Cerri [2])

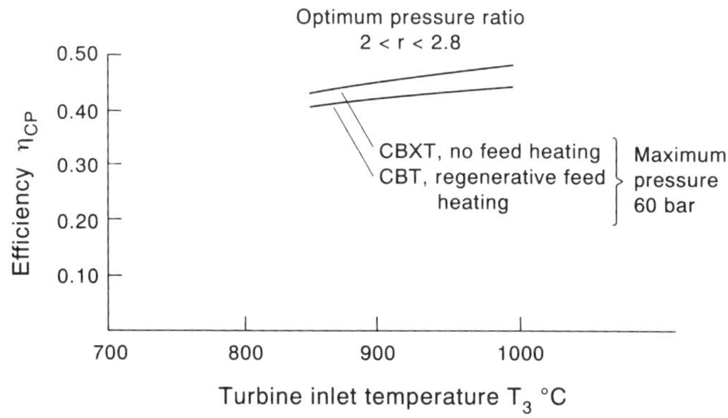

FIGURE 4.44 Effect of heat exchanger on thermal efficiency of binary "Joule"/ "Rankine" plant (after Dibelius and Ziemann [13], quoted in [11])

temperature of 900° C to 1100° C) than Buxmann and Bormann (who suggest $r = 3$–4 for 800° C for the doubly closed series plant). For the open circuit/closed cycle plant, perhaps the best guide to the relationship between optimum pressure ratio and turbine inlet temperature is provided by Cerri. A plot of the maximum possible efficiency against turbine inlet temperature, taken from Cerri's comprehensive calculations (but with no feed heating), is shown in Fig. 4.43. The corresponding pressure ratio is also shown. However, we should remember that these calculations are for a single pressure steam system, and that the plant would require a pre-heating loop to heat the feed water from

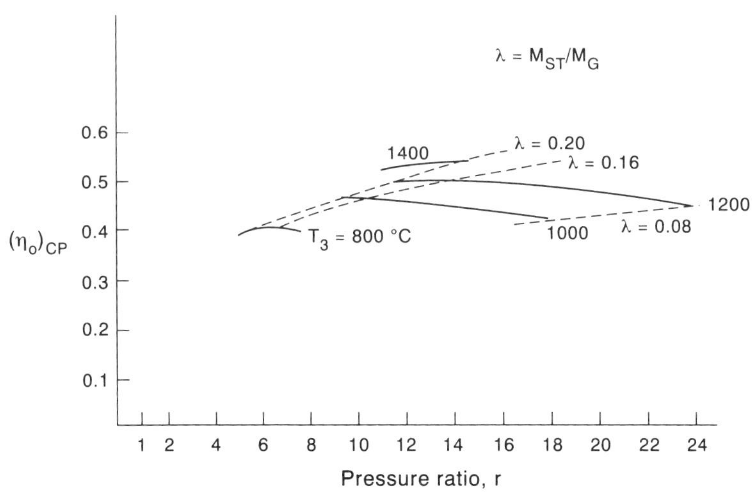

FIGURE 4.45 Overall efficiency of open circuit/closed cycle CCGT plant with supplementary firing (after Cerri [2])

condenser temperature to the minimum allowable temperature at entry to the HRSG.

4. Dibelius and Ziemann (quoted by Buxmann and Bormann) consider the use of a heat exchanger in the gas turbine side of the doubly closed combined plant. Figure 4.44 shows their plot of efficiency against turbine inlet temperature (for optimum pressure ratio). This result shows a gain of some 3–4% over the doubly closed plant without a heat exchanger, at a given turbine inlet temperature (Buxmann and Bormann quote the optimum pressure ratio for both cycles as in the range 2–2.8). The role of the gas turbine (including a heat exchanger) in the combined plant is now even more important, with an order of magnitude increase in the gas/steam power ratio.

   A more pragmatic approach to the use of a heat exchanger was taken by Pfenninger, who considered its introduction in a gas turbine plant (the CBTX plant of Section 1.5.1.2) as an alternative to the introduction of a bottoming steam turbine. That this was a marginal decision at the time (1973) may be illustrated by comparison of Pfenninger's estimate of a peak overall efficiency of 0.357 for the heat exchanger gas turbine plant (for a turbine entry temperature T_3 of 880° C) with an overall efficiency of 0.385 for a combined gas/steam plant operating with the same gas turbine inlet temperature. (We may note that Cerri's estimate of the maximum possible combined plant efficiency is just over 0.40 at an inlet temperature T_3 of 900° C, but of course his basic assumptions differ somewhat.) Moreover, Pfenninger was hesitant about the CBTX cycle on the grounds of cost, pointing out that the combined cycle could achieve comparable if not better overall efficiency coupled with increased specific power output.

5. In spite of the thermodynamic arguments presented earlier (e.g. Section 3.4.1), that combined plant efficiency will drop with supplementary heating (due to the introduction of the parameter v_S in the simple equation for plant efficiency), Pfenninger argued for the combined plant with supplementary firing, on the grounds that high efficiency can be maintained as firing is increased because of increase in the steam cycle efficiency. Pfenninger presented a convincing graph (Fig. 4.28) showing combined plant efficiency against the gas temperature reached in the supplementary heater. Plant efficiency increased somewhat with fuel addition in the supplementary heater, but then fell as the excess air in the gas turbine exhaust was used up in combustion. The gas turbine conditions were held constant in this analysis ($T_3 = 880$° C, $r = 9.56$, $(\eta_O)_H = 0.277$, $\dot{W}_H = 65.8$ MW). Steam conditions increased both in pressure and in temperature (up to a gas temperature of 600° C) but were then held constant at $p_c = 116$ bar, $T_e = 530$° C; steam mass flow was then increased as the boiler gas temperature was increased.

Wunsch [17] confirmed this small variation in overall efficiency with boiler firing temperature at a slightly higher gas turbine inlet temperature ($T_3 = 945°$ C). However, he extended Pfenninger's graph (Fig. 4.28) to both higher and lower values of T_3. For a lower value of (620° C) there was a marked increase in overall plant efficiency with HRSG firing temperature, as the steam cycle efficiency improved with raising the live steam conditions. For higher values of T_3 (1180° C and 1350° C) plant overall efficiency dropped with increase of boiler inlet temperature, up to a limiting value of excess of combustion air (5% greater than required by stoichiometric conditions).

Wunsch's results are broadly in line with Cerri's estimates of plant efficiency for supplementary firing (Fig. 4.45). He observed an increase in $(\eta_O)_{CP}$ with afterburning at low T_3 (800° C) but little change at $T_3 = 1000°$ C and 1200° C, and a small decrease in $(\eta_O)_{CP}$ with supplementary firing at $T_3 = 1400°$ C.

These conclusions have subsequently been confirmed by Davidson and Keeley [29] (see Chapter 8, the summary chapter).

6. The introduction of feed heating into the steam cycle of an open circuit (gas turbine)/closed cycle (steam turbine) combined plant is a complex matter. A summary of the thermodynamic issues is as follows.

(i) The simplest recuperative plant, with no regenerative feed heating (Fig. 4.30a) and all the feed water heated directly in the HRSG, is not feasible because of the limits that have to be placed on the temperature T_b of the feed water entering the HRSG (in order for corrosion of the metal surfaces to be avoided). However, a thermodynamic performance the same as this simplest plant (no regenerative feed heating) can be achieved by extracting from the exhaust gases the heat required to raise the feed water from condenser temperature (T_a) to the prescribed minimum temperature (T_b) by a pre-heating loop. This modified plant would give higher efficiency than a plant which uses bled steam to heat the feed water to T_b.

Each of these systems has to be optimised separately. In essence, Cerri's first comprehensive parametric calculations were made for the simplest recuperative system with no regenerative feed heating, but are valid for the first system. Rufli's single pressure calculations are for the system with feed heating, although Cerri and Colagé also studied this system subsequently.

(ii) The two pressure steam cycle using fuel containing sulphur requires the introduction of substantial feed heating and a pre-heating loop, avoiding use of an L.P. economiser. The introduction of two steam pressure levels in a plant using sulphur-free fuel, with a much lower feed-water temperature at entry to the HRSG, allows an L.P. economiser to be introduced, with a single stage of direct contact feed heating (with deaeration) providing a small temperature rise of the feed water before the L.P. economiser. In practice this appears to provide the highest overall efficiency of the combined plant.

Rufli's calculations provide comprehensive parametric analyses of these dual pressure plants, both for fuel oil and natural gas and show this higher efficiency of the plant fuelled by natural gas.

References

[1] Mangan, J. L. and Pettit, J. L. *A Highly Efficient Steam Turbine Gas Turbine Cycle*. ASME Paper 63-GT-51, 1963.

[2] Cerri, G. Parametric Analysis of Combined Gas-Steam Cycles. *Trans. ASME Journal of Engineering for Gas Turbines and Power* **109,** 46–55, 1987.

[3] Rufli, P. A Systematic Analysis of the Combined Gas/Steam Cycle. *Proc. ASME COGEN-TURBO I* 135–146, 1987.

[4] Kehlhofer, R. *Combined Cycle Gas and Steam Turbine Power Plants*. Fairmont Press, Lilburn GA, 1991.

[5] Wood, B. Alternative Fluids for Power Generation. *Proc. I.Mech.E.* **184,** 1, 40, 713, 1969–1970.

[6] Haywood, R. W. *Analysis of Engineering Cycles*, 4th edition. Pergamon Press, Oxford, 1991.

[7] Horn, G. and Norris, I. D. The Selection of Working Fluids Other Than Steam for Future Power Generation Cycles. *The Chemical Engineer*, 285–305, Nov. 1966.

[8] Colosimo, D. D., Rose, R. K. and Decker, O. *Power Generation with Binary Cycles*. Von Karman Institue for Fluid Dynamics, Lecture Series 6, Vol. 2, 1978.

[9] Kalina, A. I. Combined Cycle Steam with Novel Bottoming Cycle. *Trans. ASME Journal of Engineering for Gas Turbines and Power*, **106,** 737–742, 1984.

[10] Kalina, A. I. and Leibowitz, H. M. *Applying Kalina Technology to a Bottoming Cycle for Utility Combined Cycles*. ASME Paper 87-GT-25, 1987.

[11] Buxmann, J. and Bormann, T. *The Closed Cycle Gas Turbine-Steam Turbine*. Von Karman Institute for Fluid Dynamics, Lecture Series 6, 1976.

[12] Keenan, J. H. and Kaye, J. *Gas Tables*, Wiley, New York, 1941.

[13] Dibelius, G. and Ziemann, M. *Vergleich von Kombinierten Kraftprozessen fur Hochtemperaturreaktoren*, HHT-Program der BRD, Julich Bericht HHT 17, 1975.

[14] Seippel, C. and Bereuter, R. The Theory of Combined Steam and Gas Turbine Installations. *Brown Boveri Review*, **47,** 783–799, 1960.

[15] Cerri, G. and Colagé, A. Steam and Regeneration Influence on Combined Gas/Steam Power Plant Performance. *Trans. ASME Journal of Engineering for Power*, **107,** 574–581, 1985.

[16] Pfenninger, H. Combined Steam and Gas Turbine Power Stations. *Brown Boveri Review,* **60,** 9, 389–397, 1973.

[17] Wunsch, A. Combined Gas/Steam Turbine Power Plants—The Present State of Progress and Future Developments. *Brown Boveri Review,* **65,** 10, 646–655, 1978.

[18] Negri di Montenegro, G., Bettochi, R., Cantore, G., Borghi, M. and Naldi, G. A Steam-Gas Combined Plants Optimization, *Proc. 21st Intersociety Energy Conversion Engineering Conference*, 129–135, 1986.

[19] Korpela, T. Characteristics of Combined Gas Steam Cycles in Heat and Power Production. *Fernwarme International* **14,** 2, 41–45, 1983.

[20] Timmermans, A. R. J. *Combined Cycles and Their Possibilities*. Von Karman Institute for Fluid Dynamics, Lecture Series 6, Vol. 1, 1978.

[21] Rice, I. G. The Combined Reheat Gas Turbine/Steam Turbine Cycle. *Trans. ASME Journal of Engineering for Power* **102,** 1, Part 1, 35–41, Part 2, 42–49, 1980.

[22] Hemsley, P. D. and Roberts, R. L. Steam Turbines for Combined Cycles. *Proc. I.Mech.E. Conference on Steam Plant for the 1990's,* 1990.

[23] Mayers, M. A. *et al. Combination Steam and Gas Turbines*. ASME Paper 55-A-184, 1955.

[24] El-Masri, M. A. Exergy Analysis of Combined Cycles: Part 1—Air-Cooled Brayton-Cycle Gas Turbines. *Trans. ASME Journal of Engineering for Gas Turbines and Power*, **109,** 2, 228–236, 1987.
[25] Manfrida, G. and Mannucci, M. Exergy Analysis of Possible Solutions for Combined Gas-Steam Cycles. *Proc. 21st Intersociety Energy Conversion Engineering Conference.*
[26] Horlock, J. H. The Use of Feed Heating in the Steam Cycle of a Combined Cycle Power Plant. *Proc. I.Mech.E.* **205,** 207–215, 1991.
[27] Haywood, R. W. A Generalised Analysis of the Regenerative Steam Cycle for a Finite Number of Heaters. *Proc. I.Mech.E.* **161,** 157, 1949.
[28] Traupel, W. *Thermische Turbomaschinen*, Vol. 1. Springer Verlag, Berlin, New York, 1977.
[29] Davidson, B. J. and Keeley, K. R. The Thermodynamics of Practical Combined Cycles. *Proc. I.Mech.E. Conference on Combined Cycle Gas Turbines*, 28–50, 1991.
[30] Gyarmathy, G. On Load Control Methods for Combined Cycle Plants. *Proc. ASME COGEN-TURBO III*, 39–50, 1989.
[31] Frutschi, H. U. Control Methods for Cogeneration with Gas Turbines and Combined Cycles. *Proc. ASME COGEN-TURBO I*, 263–268, 1987.
[32] Fraize, W. E. and Kinney, C. Effects of Steam Injection on the Performance of Gas Turbine Power Cycles. *Trans. ASME Journal of Engineering for Power* **101,** 217–227, 1979.
[33] Cerri, G. and Arsuffi, G. *Calculation Procedure for Steam Injected Gas Turbine Cycle with Autonomous Distilled Water Production.* ASME Paper 86-GT-297, 1986.
[34] Horlock, J. H. Some Approximate Equations for the Properties of Steam at Moderately High Pressures. *Proc. I.Mech.E.* **173**, 33, 773–744, 1959.
[35] Brown, D. H. and Cohn, A. *An Evaluation of Steam Injected Combustion Turbine Systems.* ASME Paper 80-GT-51, 1980.
[36] Messerlie, R. L. and Strother, J. R. *Integration of the Brayton and Rankine Cycles to Maximise Gas Turbine Performance—a Cogeneration Option.* ASME Paper 84-GT-52, 1984.
[37] Kolp, D. A. and Moeller, D. J. World's First Full STIG LM 5000 Installed at Simpson Paper Company. *Trans. ASME Journal of Engineering for Gas Turbines and Power,* **111,** 2, 200–210, 1989.
[38] Frutschi, H. U. and Plancherel, A. Comparison of Combined Cycles with Steam Injection and Evaporisation Cycles. *Proc. ASME COGEN-TURBO II*, 137–145, 1988.

CHAPTER 5

Exergy Analysis

5.1 Introduction

We have defined both thermal efficiency and overall efficiency for combined plant, and have seen how those efficiencies vary with various parameters. In the first chapter we considered how the thermal efficiency of a power plant differs from that of an ideal plant (a reversible Carnot engine, receiving all its heat supply at constant maximum temperature and rejecting all its waste heat at constant minimum temperature), drawing attention to two factors:

(i) a factor ξ, which measured the failure to achieve maximum and minimum temperature;
(ii) a factor μ, which measured the "widening" of the cycle on a diagram, due to internal irreversibilities.

However, we have not as yet systematically studied how irreversibilities arise, and how potential work output is lost in the various processes throughout the plant. Here we use the concepts of availability, exergy, and rational efficiency which were outlined in Section 1.6 for that purpose.

In defining a *rational efficiency* for an *open circuit plant*, we related the work output to the maximum (reversible) work that could be achieved between the reactants supplied (each at the pressure p_0 and temperature T_0 of the environment) and the products produced, each at the same p_0 and T_0,

$$(\eta_R)=\frac{W}{(-\Delta G_0)}, \tag{5.1}$$

where $(-\Delta G_0)=(G_R)_0-(G_P)_0$ is the change in the Gibbs function from reactants to products.

It should be noted, however, that the reversible work (W_{REV}) achieved by the reactants after burning with oxygen drawn from the atmosphere and reaching a final "product" dead state in chemical equilibrium with the atmosphere is strictly

$$W_{\mathrm{REV}}=(-\Delta G_0)+T_0\left[\sum_{\mathrm{k}} M_{\mathrm{k}} R_k \ln\left(\frac{p_0}{p_{\mathrm{k}}}\right)-M_{\mathrm{O}_2} R_{\mathrm{O}_2} \ln\frac{p_0}{p_{\mathrm{O}_2}}\right], \quad (5.2)$$

where M_{k} is the mass of product k in the exhaust products, which attains an equilibrium partial pressure p_{k} in the atmosphere, and M_{O_2} is the mass of oxygen in the atmosphere, with partial pressure p_{O_2}.

This quantity may be taken as the exergy of the fuel E_{F}, the exergy of the air in the entry state and the products in the dead state being taken as zero, $(E_{\mathrm{A}})_0=0$, $(E_{\mathrm{P}})_0=0$. (Kotas [1]). Thus a stricter definition of the rational efficiency is

$$(\eta_{\mathrm{R}})=\frac{W}{(W_{\mathrm{REV}})}=\frac{W}{E_{\mathrm{F}}}, \quad (5.3)$$

but is usually approximated to $(\eta_{\mathrm{R}})=\dfrac{W}{(-\Delta G_0)}$. This approximation is made because the partial pressure terms in equation (5.2) are small.

For an open circuit plant the rational efficiency is frequently replaced by an *arbitrary overall efficiency*

$$(\eta_{\mathrm{O}})=\frac{W}{F}=\frac{W}{(-\Delta H_0)}=\frac{W}{M_{\mathrm{F}}(\mathrm{CV})_0}, \quad (5.4)$$

where M_{F} is the mass flow of fuel and $(\mathrm{CV})_0$ is its calorific value.

The definition of rational efficiency can also cover a plant comprising a closed cycle with heat supplied from a boiler with external combustion. (η_{R}) then expresses the ratio of actual work output to the maximum possible work output that could be obtained from the reactants supplied to the boiler of the plant. But again the rational efficiency is frequently replaced by an arbitrary overall efficiency

$$(\eta_{\mathrm{O}})=\frac{W}{(-\Delta H_0)}, \quad (5.5)$$

which can be written

$$(\eta_{\mathrm{O}})=\left(\frac{W}{Q_{\mathrm{B}}}\right)\left(\frac{Q_{\mathrm{B}}}{-\Delta H_0}\right)=\eta(\eta_{\mathrm{B}}), \quad (5.6)$$

where $(\eta_{\mathrm{B}})=\dfrac{Q_{\mathrm{B}}}{(-\Delta H_0)}$ is the boiler efficiency and η is the thermal efficiency of the closed cycle.

The definition of rational efficiency [equation (5.1)] has been used by many authors in discussing the performance of power plants (e.g. Kotas [1], Horlock [2], Traupel [3] and others) and we shall use it here for combined plants. Following the approach of Section 1.6, for flow through a component from state X to state Y (Fig. 5.1) we may write the exergy transfer in the form

$$E_Y = E_X - I^{CR} - I^Q - W_{CV}, \tag{5.7}$$

where W_{CV} is the work output from the control surface surrounding the component, I^{CR} is the lost work due to internal irreversibility within the control volume, I^Q is the lost work due to external irreversibility, arising from heat transfer *out* ($\int dQ$) at temperatures other than T_0.

We can then trace the exergy out of one component (e.g. a compressor or combustion chamber) into the next, as illustrated in Table 5.1.

For the complete power plant the total exergy in (E_X) will usually be the exergy of the fuel (E_F) since the exergy of the air at atmospheric conditions is zero. Thus the rational efficiency will be given by

$$(\eta_R) = \frac{\sum_{i=A,B\ldots} (W_{CV})_i}{E_F}$$

$$\approx \frac{\sum_{i=A,B\ldots} (W_{CV})_i}{(-\Delta G_0)}$$

$$= 1 - \left\{ \frac{\sum_{i=A,B\ldots} (I^Q + I^{CR})_i}{(-\Delta G_0)} + \frac{(E_Y)_Z}{(-\Delta G_0)} \right\}, \tag{5.8}$$

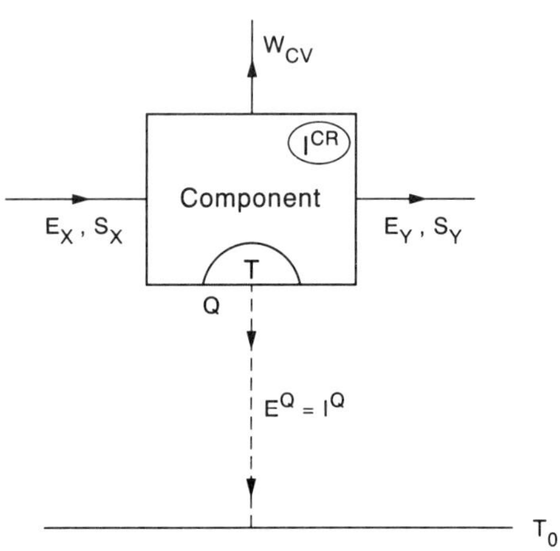

FIGURE 5.1 Exergy flows through component losing heat Q

TABLE 5.1

	Exergy in	Lost work due to heat transfer out $(I^{Q})_i$	Lost work due to internal irreversibility $(I^{CR})_i$	Work output	Exergy out
For component A	$[E_X]_A$	$\left[\int\left(\frac{T-T_0}{T}\right)dQ\right]_A$	$T_0[\Delta S_{CR}]_A$	$[W_{CV}]_A$	$[E_Y]_A$
For component B (e.g. a combustion chamber receiving fuel with exergy E_F)	$[E_X]_B = ([E_Y]_A + E_F)$	$\left[\int\left(\frac{T-T_0}{T}\right)dQ\right]_B$	$T_0[\Delta S_{CR}]_B$	$[W_{CV}]_B$	$[E_Y]_B$
For component C	$[E_X]_C = [E_Y]_B$	—	—	—	—
For component Z	—	—	—	—	$[E_Y]_Z$
For total power plant	$[E_X]_A + E_F$	$\sum_{i=A,B,C\ldots} I_i^Q$	$T_0 \sum_{i=A,B,C\ldots} (\Delta S_{CR})_i$	$\sum_{i=A,B,C\ldots} (W_{CV})_i$	$[E_Y]_Z$

where $(E_Y)_Z$ is the exergy of gases finally discharged from a combined plant (in the exhaust gases for an open circuit plant with internal combustion or in the boiler flue gases for a plant with external combustion in a boiler).

Since there is no work output from these gases and their ultimate exergy is zero, we may write

$$(E_Y)_Z = (I^Q)_Z = \left[\int\left(\frac{T-T_0}{T}\right)dQ\right]_Z$$

where $[\int dQ]_Z = (H_Y)_Z - H_0$ would be the heat transferred out in reducing the exhaust gases to atmospheric temperature T_0.

The quantity in the curly brackets in equation (5.8) is sometimes referred to as an efficiency defect (Kotas [1]), $\delta = \sum_{i=A,B\ldots} \delta_i$. Then the rational efficiency is written

$$(\eta_R) = 1 - \delta = 1 - \sum_{i=A,B\ldots} \delta_i, \tag{5.9}$$

where δ_i includes $(\delta)_Z = (I^Q)_Z/(-\Delta G_0)$.

The relative contribution of a component to the failure to achieve ideal work output is then (δ_i/δ).

However, there have been other attempts to define component performance, and to relate that performance to plant rational efficiency (see Horlock [4]).

The simplest definition of rational efficiency for a component *i* follows that for a plant

$$(\eta_R)_i = \frac{[W_{CV}]_i}{[(W_{REV})_{CV}]_i}, \tag{5.10}$$

where $[W_{CV}]_i$ is the actual work output and $[(W_{REV})_{CV}]_i = (E_X - E_Y)_i$ is the maximum possible work output between the entry and exit states in the presence of a given environment (at p_0, T_0). For a component with work *input* $[W_{CV}]_i$, compared with minimum possible work *input* $[(W_{REV})_{CV}]_i = (E_Y - E_X)_i$

$$(\eta_R)_i = \frac{[(W_{REV})_{CV}]_i}{[W_{CV}]_i}. \tag{5.11}$$

Horlock [4] shows the plant rational efficiency is then given by

$$(\eta_R) = \sum_{\substack{i \\ \text{OUTPUT}}} (\eta_R)_i \lambda_i - \sum_{\substack{i \\ \text{INPUT}}} \lambda_i/(\eta_R)_i,$$

where

$$\underset{\text{(WORK OUTPUT)}}{(\lambda_i)} = (E_X - E_Y)_i / (-\Delta G_0), \qquad \underset{\text{(WORK INPUT)}}{(\lambda_i)} = (E_Y - E_X)_i / (-\Delta G_0).$$

However, not all components have a work output or input, so that their rational efficiency is zero and $(\eta_R)_i$ is not a good measure of their performance.

There may be two or more streams flowing through a component. Fratzscher [5] then defines a degree of quality ν_i for the component as

$$\nu_i = \frac{\sum_{str} \mathrm{E_Y}}{\sum_{str} \mathrm{E_X}} = 1 - \frac{(I^{CR})_i}{\sum_{str} \mathrm{E_X}}, \tag{5.12}$$

where

$$\mathrm{E_X} = [E + E^Q + W_{CV}]_X, \; \mathrm{E_Y} = [E + E^Q + W_{CV}]_Y,$$

and $E^Q = \int \left(\frac{T - T_0}{T}\right) dQ$ is now regarded as an exergy flux due to heat transfer in or out. $\sum_{str}$ denotes the sum for several fluid streams.

Fratzscher further suggests yet another "second law" criterion of performance, the exergetical efficiency $(\eta_e)_i$ for a component where there is a desired output from one stream (b) and an expenditure from another stream (a),

$$(\eta_e)_i = \frac{\Delta \mathrm{E_b}}{\Delta \mathrm{E_a}} = 1 \frac{(I^{CR})_i}{\Delta \mathrm{E_a}}. \tag{5.13}$$

There are arguments for using ν_i and $(\eta_e)_i$ for components (see Horlock [4]) but it is difficult to relate them to the plant rational efficiency (η_R). However, it is shown in reference [4] that for a combined power plant

$$\eta_R = (\eta_R)_H + \eta_e(\eta_R)_L - \eta_e(\eta_R)_H(\eta_R)_L - \eta_e(\eta_R)_L \delta_H \tag{5.9a}$$

where η_e is the exergetical efficiency of the HRSG linking the higher and lower plants, and δ_H is the sum of the fractional irreversibilites in the former (excluding the HRSG). This equation has somewhat similar form to equation (3.33) for combined plant overall efficiency.

We shall not follow any of these alternative approaches here but use the expression given by Kotas and employed by most authors who have uced exergy analyses,

$$(\eta_R) = 1 - \delta, \tag{5.9b}$$

where

$$\delta = \sum_{i=A,B\ldots} \delta_i$$

$$= \frac{\sum_i [I^{CR} + I^Q]_i}{(-\Delta G_0)}.$$

We illustrate this approach by first deriving $(I^{CR})_i$ and $(I^Q)_i$ for several components and then applying exergy analysis to two examples:

(i) a binary (closed cycle) plant with heat supplied from a boiler to the closed cycles;
(ii) an open circuit gas turbine/closed steam turbine plant.

We subsequently refer to some parametric studies based on exergy analysis.

In Chapter 6 we show how exergy analysis can be extended to provide a "thermo-economic" analysis of a combined power plant.

5.2 Lost Work Due to Irreversibility in Components

Before giving examples of exergy analyses of two combined plants it is useful to derive expressions for $(I^{CR})_i$ and $(I^Q)_i$, for various components. We use a standard notation for each component, X denoting the inlet state and Y the outlet state.

5.2.1 Compressor

For an adiabatic compressor, $(I^Q)_i = 0$ and

$$(I^{CR})_i = T_0(\Delta S^{CR})_i$$

$$= M_A T_0(s_Y - s_X), \tag{5.14}$$

where M_A is the mass flow. If the fluid is a perfect gas with specific heat (at constant pressure) $(c_p)_A$ and gas constant R_A, then

$$(I^{CR})_i = M_A T_0[(c_p)_A \ln(T_Y/T_X) - R_A \ln(p_Y/p_X)]. \tag{5.15}$$

5.2.2 Turbine

For an adiabatic turbine, $(I^{Q})_i = 0$ and

$$(I^{CR})_i = T_0(\Delta S_{CR})_i$$
$$= M_G T_0(s_Y - s_X), \qquad (5.16)$$

where M_G is the mass flow.

For a steam or mercury turbine, values of entropy s_X and s_Y are obtained from tables of properties. For a gas turbine

$$(I^{CR})_i = M_G T_0[(c_p)_G \ln(T_Y/T_X) - R_G \ln(p_Y/p_X)]$$
$$= M_G T_0[R_G \ln(p_X/p_Y) - (c_p)_G \ln(T_X/T_Y)], \qquad (5.17)$$

where $(c_p)_G$, R_G are specific heat (at constant pressure) and gas constant respectively.

5.2.3 Combustion Chamber

For a combustion chamber, to which liquid fuel and air are supplied, the mass flows being M_F and M_A respectively, we may write, for adiabatic combustion,

$$(I^{CR})_i = T_0[(S_P)_Y - (S_R)_X] \qquad (5.18)$$

and

$$(H_R)_X = (H_P)_Y. \qquad (5.19)$$

In a "calorific value" process, reactants would be supplied at temperature T_0 and pressure p_0, and heat $(-\Delta H_0) = (H_R)_0 - (H_P)_0$ would be abstracted in order to allow the products to leave at the temperature T_0. For this process

$$T_0[(S_P)_0 - (S_R)_0] = T_0 \Delta S_0$$
$$= (-\Delta G_0) - (-\Delta H_0). \qquad (5.20)$$

Equations (5.18) and (5.20) yield

$$(I^{CR})_i = T_0[(S_P)_Y - (S_P)_0] - T_0[(S_R)_X - (S_R)_0] + T_0 \Delta S_0. \qquad (5.21)$$

If the reactants consist of liquid fuel (usually at p_0, T_0) and air at p_X, T_X then

$$(S_R)_X - (S_R)_0 = [(S_F)_0 + (S_A)_X] - [(S_F)_0 + (S_A)_0]$$
$$= M_A(\phi_A)_X - M_A R_A \ln(p_X/p_0), \qquad (5.22)$$

where M_A is the mass flow of air and $(\phi_A)_X = \int_{T_0}^{T_X} (c_p)_A \, dT/T$.

For the products

$$(S_P)_Y - (S_P)_0 = M_G(\phi_G)_Y - M_G R_G \ln(p_Y/p_0), \tag{5.23}$$

where

$$(\phi_G)_Y = \int_{T_0}^{T_Y} (c_p)_G \, dT/T,$$

$$M_G = M_A + M_F.$$

It therefore follows from (5.21), (5.22) and (5.23) that

$$\begin{aligned}(I^{CR})_i = {} & T_0 \Delta S_0 \\ & + M_G T_0[(\phi_G)_Y - R_G \ln(p_Y/p_0)] \\ & - M_A T_0[(\phi_A)_X - R_A \ln(p_X/p_0)].\end{aligned} \tag{5.24}$$

El-Masri [6] provides an interesting approximation to this expression for $(I^{CR})_i$ by

(i) pointing out that ΔG_0 is little different from ΔH_0 for most fuels, i.e. that the term $T_0 \Delta S_0 = (-\Delta G_0) - (-\Delta H_0)$ is small and cannot be varied;
(ii) assuming that (M_F/M_A) is small for most practical cycles (so that $M_G \approx M_A = M$) and that $(c_p)_G = (c_p)_A = c_p$;
(iii) assuming that pressure changes are small ($p_Y/p_X \approx 1$).

He therefore writes

$$\begin{aligned}(I^{CR})_i & \approx M T_0(\phi_Y - \phi_X) \\ & = M c_p T_0 \ln(T_Y/T_X) \\ & = M c_p T_0 \frac{(T_Y - T_X)}{T_{LM}},\end{aligned} \tag{5.25}$$

where T_{LM} is the logarithmic mean temperature

$$T_{LM} = \frac{T_Y - T_X}{\ln(T_Y/T_X)}. \tag{5.26}$$

With similar assumptions

$$(-\Delta H_0) \approx M c_p (T_Y - T_X), \tag{5.27}$$

so that

$$\frac{(I^{\mathrm{CR}})_{\mathrm{i}}}{(-\Delta H_0)} \approx \frac{T_0}{T_{\mathrm{LM}}}. \tag{5.28}$$

El-Masri points out that (T_0/T_{LM}) is the complement of the Carnot efficiency based on a heat supply temperature of T_{LM},

$$\eta_{\mathrm{CARNOT}} = 1 - \frac{T_0}{T_{\mathrm{LM}}}. \tag{5.29}$$

He therefore takes the major irreversibility in the combustion process as

$$(I^{\mathrm{CR}})_{\mathrm{i}} = \left(\frac{T_0}{T_{\mathrm{LM}}}\right) Q_{\mathrm{B}} = Q_{\mathrm{B}}(1 - \eta_{\mathrm{CARNOT}}), \tag{5.30}$$

where $Q_{\mathrm{B}} = (-\Delta H_0)$ is the "equivalent heat supply" to the cycle.

It is of interest to compare $(I^{\mathrm{CR}})_{\mathrm{i}}$ for combustion with the irreversibility involved in heat transfer from a large reservoir at the top temperature T_{Y} to a gas flow initially at T_{X}. Then the lost work due to irreversibility is

$$I^{\mathrm{Q}} = \int_{\mathrm{X}}^{\mathrm{Y}} \left(\frac{T_{\mathrm{Y}} - T_0}{T_{\mathrm{Y}}}\right) dQ - \int_{\mathrm{X}}^{\mathrm{Y}} \left(\frac{T - T_0}{T}\right) dQ, \tag{5.31}$$

where $dQ = Mc_{\mathrm{p}} dT$, so that

$$\frac{I^{\mathrm{Q}}}{Mc_{\mathrm{p}}(T_{\mathrm{Y}} - T_{\mathrm{X}})} = \frac{T_0 \ln(T_{\mathrm{Y}}/T_{\mathrm{X}})}{(T_{\mathrm{Y}} - T_{\mathrm{X}})} - \frac{T_0}{T_{\mathrm{Y}}}. \tag{5.32}$$

Hence from equation (5.25)

$$\frac{(I^{\mathrm{CR}})_{\mathrm{i}}}{Mc_{\mathrm{p}}(T_{\mathrm{Y}} - T_{\mathrm{X}})} = \frac{I^{\mathrm{Q}}}{Mc_{\mathrm{p}}(T_{\mathrm{Y}} - T_{\mathrm{X}})} + \frac{T_0}{T_{\mathrm{Y}}}. \tag{5.33}$$

El-Masri also uses this form of presenting the lost work due to irreversibility in combustion—the sum of the "Carnot unavailability" (T_0/T_{Y}) and the term $I^{\mathrm{Q}}/Mc_{\mathrm{p}}(T_{\mathrm{Y}} - T_{\mathrm{X}})$; we shall refer to this form later in Section 5.6.

If the combustion is of a gaseous fuel which is compressed to state X', and air, fuel and products are all treated as semi-perfect gases,

$$\begin{aligned}(I^{\mathrm{CR}})_{\mathrm{i}} &= T_0 \Delta S_0 + T_0[(S_{\mathrm{P}})_{\mathrm{Y}} - (S_{\mathrm{P}})_0] \\ &\quad - T_0[(S_{\mathrm{A}})_{\mathrm{X}} - (S_{\mathrm{A}})_0] - T_0[(S_{\mathrm{F}})_{\mathrm{X}'} - (S_{\mathrm{F}})_0] \\ &= T_0 \Delta S_0 + M_{\mathrm{G}} T_0[(\phi_{\mathrm{G}})_{\mathrm{Y}} - R_{\mathrm{G}} \ln(p_{\mathrm{Y}}/p_0)] \\ &\quad - M_{\mathrm{A}} T_0[(\phi_{\mathrm{A}})_{\mathrm{X}} - R_{\mathrm{A}} \ln(p_{\mathrm{X}}/p_0)] \\ &\quad - M_{\mathrm{F}} T_0[(\phi_{\mathrm{F}})_{\mathrm{X}'} - R_{\mathrm{F}} \ln(p_{\mathrm{X}'}/p_0)].\end{aligned} \tag{5.34}$$

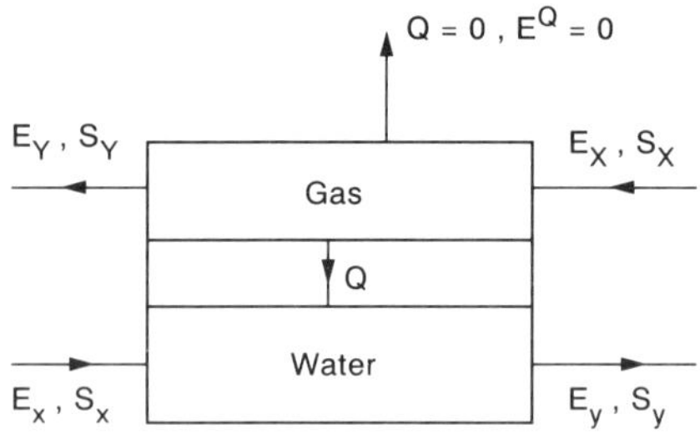

FIGURE 5.2 Exergy flows through boiler or HRSG

If the combustion is of a solid fuel then the expression given earlier for combustion of liquid fuel remains valid [equation (5.24)],

$$(I^{CR})_i = T_0 \Delta S_0 + M_G T_0[(\phi_G)_Y - R_G \ln(p_Y/p_0)] - M_A T_0[(\phi_A)_X - R_A \ln(p_X p_0)], \tag{5.24}$$

assuming adiabatic combustion, where

$$T_0 \Delta S_0 = (-\Delta G_0) - (-\Delta H_0) = (\psi - 1)\Delta H_0, \tag{5.20}$$

with $\psi = (-\Delta G_0)/(-\Delta H_0)$.

Kotas [1] defines the exergy of the fuel (the truly maximum reversible work obtainable) as

$$E_F = (-\Delta G_0) - T_0\left[\sum_k M_k R_k \ln(p_k/p_0) - M_{O_2} R_{O_2} \ln(p_{O_2}/p_0)\right], \tag{5.2}$$

as explained above. He gives tables of E_F for various fuels, with expressions for ψ, and bases his calculations on these tabled values. Such tables are not given here, but the reader is referred to Kotas [1] who gives a detailed example of combustion in a coal fired steam boiler the results of which we use below. The irreversibility in combustion is shown to be substantial.

5.2.4 Heat Transfer in a Boiler or HRSG

For a boiler or an HRSG we consider the heat transfer between the hot gas and the colder fluid (water/steam), i.e. we assume adiabatic combustion has taken place in an earlier (first) stage, so that the analysis given in the last section for a combustion chamber can be used.

We suppose that the gas enters in state X and leaves in state Y; that the water enters in state x and steam leaves in state y; that there is no heat loss to the atmosphere, so that $(I^Q)_i = 0$ (Fig. 5.2).

Then

$$H_X + H_x = H_Y + H_y, \tag{5.35}$$

and the total lost work due to irreversibility is

$$\begin{aligned}(I^{CR})_i &= (E_X + E_x) - (E_Y + E_y)\\ &= T_0[(S_Y + S_y) - (S_X + S_x)].\end{aligned} \tag{5.36}$$

For the gas flow

$$(S_Y - S_X) = M_G[(\phi_G)_Y - (\phi_G)_X - R_G \ln(p_Y/p_X)] \tag{5.37}$$

and for the water/steam, flow

$$(S_y - S_x) = M_{ST}(s_y - s_x) \tag{5.38}$$

is obtained from tables of properties.

If the heat transfer is to two fluids consecutively, as in a dual pressure HRSG, then the term $(S_y - S_x)$ becomes $[M_{ST}(s_y - s_x) + M'_{ST}(s_{y'} - s_{x'})]$, where the dash refers to the second fluid.

5.2.5 Steam Turbine with Bleed for Regenerative Feed Heating

For a steam turbine, with a steam mass flow fraction m bled for regenerative feed heating, the work output is

$$W_T = M_{ST}(h_X - h_Y) - mM_{ST}(h_{bled} - h_Y), \tag{5.39}$$

where h_{bled} is the enthalpy of the bled steam and the flow is assumed adiabatic.

The lost work due to irreversibility is then

$$\begin{aligned}(I^{CR})_i &= T_0(\Delta S_{CR})_i\\ (I^{CR})_i/T_0 &= M_{ST}(1-m)s_Y + mM_{ST}s_{bled} - M_{ST}s_X\\ &= M_{ST}[(s_Y - s_X) - m(s_Y - s_{bled})].\end{aligned} \tag{5.40}$$

The analysis may be simply developed for more than one stage of bled steam.

5.2.6 Feed Heating Train

Consider a single direct contact feed heater (Fig. 5.3) receiving a mass flow M_{CON} of water from the condenser (state X) and a mass flow mM_{ST}

of bled steam (state "bled") from the turbine, and delivering a mass flow M_{ST} of water in state Y [i.e. $M_{CON} = M_{ST}(1-m)$].

Then if the mixing process is adiabatic, $I^{Q}=0$ and the lost work due to internal irreversibility is

$$(I^{CR})_i = T_0[M_{ST}s_Y - M_{ST}(1-m)s_X - mM_{ST}s_{bled}]$$
$$= M_{ST}T_0[(s_Y - s_X) - m(s_{bled} - s_X)]. \quad (5.41)$$

But from the first law

$$mM_{ST}(h_{bled} - h_Y) = mM_{ST}T_{bled}(s_{bled} - s_Y)$$
$$= M_{CON}(h_Y - h_X)$$
$$= M_{ST}(1-m)T_{XY}(s_Y - s_X), \quad (5.42)$$

where T_{XY} is the mean water temperature between states X and Y. Substituting for m in equation (5.41), it follows that

$$(I^{CR})_i = M_{CON}T_0\left[(s_Y - s_X)\left(1 - \frac{T_{XY}}{T_{bled}}\right)\right]. \quad (5.43)$$

The analysis may be developed further for a chain of heaters. We may note that the mass flow mM_{ST} is determined from the heat balance on the individual heater. For the first heater

$$\frac{mM_{ST}}{M_{CON}} = \frac{m}{(1-m)} = \frac{h_Y - h_X}{h_{bled} - h_Y} = \frac{\gamma'}{\beta},$$

$$m = \frac{\gamma'}{\gamma' + \beta}, \quad (5.44)$$

$$\frac{M_{CON} + mM_{ST}}{M_{CON}} = \left(1 + \frac{\gamma'}{\beta}\right), \quad (5.45)$$

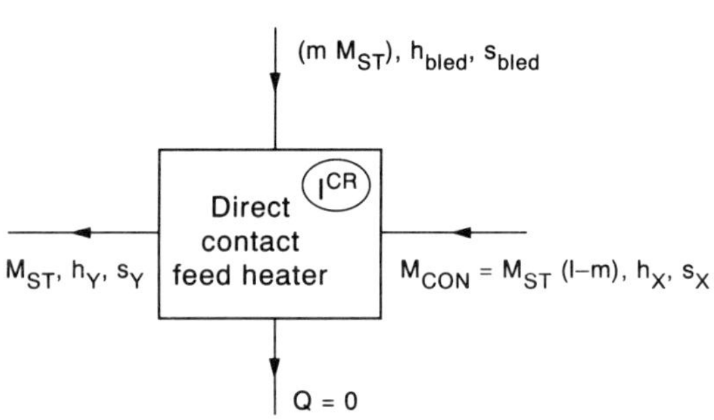

FIGURE 5.3 Flow through direct contact feed heater

where $\gamma' = (h_Y - h_X)$ is the enthalpy rise of the water, and β is the difference between the enthalpy along the turbine expansion line and the enthalpy of saturated water at that pressure.

Haywood [7] has developed analyses of regenerative feed heating on the basis that β is constant along the expansion line and we shall use that approach in the examples below. The ratio of the mass flow leaving *any* heater to that entering is then $\left(1+\frac{\gamma'}{\beta}\right)$. Thus the mass flow to the boiler M_{ST} after say three heaters is given by

$$\frac{M_{ST}}{M_{CON}} = \left(1+\frac{\gamma'}{\beta}\right)^3 = \left(1+\frac{\gamma}{3\beta}\right)^3, \tag{5.46}$$

where $\gamma = 3\gamma'$ is the total enthalpy rise distributed equally in the three heaters. Expressions similar to equation (5.43) may be derived from each heater.

5.2.7 Condenser

For a condenser, we assume the heat is transferred from the vapour at temperature T_{XY} to cooling water at ambient temperature T_0. If Q is the heat rejected, then

$$\begin{aligned} Q &= M_{CON}(h_Y - h_X) \\ &= M_{CON}\, T_{XY}(s_Y - s_X). \end{aligned}$$

The lost work due to irreversibility is then

$$\begin{aligned} (I^Q)_i &= \int \left(\frac{T-T_0}{T}\right) dQ \\ &\approx \left(1-\frac{T_0}{T_{XY}}\right) M_{CON}\, T_{XY}(s_Y - s_X) \\ &= M_{CON}(T_{XY} - T_0)(s_Y - s_X) \\ &= \frac{Q(T_{XY} - T_0)}{T_{XY}}. \end{aligned} \tag{5.47}$$

5.3 Exergy Analysis of a Binary Vapour Plant

As a first example of the use of exergy analysis we consider a mercury/steam binary vapour plant, for which the (T, s) diagram and the various state points are shown in Fig. 5.4.

Heat is supplied from the boiler, jointly to the mercury plant (to heat

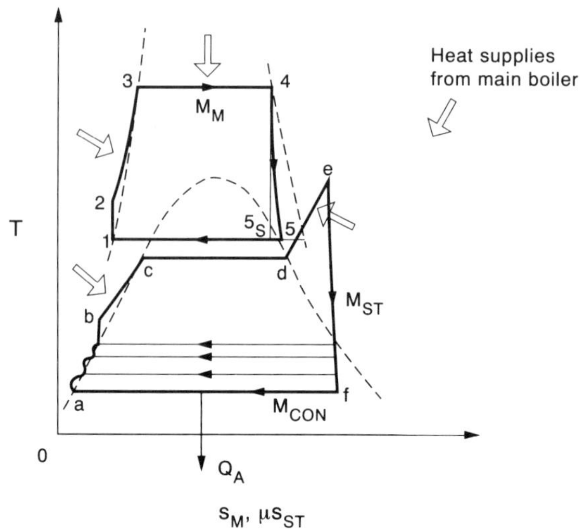

FIGURE 5.4 Mercury/steam binary plant

TABLE 5.2

Pressure	T, K	h_{liquid} (kJ/kg)	h_{vapour} (kJ/kg)	s_{liquid} (kJ/kg K)	s_{vapour} (kJ/kg K)
Evaporator, 900 kN/m^2	781.7	68.4	358.1	0.1415	0.5121
Condenser, 17 kN/m^2	542.6	36.9	331.1	0.0940	0.6362

the liquid mercury and then to evaporate it) and to the steam plant (to heat the feed water to boiler saturation temperature and to superheat the saturated steam).

We take the cycle conditions as specified by Haywood [8] in Example 9.4 of *Analysis of Engineering Cycles*, with three direct contact feed heaters heating the feed water to 170.4° C from a condenser temperature of 29° C.

The properties of saturated mercury corresponding to states in the mercury cycle are given by Haywood as in Table 5.2.

The mercury turbine efficiency is 0.75.

For the steam cycle, dry saturated steam leaves the mercury/steam boiler at 4.0 MN/m^2, is superheated to 425° C by the main boiler flue gases, passes through a turbine with a "constant β" expansion, and condenses at 4.0 kN/m^2. After the direct contact feed heating to 170.4° C (at $0.8\,MN/m^2$), the feed water is heated to the steam saturation temperature of 250.3° C (at $4.0\,MN/m^2$) (by the main boiler flue gases) before entering the mercury/steam boiler.

The mercury/steam conditions are very similar to those of the Schiller plant in New Hampshire (although there is not enough detail in the literature to show the feed heating system of the Schiller plant).

For the main boiler we assume conditions are as given in the example for a coal fired steam boiler as given by Kotas [1, p. 157]. We assume the boiler is scaled to give a mercury flow similar to that of the Schiller station (200 kg/s). This requires a supply of anthracite of 3.403 kg/s. Conditions for combustion are as given by Kotas, with temperatures of 1558° C after combustion to state P (before heat transfer to mercury and water/steam) and 200° C in state P_S (after heat transfer before discharge in the exhaust stack).

In analysing the mercury/steam cycles we make the following assumptions:

(i) The expansion in the steam turbine is such that the local enthalpy is given by $h = h_w + \beta = (h_w + 2187)$ kJ/kg, where h_w is the enthalpy of saturated water at the same pressure.

(ii) The enthalpy rises in the three direct contact heaters are equal, i.e.

$$\gamma' = \frac{720.9 - 121.8}{3} = 199.8\,\text{kJ/kg}$$

[The ratio of the mass flow through the boiler (M_{ST}) to that through the condenser (M_{CON}) is then given by

$$\frac{M_{ST}}{M_{CON}} = \left(1 + \frac{\gamma'}{\beta}\right)^3 = \left(1 + \frac{199.8}{2187}\right)^3 = 1.3.] \tag{5.46}$$

(iii) The feed pump work term in the mercury cycle is negligible (less than 0.1 kJ/kg).

(iv) The feed pump work terms in the lower pressure steam cycle are negligible (less than 1 kJ/kg total).

(v) The boiler feed pump has an efficiency of 0.8, so the boiler feed pump work is then given by

$$h_b - h_{a'''} = \frac{3.2}{850 \times 0.8} = 4.7\,\text{kJ/kg}.$$

Further, in analysing the combustion process and the heat transfer in the main boiler, we follow Kotas in assuming that the maximum possible reversible power from the reactants is the exergy flux of the fuel,

$$\begin{aligned}\dot{E}_F &= \dot{M}_F \varepsilon_F \\ &= \dot{M}_F\left[-(\Delta h_0 - T_0 \Delta S_0) + R T_0 [x_{O_2} \ln(p_{O_2}/p_0) - \sum_k x_k \ln(p_k/p_0)]\right],\end{aligned} \tag{5.2}$$

i.e. that final exhaust is to the dead state with atmospheric partial presures $(E_P)_0 = 0$. We use the tables given by Kotas for E_F and his other values of exergy E_P, $(E_P)_S$.

5.3.1 Cycle Calculations (for Unit Flow Rate of Mercury) and Plant Overall Efficiency

Main boiler

$$\dot{M}_M = 1 \text{ kg/s}$$

$$(\dot{Q}_M)_B = \text{Heat supplied to mercury} = 1(h_4 - h_3)$$
$$= 358.1 - 36.9$$
$$= 321.2 \text{ kW}.$$

$$(\dot{Q}_{ST})_B = \text{Heat supplied to water and steam}$$
$$= \dot{M}_{ST}[(h_c - h_b) + (h_e - h_d)]$$
$$= \dot{M}_{ST}[1087.4 - (720.9 + 4.7) + (3274 - 2800.3)]$$
$$= \dot{M}_{ST}(361.8 + 473.7)$$
$$= \dot{M}_{ST}(835.5) \text{ kW}.$$

Mercury/steam boiler

An energy balance gives

$$(h_5 - h_1) = \dot{M}_{ST}(h_d - h_c)$$
$$(287.3 - 36.9) = M_{ST}(2800.3 - 1087.4)$$
$$M_{ST} = \frac{250.4}{1712.9} = \frac{1}{6.841}.$$

Note that h_5 is obtained by taking 0.75 of the isentropic enthalpy drop between 900 kN/m^2 and 17 kN/m^2, to point 5_S. Thus

$$x_{5S} = \frac{0.5121 - 0.0940}{0.6362 - 0.0940} = \frac{0{\cdot}4181}{0.5422} = 0.771,$$

$$h_4 - h_{5S} = 358.1 - [36.9 + 0.771\ (331.1 - 36.9)]$$
$$= 358.1 - 263.7$$
$$= 94.4 \text{ kJ/kg},$$

$$h_4 - h_5 = 0.75 \times 94.4 = 70.8 \text{ kJ/kg},$$

$$h_5 = 358.1 - 70.8 = 287.3 \text{ kJ/kg}.$$

Total heat supplied in main boiler

$$\dot{Q}_{\mathrm{B}}=(\dot{Q}_{\mathrm{M}})_{\mathrm{B}}+(\dot{Q}_{\mathrm{ST}})_{\mathrm{B}}$$

$$=321.2+\frac{835.5}{6.841}=321.2+122.1=443.3\,\mathrm{kW}.$$

Heat rejected in steam condenser

$$\dot{Q}_{\mathrm{A}}=\dot{M}_{\mathrm{CON}}(h_f-h_a)$$

$$=\dot{M}_{\mathrm{CON}}(2187)\mathrm{kW},$$

where $\dot{M}_{\mathrm{CON}}$ = condenser mass flow $=\dfrac{\dot{M}_{\mathrm{ST}}}{1.3}$,

$$\dot{Q}_{\mathrm{A}}=\frac{1}{6.841\times 1.3}(2187)=245.9\,\mathrm{kW}.$$

Combined plant thermal efficiency

$$\eta_{\mathrm{CP}}=1-\frac{\dot{Q}_{\mathrm{A}}}{\dot{Q}_{\mathrm{B}}}$$

$$=1-\frac{245.9}{443.3}$$

$$=0.4453.$$

Mercury cycle thermal efficiency

$$\eta_{\mathrm{M}}=1-\frac{(h_5-h_1)}{(h_4-h_2)}$$

$$=1-\frac{(287.3-36.9)}{(358.1-36.9)}=0.2204.$$

Steam cycle thermal efficiency

$$\eta_{\mathrm{ST}}=1-\frac{\dot{M}_{\mathrm{CON}}(h_f-h_a)}{\dot{M}_{\mathrm{ST}}(h_e-h_b)}$$

$$=1-\frac{1\,(2187)}{1.3\,(3274-725.6)}=0.3399.$$

[Note that combined plant thermal efficiency should be given by

$$(1-\eta_{\mathrm{CP}})=(1-q\eta_{\mathrm{M}})(1-\eta_{\mathrm{ST}}), \qquad (3.38\mathrm{b})$$

where

$$q = \frac{(\dot{Q}_M)_B}{\dot{Q}_B} = \frac{321.2}{443.3} = 0.7246$$

i.e.

$$\begin{aligned}(1 - \eta_{CP}) &= (1 - 0.7246 \times 0.2204)(0.6601)\\ &= 0.8403 \times 0.6601\\ &= 0.5546,\\ \eta_{CP} &= 0.4454, \text{ cf. } 0.4453 \text{ above.]}\end{aligned}$$

5.3.2 Power Output

For a mercury flow of 200 kg/s, and a steam flow of $\frac{200}{6.841} = 29.235$ kg/s, the power outputs are as follows.

Mercury cycle

$$\begin{aligned}W_M &= \eta_M(\dot{Q}_M)_B\\ &= 0.2204 \times 200 \times 321.2\\ &= 14{,}158 \text{ kW}.\end{aligned}$$

Steam cycle

$$\begin{aligned}W_{ST} &= \eta_{ST} Q_{ST}\\ &= 0.3399 \times 29.235 \times 2548.4\\ &= 25{,}323 \text{ kW}.\end{aligned}$$

Combined plant

$$\begin{aligned}W_{CP} &= W_M + W_{ST}\\ &= 39{,}481 \text{ kW}.\end{aligned}$$

$$\begin{aligned}(\text{cf. } W_{CP} &= \eta_{CP}\dot{Q}_B\\ &= 0.4453 \times 200 \times 443.3\\ &= 39{,}480 \text{ kW.)}\end{aligned}$$

5.3.3 Coal Supply

Kotas [1] calculates the enthalpy change $[H_P-(H_P)_S]$ for 1 kg of coal as

$$H_P-(H_P)_S=(2.8932\times 10^4-0.2879\times 10^4)$$
$$=2.6053\times 10^4\ \text{kJ}.$$

For the current example, the coal flow rate is therefore, for 200 kg/s of mercury

$$\dot{M}_{COAL}=\frac{200\times 443.3}{2.6053\times 10^4}=3.403\ \text{kg/s}.$$

5.3.4 Exergy Fluxes and "Lost Work" Terms

Entry fuel exergy and combustion loss (from Kotas)

$$\dot{E}_F=3.403\times 30.705\times 10^4=104{,}492\ \text{kW},$$
$$\dot{E}_P=3.403\times 2.009\times 10^4=68{,}366\ \text{kW}.$$

Irreversibility in combustion =

$$\dot{E}_F-\dot{E}_P=36{,}126\ \text{kW},$$
$$(\dot{E}_P)_S=3.403\times 0.16514\times 10^4=5620\ \text{kW},$$
$$\dot{E}_P-(\dot{E}_P)_S=68{,}366-5620=62{,}746\ \text{kW}.$$

Main boiler

With no external heat loss the "lost work" is

$$(\dot{I}^{CR})_i=T_0(\Delta S_{CR})_i$$
$$=T_0[(\dot{S}_P)_S-(\dot{S}_P)+\dot{M}_M(s_4-s_3)$$
$$+\dot{M}_{ST}(s_e-s_d)+\dot{M}_{ST}(s_c-s_b)],$$
$$T_0[(\dot{S}_P)_S-(\dot{S}_P)]=[\dot{E}_P-(\dot{E}_P)_S]-(\dot{H}_P-(\dot{H}_P)_S]$$
$$=62{,}746-\left[\frac{2.6053\times 10^6\times 3.403}{100}\right]$$
$$=62{,}746-88{,}660$$
$$=-25{,}914\ \text{kW}.$$

$$T_0\dot{M}_M(s_4-s_3)=298.16\times200\times(0.5121-0.0940)=24{,}932\ \text{kW}$$
$$T_0\dot{M}_{ST}(s_e-s_d)=298.16\times29.235(6.858-6.069)\ =\ 6877\ \text{kW}$$
$$T_0\dot{M}_{ST}(s_c-s_b)=298.16\times29.235(2.797-2.046)\ =\ 6546\ \text{kW}$$

Total 38,355 kW

$$(\dot{I}^{CR})_i=38{,}355-25{,}914$$
$$=12{,}441\ \text{kW}.$$

Mercury turbine

The "lost work" is

$$(\dot{I}^{CR})_i=T_0(\Delta\dot{S}_{CR})_i=\dot{M}_M T_0(s_5-s_4) \quad (5.16)$$
$$=200\times298.16\ (0.5554-0.5121)$$
$$=2643\ \text{kW}.$$

[Note:

$$s_5=0.0940+x_5(0.6362-0.0940)$$
$$x_5=\frac{(287.3-36.9)}{(331.1-36.9)}=\frac{250.4}{294.2}=0.851$$
$$s_5=0.5554\ \text{kJ/kg K}.]$$

Mercury/steam boiler

The "lost work" is

$$(\dot{I}^{CR})_i=T_0(\Delta\dot{S}_{CR})_i$$
$$=T_0[\dot{M}_{ST}(s_d-s_c)-\dot{M}_M(s_5-s_1)] \quad (5.36)$$
$$=298.16[29.235(6.069-2.797)-200(0.5554-0.0940)]$$
$$=28{,}521-27{,}514$$
$$=1007\ \text{kW}.$$

Steam turbine

From a Mollier chart, points on expansion line are as given in Table 5.3 The "lost work" is, from equation (5.40),

$$(\dot{I}^{CR})_i=298.16\times29.235\ (1\times0.11+0.916\times0.12$$
$$+0.84\times0.20+0.77\times0.38)$$
$$=5924\ \text{kW}.$$

TABLE 5.3

Pressure (MN/m^2)	Enthalpy (kJ/kg)	Entropy (kJ/kg K)	Mass flow
4	3274	6.85	1.0
0.8	2908	6.96	0.916
0.236	2708	7.08	0.840
0.041	2508	7.28	0.770
0.004	2308	7.66	

Condenser

The "lost work" is

$$(\dot{I}^{Q})_{i} = Q_{A}\left(\frac{T_{CON} - T_{0}}{T_{CON}}\right) \tag{5.47}$$

$$= \frac{29.235 \times 2187 \times 4}{1.3 \times 302.16}$$

$$= 651\ \text{kW}.$$

Feed heaters

For a feed heater receiving feed water $\dot{M}_{IN}$

$$(\Delta\dot{S}_{CR})_{i} = \dot{M}_{IN}\, T_{0}\, \Delta s_{water}\left[\frac{T_{bled} - \bar{T}_{water}}{T_{bled}}\right]. \tag{5.43}$$

Hence the total "lost work" in the three feed heaters is

$$(\dot{I}^{CR})_{i} = T_{0}(\Delta\dot{S}_{CR})_{i}$$

$$= 298.16 \times 29.235 \times \left[0.770 \times 0.614 \times \frac{23.2}{350} + 0.840 \times 0.535 \times \frac{23.8}{397.3} + 0.916 \times 0.475 \times \frac{23.2}{443.6}\right]$$

$$= 679\ \text{kW}.$$

5.3.5 Summary

TABLE 5.4

	kW
	$\dot{E}_F = 104{,}492$
Power output	$W_{CP} = 39{,}481$
	$(\dot{E}_F - W_{CP}) = 65{,}011$
cf "Lost Work" $\left(\sum_i \dot{I}_i\right)$	
Combustion	36,126
Exhaust gases	5620
Main boiler	12,441
Mercury turbine	2643
Mercury/steam boiler	1007
Steam turbine	5924
Condenser	651
Feed heaters	679
	65,091

$$[(\eta)_R]_{CP} = \frac{39{,}481}{104{,}492} = 0.3778$$

The small difference between $\sum_i \dot{I}_i$ and $(\dot{E}_F - W_{CP})$ is due to numerical rounding errors and neglect of some feed pump terms. (The lost work in the mercury pump is of the order of 100 kW.)

5.4 Exergy Analysis of a CCGT Plant (with Feed Heating in the Steam Cycle)

We next consider a combined power plant consisting of a simple open circuit gas turbine plant, fired by liquid n-octane, and rejecting heat to a closed cycle steam plant.

The latter is identical to the steam plant described in the earlier example of the binary cycle plant (i.e. with three direct contact feed heaters) but we shall also consider the steam plant with no feed heating, in Section 5.5 subsequently.

Figure 5.5 shows the temperature–entropy diagrams for the component plants.

The following assumptions are made for the gas turbine plant:

Entry pressure, $p_1 = 10^2\,\mathrm{kN/m^2}$
Entry temperature, $T_1 = 25°\,\mathrm{C}\ (= T_0)$
Polytropic efficiency of compressor, $(\eta_p)_C = 0.88$
Polytropic efficiency of gas turbine, $(\eta_p)_T = 0.88$
Pressure ratio $= 8$
Maximum temperature at entry to gas turbine, $T_3 = 900°\,\mathrm{C}$

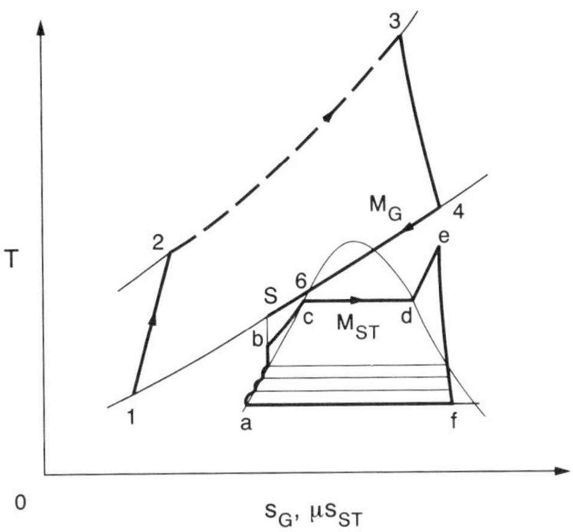

FIGURE 5.5 Open circuit (gas turbine)/closed cycle (steam turbine) combined plant, with regenerative feed heating in steam cycle

Calorific value of fuel $(CV)_0 = 44{,}430$ kJ/kg
(liquid n-octane)
Pressure losses (i) combustion chamber, $(p_2 - p_3)/p_2 = 0.03$
(ii) HRSG, 5 kN/m^2

The fuel–air ratio to produce a turbine inlet temperature of 900°C is shown to be 0.0162 (i.e. combustion with about 270% excess air). Specific heats for air and gas are taken as 1.006 kJ/kg K and 1.148 kJ/kg K respectively, with specific heat ratios of 1.4 and 1.333 respectively.

Steam turbine conditions are as specified in the binary/mercury plant described in Section 5.3.

These assumptions are similar to those made by Cerri [9] whose calculations we have described in Chapter 4.

5.4.1 Calculations for the Gas Turbine Plant (for Unit Flow Rate of Gas Products)

Compressor

$$p_1 = 10^2 \text{ kN/m}^2$$

$$p_2 = 8 \times 10^2 \text{ kN/m}^2$$

$$T_1 = 298 \text{ K}$$

$$\left(\frac{T_2}{T_1}\right) = \left(\frac{p_2}{p_1}\right)^{\frac{\kappa - 1}{\kappa(\eta_p)_C}} = 8^{\left(\frac{0.4}{1.4 \times 0.88}\right)} = 8^{0.325} = 1.965$$

$$T_2 = 586 \text{ K} \equiv 313^\circ \text{C}$$

Combustion

$$\text{Pressure loss} = 0.03 \times 8 \times 10^2$$
$$= 24\,\text{kN/m}^2$$
$$p_3 = 776\,\text{kN/m}^2$$

We consider unit flow rate of combustion gas (1 kg/s) and a flow of fuel $= f'$ kg/s. The flow of air is therefore $(1 - f')$ kg/s. Then T_3 is given by

$$f'(\text{CV})_0 = (c_p)_G(T_3 - T_0) - (1 - f')(c_p)_A(T_2 - T_0),$$

i.e.

$$f' \times 44{,}430 = 1.148(900 - 25) - (1 - f')1.006 \times (313 - 25)$$
$$= 1004.5 - 289.7(1 - f'),$$

so that

$$f' = \frac{714.8}{44{,}140} \approx 0.0162.$$

$$\text{Air-fuel ratio} = \frac{(1 - f')}{f'} = 60.7.$$

Air-fuel ratio for stoichiometric combustion of C_8H_{18}

$$= \frac{(12.5 \times 28 + 47.32)}{(96 + 18)} = 16.3.$$

$$\text{Excess air } \frac{(60.7 - 16.3)}{(16.3)} = 2.72\ (272\%).$$

Turbine

$$p_4 = 100 + 5 = 105\,\text{kN/m}^2$$
$$T_3 = 900^\circ\,\text{C} = 1173\,\text{K}$$

$$\left(\frac{T_3}{T_4}\right) = \left(\frac{p_3}{p_4}\right)^{\frac{(\kappa - 1)(\eta_p)_T}{\kappa}} = \left(\frac{776}{105}\right)^{\frac{0.333 \times 0.88}{1.333}}$$
$$= (7.39)^{0.22}$$
$$= 1.553$$

$$T_4 = \frac{1173}{1.553} = 755\,\text{K} = 482^\circ\,\text{C}$$

HRSG

If the pinch-point temperature difference is 30° C, then

$$T_6 = T_c + 30 = 250.3 + 30 \approx 280^\circ\text{ C}.$$

The steam mass flow rate corresponding to unit gas flow rate is then given by

$$\dot{M}_{ST} = \frac{(c_p)_G(T_4 - T_6)}{(h_e - h_c)} = \frac{1.148(482 - 280)}{(3274 - 1087)}$$
$$= 0.106 \text{ kg/s}.$$

The total heat exchange in the HRSG yields the stack temperature T_S,

$$1.148(482 - T_S) = 0.106(3274 - 725.6).$$

Hence

$$T_S = 247^\circ\text{ C}.$$

5.4.2 Power Output

We assume that the actual steam mass flow rate through the HRSG is that given in the previous example of the binary plant, 29.235 kg/s, and the power output from the steam turbine is 25,323 kW, as calculated there.

The gas turbine mass flow rate is therefore 29.235/0.106 = 275.8 kg/s and the air flow entering the compressor is $(1 - f')275.8 = (0.9838)275.8 = 271.3$ kg/s. The power output from the gas turbine plant is then

$$W_G = 275.8 \times 1.148(900 - 482) - 271.3 \times 1.006(313 - 25)$$
$$= 132{,}347 - 78{,}603$$
$$= 53{,}744 \text{ kW}.$$

The total power output is

$$W_{CP} = W_G + W_{ST} = 79{,}067 \text{ kW}.$$

The fuel mass flow rate is

$$0.0162 \times 275.8 = 4.466 \text{ kg/s}.$$

5.4.3 Plant Overall Efficiency

The overall efficiency of the combined plant is

$$(\eta_O)_{CP} = \frac{79{,}067}{4.466 \times 44{,}430} = 0.3985.$$

The thermal efficiency of the steam plant was calculated as $\eta_{ST} = 0.3399$.

The gas turbine plant overall efficiency is

$$(\eta_O)_G = \frac{53{,}744}{4.466 \times 44{,}430} = 0.2709.$$

The "lost heat" coefficient in the exhaust stack is

$$\nu_{UN} = \frac{275.8 \times 1.148 \times (247 - 25)}{4.466 \times 44{,}430} = 0.354.$$

A check on the overall efficiency of the combined plant is then given by

$$\begin{aligned}(\eta_O)_{CP} &= (\eta_O)_G + \eta_{ST} - (\eta_O)_G \eta_{ST} - \nu_{UN} \eta_{ST} \\ &= 0.271 + 0.340 - 0.271 \times 0.340 - 0.354 \times 0.340 \\ &= 0.611 - 0.092 - 0.120 \\ &= 0.399. \qquad (3.51a)\end{aligned}$$

The results for this combined power plant are very close to the optimum conditions calculated by Cerri [9]. For $r = 8$ and $T_3 = 900°$ C Cerri gives $(\eta_O)_{CP} = 0.392$ and $(\dot{M}_{ST}/\dot{M}_G) = 0.1$.

5.4.4 Exergy Fluxes and "Lost Work" Terms

We shall calculate the exergy balance assuming that the maximum (reversible) work from the fuel is given by $(-\Delta G_0)$, i.e. neglecting the exergies of extraction and delivery. We shall further assume that

$$\psi = \frac{(-\Delta G_0)}{(-\Delta H_0)} = 1.0401 + 0.1728 \left(\frac{h}{c}\right),$$

where (h/c) is the mass ratio of hydrogen to carbon in the fuel (after Kotas, but neglecting partial pressure terms). Hence $\psi = 1.0725$.

Thus

$$\begin{aligned}(-\Delta \dot{G}_0) &= 1.0725 \times 44{,}430 \times 4.466 \\ &= 212{,}810\,\text{kW}, \\ (T_0 \Delta \dot{S}_0) &= 212{,}810 - 198{,}424 = 14{,}386\,\text{kW}.\end{aligned}$$

The various amounts of lost work due to irreversibility are as follows.

Compressor

The "lost work" is

$$(I^{\mathrm{CR}})_{\mathrm{i}} = \dot{M}_{\mathrm{A}} T_0 (s_2 - s_1), \tag{5.14}$$

where

$$\begin{aligned}
(s_2 - s_1) &= (c_{\mathrm{p}})_{\mathrm{A}} \ln\left(\frac{T_2}{T_1}\right) - R_{\mathrm{A}} \ln\left(\frac{p_2}{p_1}\right) \\
&= 1.006 \ln 1.965 - \left(\frac{0.4 \times 1.006}{1.4}\right) \ln 8 \\
&= 0.6795 - 0.5977, \\
&= 0.0818\ \mathrm{kJ/kg\,K}, \\
(\dot{I}^{\mathrm{CR}})_{\mathrm{i}} &= 271.3 \times 298 \times 0.0818 \\
&= 6613\ \mathrm{kW}.
\end{aligned}$$

Combustion

The "lost work" is

$$(\dot{I}^{\mathrm{CR}})_{\mathrm{i}} = T_0[(\dot{S}_{\mathrm{P}})_3 - (\dot{S}_{\mathrm{R}})_2], \tag{5.18}$$

where

$$(\dot{S}_{\mathrm{R}})_2 = (\dot{S}_{\mathrm{A}})_2 + (\dot{S}_{\mathrm{F}})_0 .$$

Hence

$$\begin{aligned}
(\dot{I}^{\mathrm{CR}})_{\mathrm{i}} = {} & T_0\{[(\dot{S}_{\mathrm{P}})_3 - (\dot{S}_{\mathrm{P}})_0] + (\dot{S}_{\mathrm{P}})_0 \\
& - [\langle(\dot{S}_{\mathrm{A}})_2 - (\dot{S}_{\mathrm{A}})_0\rangle + (\dot{S}_{\mathrm{F}})_0] - (\dot{S}_{\mathrm{A}})_0\} \\
= {} & T_0\left\{\Delta\dot{S}_0 + \dot{M}_{\mathrm{G}}\left[(c_{\mathrm{p}})_{\mathrm{G}} \ln\left(\frac{T_3}{T_0}\right) - R_{\mathrm{G}} \ln\left(\frac{p_3}{p_0}\right)\right]\right. \\
& \left. - \dot{M}_{\mathrm{A}}\left[(c_{\mathrm{p}})_{\mathrm{A}} \ln\left(\frac{T_2}{T_0}\right) - R_{\mathrm{A}} \ln\left(\frac{p_2}{p_0}\right)\right]\right\} \\
= {} & 14{,}386 + 275.8 \times 298\left[1.148 \ln\left(\frac{1173}{298}\right) - \frac{1.148}{4} \ln(7.76)\right] \\
& - 271.4 \times 298\left[1.006 \ln\left(\frac{586}{298}\right) - \frac{1.006}{3.5} \ln 8\right] \\
= {} & 14{,}386 + 80{,}947 - 6672 \\
= {} & 88{,}661\ \mathrm{kW}.
\end{aligned}$$

[*Note*: This may be compared with El-Masri's approximation to $(\dot{I}^{\mathrm{CR}})_{\mathrm{i}}$, which neglects $T_0 \Delta \dot{S}_0$ and pressure loss,

$$(\dot{I}^{CR})_i \approx \left(\frac{T_0}{T_{LM}}\right) \Delta \dot{H}_0, \tag{5.28}$$

where

$$\begin{aligned} T_{LM} &= (T_3 - T_2)/\ln(T_3/T_2) \\ &= (1173 - 586)/\ln(1173/586) \\ &= 587/0.694 \\ &= 846\,\mathrm{K}, \end{aligned} \tag{5.26}$$

i.e.

$$\begin{aligned} (\dot{I}^{CR})_i &\approx \frac{4.466 \times 44{,}430 \times 298}{846} \\ &= 69{,}894\,\mathrm{kW}.] \end{aligned}$$

Turbine

The "lost work" is

$$(\dot{I}^{CR})_i = \dot{M}_G T_0 (s_4 - s_3), \tag{5.16}$$

$$\begin{aligned} s_4 - s_3 &= (c_p)_G \ln\left(\frac{T_4}{T_3}\right) - R_G \ln\left(\frac{p_4}{p_3}\right) \\ &= \frac{1.148}{4} \ln(7.39) - 1.148 \ln(1.553) \\ &= 0.0687\,\mathrm{kJ/kg\,K}, \end{aligned}$$

$$\begin{aligned} (\dot{I}^{CR})_i &= 275.8 \times 298 \times 0.0687 \\ &= 5646\,\mathrm{kW}. \end{aligned}$$

HRSG

The "lost work" is

$$\begin{aligned} (\dot{I}^{CR})_i &= T_0 (\Delta \dot{S}_{CR})_i \\ &= T_0 [\dot{M}_{ST}(s_e - s_b) + \dot{M}_G (s_S - s_4)]. \end{aligned} \tag{5.36}$$

Here

$$\begin{aligned} s_S - s_4 &= (c_p)_G \ln\left(\frac{T_S}{T_4}\right) - R_G \ln\left(\frac{p_S}{p_4}\right) \\ &= \frac{1.148}{4} \ln\left(\frac{105}{100}\right) - 1.148 \ln\left(\frac{755}{520}\right) \\ &= 0.014 - 0.428 \\ &= -0.414. \end{aligned}$$

Hence

$$\dot{I}^{CR} = 29.235 \times 298(6.858 - 2.046) - 275.8 \times 298 \times 0.414$$
$$= 41{,}922 - 34{,}026$$
$$= 7896\,\text{kW}.$$

Exhaust "loss"

The "lost work" is

$$\dot{I}^{Q} = \int_{T_S}^{T_0} \left(1 - \frac{T_0}{T}\right) dQ$$
$$= \dot{M}_G (c_p)_G (T_S - T_0) - \dot{M}_G T_0 (c_p)_G \ln (T_S/T_0) \qquad (5.48)$$
$$= 275.8 \times 1.148[(247 - 25) - 298 \ln 1.74]$$
$$= 17{,}760\,\text{kW}.$$

5.4.5 Summary

TABLE 5.5

Input (kW)	Power output (kW)	$(\dot{I}^{CR})_i$, $(\dot{I}^{Q})_i$ (kW)	
$\dot{E}_F = (-\Delta \dot{G}_0) = 212{,}810$			
	$W_G = 53{,}744$		
	$W_{ST} = 25{,}323$		
		Compressor	$\dot{I}^{CR} = 6613$
		Combustion	$\dot{I}^{CR} = 88{,}661$
		Gas turbine	$\dot{I}^{CR} = 5646$
		HRSG	$\dot{I}^{CR} = 7896$
		Exhaust	$\dot{I}^{Q} = 17{,}760$
		from previous example: Steam turbine	$\dot{I}^{CR} = 5866$
		from previous example: Condenser	$\dot{I}^{Q} = 651$
		from previous example: Feed water	$\dot{I}^{CR} = 679$
$\dot{E}_F = 212{,}810$	$W_{CP} = 79{,}067$		$\sum_i (\dot{I}^{CR} + \dot{I}^{Q})_i = 133{,}772$

$$(\eta_R)_{CP} = \frac{79{,}067}{212{,}810} = 0.3715, \text{ cf. } (\eta_O)_{CP} = 0.3985$$

The small difference between $\dot{E}_F$ and $W_{CP} + \sum_i (\dot{I}^{CR} + \dot{I}^{Q})_i$ is due partly to numerical rounding errors and partly due to the neglect of feed pump irreversibility ($\dot{I}^{CR} \approx 30\,\text{kW}$).

5.5 Exergy Analysis of a CCGT Plant (with No Feed Heating in the Steam Cycle)

It was argued in Section 4.3.2 that feed heating usually lowers combined plant overall efficiency (although it increases the thermal efficiency of the steam cycle). It is instructive to repeat the example of exergy analysis given above with no feed heating in the steam cycle, but with the same boiler and condenser pressures. The gas turbine conditions and performance remain unchanged (see Fig. 5.6.)

The analysis of the pinch-point condition remains valid, so that for *unit* gas flow rate $\dot{M}_{ST}=0.106\,kg/s$. We assume the gas flow remains unchanged at 275.8 kg/s so that the steam flow also remains unchanged at 29.235 kg/s.

However, it is necessary to calculate a new value of gas temperature at exit from the HRSG, T_S.

The enthalpy of the water at exit from the feed pump is now

$$\begin{aligned} h_b &= h_a + \frac{(p_b - p_a)}{\rho \eta_P} \\ &\approx 121.4 + \frac{(4-0.004)1000}{0.8} \\ &= 126.4\,kJ/kg. \end{aligned}$$

The overall energy balance in the HRSG is then

$$\dot{M}_G(h_4 - h_S) = \dot{M}_{ST}(h_e - h_b),$$

$$1.148(482 - T_S) = 0.106(3274 - 126.4),$$

$$T_S = 482 - 290.7 \approx 191°\,C.$$

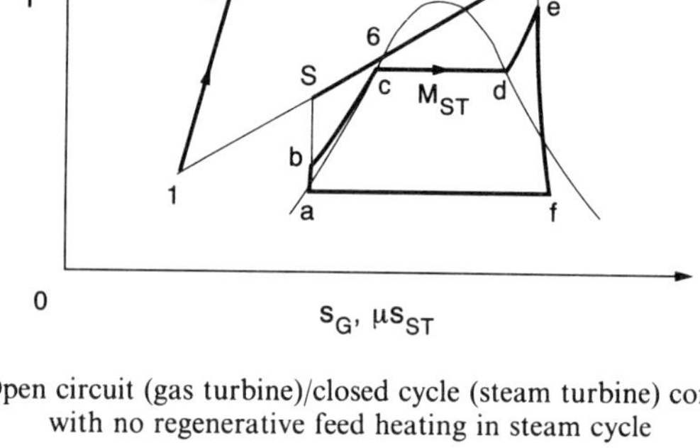

FIGURE 5.6 Open circuit (gas turbine)/closed cycle (steam turbine) combined plant, with no regenerative feed heating in steam cycle

5.5.1 Power Output

The power output from the gas turbine is unchanged at 53,744 kW. The power output from the steam plant is now

$$
\begin{aligned}
W_{ST} &= \dot{M}_{ST}[(h_e-h_f)-(h_b-h_a)] \\
&= \dot{M}_{ST}[(h_c+\beta)-(h_a+\beta)-(h_b-h_a)] \\
&= 29.235[1087.4-121.4-5.0] \\
&= 28{,}095\,\text{kW},
\end{aligned}
$$

and

$$W_{CP}=53{,}744+28{,}095=81{,}839\,\text{kW}.$$

5.5.2 Plant Overall Efficiency

The overall efficiency of the combined plant is

$$(\eta_O)_{CP}=\frac{81{,}839}{4.466\times 44{,}430}=0.4124.$$

The thermal efficiency of the steam plant is now

$$
\begin{aligned}
\eta_{ST} &= 1-\frac{(h_f-h_a)}{(h_e-h_b)} \\
&= 1-\frac{2187}{(3274-126.4)} \\
&= 0.3052.
\end{aligned}
$$

The gas turbine overall efficiency is unchanged at $(\eta_O)_G=0.2709$.

The "lost heat" coefficient in the exhaust stack is now

$$\nu_{UN}=\frac{275.8\times 1.148\times(191-25)}{4.466\times 44{,}430}=0.265.$$

Hence a check on $(\eta_O)_{CP}$ is given by

$$
\begin{aligned}
(\eta_O)_{CP} &= (\eta_O)_G+\eta_{ST}-(\eta_O)_G\eta_{ST}-\nu_{UN}\eta_{ST} \qquad (3.51a) \\
&= 0.3052+0.2709-0.3052\times 0.2709-0.265\times 0.3052 \\
&= 0.5761-0.0827-0.0808 \\
&= 0.4126.
\end{aligned}
$$

As anticipated, the overall efficiency of the combined plant with no feed heating is higher than that with feed heating (0.3895). The steam cycle efficiency η_{ST} is higher, but the "heat loss" coefficient ν_{UN} is less because the "stack" temperature is lower at 191° C (cf 247° C).

5.5.3 Exergy Fluxes and "Lost Work" Terms

The value of the maximum reversible work $(-\Delta\dot{G}_0)$ is unchanged. The lost work due to irreversibility in the compressor, in the combustion chamber and the gas turbine are also unchanged.

In the steam cycle, the power output is increased to 28,095 kW, and there is of course no lost work due to feed heating.

The irreversibility in the steam turbine, the condenser, the HRSG and the exhaust stack must all be recalculated, however.

Steam turbine

The "lost work" is

$$(\dot{I}^{CR})_i = T_0 \dot{M}_{ST}(s_e - s_f) \tag{5.16}$$
$$= 298 \times 29.235(7.663 - 6.858)$$
$$= 7013 \text{ kW}.$$

Condenser

The mass flow through the condenser is now the full steam flow $\dot{M}_{ST}$, so that the "lost work" is

$$(\dot{I}^Q)_i = \dot{M}_{ST}(h_f - h_a)\left(\frac{T_{CON} - T_0}{T_{CON}}\right) \tag{5.47}$$
$$= 29.235 \times 2187 \times \left(\frac{4}{302}\right)$$
$$= 847 \text{ kW}.$$

HRSG

The "lost work" is

$$(\dot{I}^{CR})_i = T_0(\Delta\dot{S}_{CR})_i \tag{5.36}$$
$$= T_0 \dot{M}_G(s_5 - s_4) + T_0 \dot{M}_{ST}(s_e - s_b)$$
$$= 298 \times 275.8\left[\frac{1.148}{4}\ln\left(\frac{105}{100}\right) - 1.148\ln\left(\frac{755}{464}\right)\right]$$
$$+ 298 \times 29.235[6.858 - 0.422]$$
$$= -44{,}710 + 56{,}071$$
$$= 11{,}361 \text{ kW}.$$

Exhaust stack

The "lost work" is

$$\dot{I}^{Q} = \dot{M}_G(c_p)_G[(T_S - T_0) - T_0 \ln (T_S/T_0)] \tag{5.48}$$

$$= 275.8 \times 1.148\left[(191 - 25) - 298 \ln\left(\frac{464}{298}\right)\right]$$

$$= 10{,}828\ \text{kW}.$$

5.5.4 Summary

TABLE 5.6

Input (kW)	Power output (kW)		$(\dot{I}^{CR})_i$, $(\dot{I}^{Q})_i$
$\dot{E}_F = (-\Delta\dot{G}_0)$ $= 212{,}810$			
	$W_G = 53{,}744$		
	$W_{ST} = 28{,}095$		
		Compressor	$\dot{I}^{CR} = 6613$
		Combustion	$\dot{I}^{CR} = 88{,}661$
		Turbine	$\dot{I}^{CR} = 5646$
		HRSG	$\dot{I}^{CR} = 11{,}361$
		Exhaust	$\dot{I}^{Q} = 10{,}828$
		Steam turbine	$\dot{I}^{CR} = 7013$
		Condenser	$\dot{I}^{Q} = 847$
$(-\Delta\dot{G}_0) = 212{,}810$	$W_{CP} = 81{,}839$		$\sum_i(\dot{I}^{CR} + \dot{I}^{Q})_i = 130{,}969$

$\dot{E}_F - W_{CP} = 130{,}971$ cf. $\sum_i(\dot{I}^{CR} + \dot{I}^{Q})_i = 130{,}969$. $(\eta_R)_{CP} = \dfrac{81{,}839}{212{,}810} = 0.3846$

Note: Again the small difference between $\dot{E}_F$ and $W_{CP} + \sum(\dot{I}^{CR} + \dot{I}^{Q})$ is due partly to numerical rounding errors and partly to the neglect of feed pump irreversibility ($\dot{I}^{CR} \approx 30$ kW).

It is of interest to observe the changes in the various items in the exergy balances, for feed heating and for no feed heating. Originally $(\dot{I}^{CR})_i$ for the feed heaters was quite small (679 kW). But its disappearance when feed heating is dispensed with has little effect compared with the major reductions in the "lost work" in the exhaust stack (from 17,760 kW to 10,828 kW). The lost work due to irreversibility in the steam turbine is increased (full mass flow through the turbine) as is that in the condenser (more mass flow), and in the HRSG (bigger temperature differences). But all together they are less than the reduced "loss" in the exhaust stack, and the rational efficiency of the combined plant $(\eta_R)_{CP}$ increases.

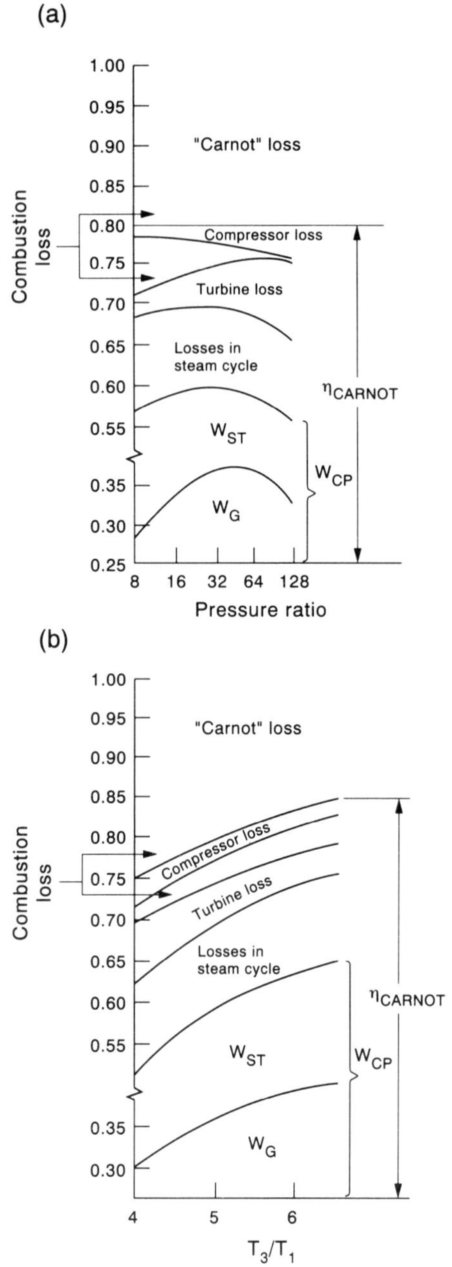

FIGURE 5.7 Work output and lost work due to irreversibility
(a) for $T_3/T_1 = 5.0$
(b) for $r = 32$
(after El-Masri [10])

5.6 Parametric Exergy Studies of CCGT Plants

Parametric studies of a CCGT plant (open circuit (gas turbine)/closed cycle (steam turbine)) reported in Chapter 4 may be repeated to calculate the rational efficiency. The irreversibilities in various parts of the plant may be estimated to show how they vary with such design parameters as gas turbine pressure ratio and top temperature, steam boiler and condenser pressures, component efficiencies, etc. Such calculations have been carried out by several authors, particularly El-Masri and Manfrida *et al.*

El-Masri has provided "second law" analyses for

(i) the basic open gas turbine/steam turbine plant [10],
(ii) a similar plant with air cooling of the gas turbine [6],
(iii) the addition of reheat and/or intercooling to the gas turbine plant [11].

The reader is referred to the original references for the results of these calculations which are extremely comprehensive. Here, by way of illustration, we show two of El-Masri's calculations for the first plant but with reheat between the turbine stages, which illustrate how the magnitudes of the irreversibilities in components vary with two basic design parameters:

(i) pressure ratio, for a given ratio of maximum temperature to ambient temperature [$(T_3/T_1)=5$, Fig. 5.7a];
(ii) the ratio of maximum temperature to ambient temperature, for a given pressure ratio ($r=32$, Fig. 5.7b).

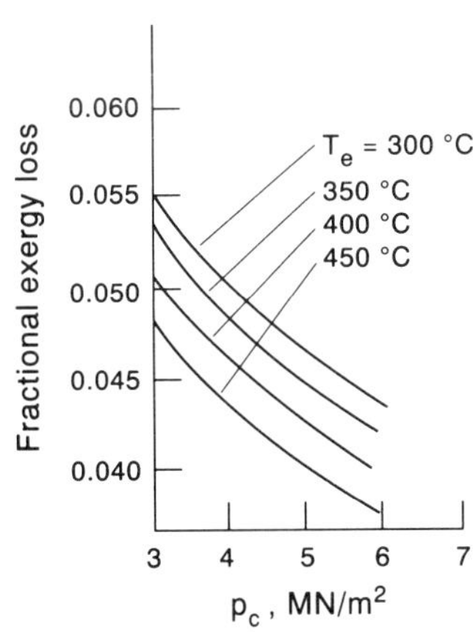

FIGURE 5.8 Lost work in HRSG, $T_3=1200$ K, $r=10$ (single pressure steam cycle) (after Manfrida and Manucci [12])

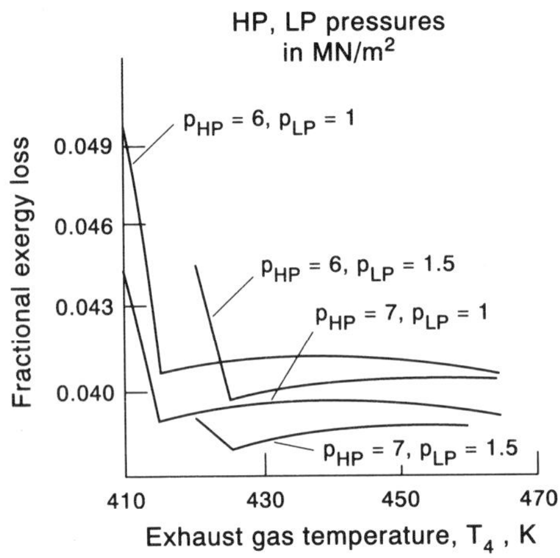

FIGURE 5.9 Lost work in HRSG, $T_3 = 1200$ K, $r = 10$ (dual pressure steam cycle, $T_e = 400°$ C (after Manfrida and Manucci [12]))

It should be noted that:

(i) three gas turbines are assumed (two reheat stages);
(ii) an optimised "effectiveness" of the steam is used;
(iii) the combustion "loss" is split into a "Carnot loss",

$$\frac{I^{Q}}{(-\Delta H_0)} = \left(\frac{T_0}{T_3}\right) = (1 - \eta_{CARNOT})$$

and the remaining loss

$$\left[\frac{T_0 \Delta S_0}{(-\Delta H_0)} - \left(\frac{T_0}{T_3}\right)\right],$$

as explained in Section 5.3.2.

The breakdown in work output between the two component plants, and the various amounts of lost work $(\dot{I}^{CR})_i$, $(\dot{I}^{Q})_i$ illustrate that the optimum pressure ratio (for a given ratio of maximum to minimum temperature) is less for the combined plant than for the gas turbine plant alone (Fig. 5.7a). The compressor and turbine "losses" increase with pressure ratio r but the combustion "loss" decreases with r, as the temperature T_2 approaches the maximum temperature T_3. The dominant effect of (T_3/T_1) on plant efficiency $(\eta_R)_{CP}$ (and on specific work output W_G) is illustrated in Fig. 5.7b, $(\eta_R)_{CP}$ approaching Carnot efficiency at highest temperature.

Manfrida and Manucci [12] have carried out similar efficiency calculations, which we shall not reproduce here. However, they have provided a useful summary of the differences in performance between a single pressure and a dual pressure steam cycle, receiving heat from the gas turbine exhaust gases in an HRSG. For example, Fig. 5.8 shows their calculation, for a single pressure cycle, of the fractional exergy losses, $\delta_i = \frac{(I^{CR})_i}{\Delta H_0}$ in the HRSG, for selected gas turbine parameters ($T_3 = 1200$ K, $r = 10$). δ_i drops sharply with boiler pressure, and also decreases with the maximum temperature of the (superheated) stream (T_e). For the same gas turbine conditions, Fig. 5.9 shows how δ_i in the HRSG varies with exhaust gas temperature (T_S), with various "pairs" of evaporation pressures, and a fixed value of maximum steam temperature (T_e). Rather surprisingly, these two figures suggest relatively little gain in rational efficiency $(\eta_R)_{CP}$ as a result of introducing dual pressure evaporation in the steam cycle.

Most of Manfrida's original calculations of $(\eta_R)_{CP}$ are for selected r and T_3, but varying steam evaporation pressure, and are not as comprehensive as the Cerri calculations described in Chapter 4. In later calculations Manfrida *et al.* [13] apply exergy analysis to combined power plants using existing or "next-generation" gas turbines, matching them with optimised single pressure and dual pressure steam cycles and estimating the HRSG heat transfer area (per unit steam turbine power output) which is an important feature of plant capital cost.

References

[1] Kotas, T. J. *The Exergy Method of Thermal Plant Analysis.* Butterworths, London, 1985.

[2] Horlock, J. H. The Rational Efficiency of Power Plants with External Combustion. *Proc. I.Mech.E.*, **178,** 31, 43, 1963/1964.

[3] Traupel, W. Reciprocating Engine and Turbine in Internal Combustion Engineering. *Proc. CIMAC [Int. Cong. Combustion Engineering*] 37, 1957.

[4] Horlock, J. H. The Rational Efficiency of Power Plants and Their Components. *Proc. ASME COGEN V*, 103–114, 1991.

[5] Fratzcher, W. Exergetical Efficiency. *B.W.K.*, **13,** 11, 486–493 [*CEGB Trans.* 2399].

[6] El-Masri, M. A. Exergy Analysis of Combined Cycles: Part 1—Air-Cooled Brayton-Cycle Gas Turbines. *Trans. ASME Journal of Engineering for Gas Turbines and Power*, **109,** 2, 228–236, 1987.

[7] Haywood, R. W. A Generalised Analysis of the Regenerative Steam Cycle for a Finite Number of Heaters. *Proc. I.Mech.E.*, **161,** 157, 1949.

[8] Haywood, R. W. *Analysis of Engineering Cycles*, 4th edition. Pergamon Press, Oxford, 1991.

[9] Cerri, G. Parametric Analysis of Combined Gas-Steam Cycles. *Trans. ASME Journal of Engineering for Gas Turbines and Power*, **109,** 1, 46–55, 1987.

[10] El-Masri, M. A. On Thermodynamics of Gas-Turbine Cycles: Part 1—Second Law Analysis of Combined Cycles. *Trans. ASME Journal of Engineering for Gas Turbines and Power*, **107,** 4, 880–889, 1989.

[11] El-Masri, M. A. Exergy Balance of the Reheat Gas Turbine Combined Cycle. *Proc. ASME/JSME Thermal Engineering Joint Conference*, 117–123, 1987.

[12] Manfrida, G. and Manucci, M. Exergy Analysis of Possible Solutions for Combined Gas-Steam Cycles. *Proc. 21st Intersociety Energy Conversion Engineering Conference*, 123–128, 1986.

[13] Manfrida, G., Bosio, A. and Bidini, G. Second Law Analysis of Combined Gas-Steam Power Plants. *Proc. 23rd Intersociety Energy Conversion Engineering Conference*, 391–397, 1988.

CHAPTER 6

Economics of Combined Plants

6.1 Introduction

The simplest way of assessing the economics of a power plant is to calculate the unit price of electricity produced by the plant (e.g. in dollars per kilowatt hour) and compare it with that from a conventional plant. This is the method adopted by many authors (e.g. Williams [1] for cogeneration plant and Wunsch [2] in arguing the case for combined plant). Alternatively, the change in costs associated with a plant modification may be assessed. Other methods involving net present values may be used (see Horlock [3] for their use in the assessment of cogeneration schemes), but all essentially involve the time value of money.

6.2 Electricity Pricing

The method is based on relating electricity price to both capital related cost and the recurrent cost of production (fuel and maintenance of plant):

$$P_E = \beta C_0 + M + (\mathrm{OM}), \tag{6.1}$$

where P_E is the annual cost of the electricity produced (e.g. \$ per annum);
C_0 is the capital cost of plant (e.g. \$);
$\beta(i, N)$ is the capital charge factor which is related to the discount rate (i) on capital and the life of the plant (N years) (see Section 6.3);
M is the annual cost of fuel supplied (e.g. \$ p.a.);
(OM) is the annual cost of operation and maintenance (e.g. \$ p.a.).

The "unitised" production cost (say \$ per kilowatt hour) for the plant is

$$Y_E = \frac{P_E}{\dot{W}H} = \frac{\beta C_0}{\dot{W}H} + \frac{M}{\dot{W}H} + \frac{(\mathrm{OM})}{\dot{W}H}, \tag{6.2}$$

where $\dot{W}$ is the rating of the plant (kW) and H is the plant utilisation (hours per annum).

The cost of fuel per annum, M, may be written as the product of the unit cost of fuel (ζ, \$/kW h), the rate of supply of energy in the fuel ($\dot{F}$, kW) and the utilisation, H:

$$M=\zeta \dot{F} H. \tag{6.3}$$

Thus the unitised production cost is then

$$Y_{\mathrm{E}}=\frac{P_{\mathrm{E}}}{\dot{W}H}=\frac{\beta C_0}{\dot{W}H}+\frac{\zeta \dot{F} H}{\dot{W}H}+\frac{(\mathrm{OM})}{\dot{W}H},$$

$$=\frac{\beta C_0}{\dot{W}H}+\frac{\zeta}{(\eta_{\mathrm{O}})}+\frac{(\mathrm{OM})}{\dot{W}H}, \tag{6.4}$$

where $(\eta_{\mathrm{O}})=\dfrac{\dot{W}}{\dot{F}}$ is the overall efficiency of the plant. The unit cost of fuel ζ may be written as the cost per unit mass S (say \$/kg) divided by the calorific value CV (kW h/kg), so that

$$Y_{\mathrm{E}}=\frac{\beta C_0}{\dot{W}H}+\frac{S}{(\mathrm{CV})(\eta_{\mathrm{O}})}+\frac{(\mathrm{OM})}{\dot{W}H}. \tag{6.5}$$

We may also note that Wunsch [2] expresses the operating and maintenance costs (OM) as the sum of constant operating costs U (for personnel etc, in \$/p.a.) and variable operating costs V (\$/kW h) so that

$$Y_{\mathrm{E}}=\frac{\beta C_0}{\dot{W}H}+\frac{\zeta}{(\eta_{\mathrm{O}})}+\frac{U}{\dot{W}H}+V. \tag{6.6}$$

Thus the production cost for one plant

$$(Y_{\mathrm{E}})_1=\frac{\beta_1 (C_0)_2}{\dot{W}_1 H}+\frac{\zeta_1}{(\eta_{\mathrm{O}})_1}+\frac{U_1}{\dot{W}_1 H}+V_1, \tag{6.7}$$

may be compared directly with that of a second plant

$$(Y_{\mathrm{E}})_2=\frac{\beta_2 (C_0)_2}{\dot{W}_2 H}+\frac{\zeta_2}{(\eta_{\mathrm{O}})_2}+\frac{U_2}{\dot{W}_2 H}+V_2, \tag{6.8}$$

where the utilisation is assumed the same $H_1=H_2=H$.

A new plant (2) with a higher efficiency $[(\eta_{\mathrm{O}})_2]$ than the first plant (1) $[(\eta_{\mathrm{O}})_1]$ may have increased capital cost $[(C_0)_2>(C_0)_1]$ and these two effects have to be balanced against each other in assessing the relative merits of the two plants.

Alternatively, if a basic plant (1) is to be modified to form plant (2) then the effect of the possible modifications may be assessed. If additional capital $\Delta C_0=(C_0)_2-(C_0)_1$ produces a change in efficiency $\Delta\eta_{\mathrm{O}}=(\eta_{\mathrm{O}})_2-(\eta_{\mathrm{O}})_1$ then subtracting equation (6.7) from (6.8) yields

$$\Delta Y_E=(Y_E)_2-(Y_E)_1=\left[\frac{\beta(C_O)_2}{\dot{W}_2 H}-\frac{\beta(C_O)_1}{\dot{W}_1 H}\right]+\zeta\left[\frac{1}{(\eta_O)_2}-\frac{1}{(\eta_O)_1}\right]$$
$$=\frac{\beta}{H}\left[\frac{(C_O)_2}{\dot{W}_2}-\frac{(C_O)_1}{\dot{W}_1}\right]-\frac{\zeta\Delta\eta_O}{(\eta_O)_2(\eta_O)_1}, \tag{6.9}$$

if β, ζ, $U/\dot{W}$ and V remain unchanged from condition (1). If the second (negative) term is greater than the first, an increased economic performance is obtained.

If an existing power plant (1) is "re-powered" to provide increased power output in plant (2), this may be done by installing additional identical plant, in which case the unitised cost $(Y_E)_1$, is approximately unchanged and $\Delta Y_E=0$. But if an alternative additional plant costing $\Delta C_0=(C_0)_2-(C_0)_1$ is installed to provide the additional power $\Delta\dot{W}=(\dot{W}_2-\dot{W}_1)$ the unitised cost is then

$$(Y_E)_2=\frac{(P_E)_2}{\dot{W}_2 H}$$
$$=\frac{(P_E)_1+\Delta P_E}{(\dot{W}_1+\Delta\dot{W})H}$$
$$=\frac{\beta[(C_O)_1+\Delta C_O]}{(\dot{W}_1+\Delta\dot{W})H}+\frac{\zeta_1\dot{F}_1+\Delta(\zeta\dot{F})}{(\dot{W}_1+\Delta\dot{W})H}$$
$$+\frac{\mathrm{OM}+\Delta(\mathrm{OM})}{(\dot{W}_1+\Delta\dot{W})H}. \tag{6.10}$$

If the modification involves adding a lower cycle steam turbine plant to an existing gas turbine plant with a HRSG then $\Delta(\zeta\dot{F})=0$, if there is no supplementary heating. If the unitised maintenance costs are unchanged, then the overall unitised production costs will almost certainly drop. (This argument is used by Wunsch [2]—see Section 6.6.1 below.)

6.3 The Capital Charge Factor

The capital charge factor (β) multiplied by the capital cost of the plant (C_0) gives the cost of servicing the capital required.

Suppose the capital cost of a plant at the beginning of the first year is C_0, and the plant has a life of N years. An annual amount must be provided which is $[C_0 i+B]$. The first term $(C_0 i)$ is the simple interest payment and the second (B) matures into the capital repayment after N years (i.e. interest is added to the accumulated sum at the end of each year).

Thus

$$B[1+(1+i)+(1+i)^2+\ldots+(1+i)^{N-1}]=C_0$$

so that

$$B=\frac{C_0 i}{(1+i)^N-1} \tag{6.11}$$

where it has been assumed that annual payments are made at the end of each year.

Hence the total annual payment is

$$\begin{aligned} C_0 i+B &= C_0\left[i+\frac{i}{(1+i)^N-1}\right] \\ &= C_0 i\left[\frac{(1+i)^N}{(1+i)^N-1}\right] \\ &= C_0\beta. \end{aligned} \tag{6.12a}$$

The capital charge factor $\beta=\left[\frac{i(1+i)^N}{(1+i)^N-1}\right]$ is sometimes referred to as the annuity present worth factor (${}_{N}f_{AP}$).

An alternative way of looking at the capital charge factor is to invoke a "capitalised cost" (C). C is greater than C_0 by an amount which will mature after N years to provide the renewal cost of the plant. Assuming this has no salvage value,

$$(C-C_0)(1+i)^N=C \tag{6.13a}$$

so that

$$C=\frac{(1+i)^N C_0}{(1+i)^N-1}. \tag{6.13b}$$

The annual interest payment must be made on the capitalised cost so that the annual capital charge rate is

$$\begin{aligned} Ci &= \frac{C_0(1+i)^N i}{(1+i)^N-1} \\ &= \beta C_0. \end{aligned} \tag{6.12b}$$

This annual capital charge rate βC_0 is added on to other cost of production (fuel, operation and maintenance charges), as in equation (6.1).

In arriving at an appropriate value of β, the choice of the interest or discount rate (i) is crucial, but a complex matter. It depends on the relative values of equity and debt financing, on whether the debt financing is less than the life of the plant, on tax rates and tax allowances (which vary from one country to another) and on inflation rates. In comparing two engineering projects the practice is often to use a "test discount rate", applicable to both projects.

An American approach has been outlined by Williams [1]. He elaborates the simple expression for β to take account of many other factors beyond a simple single interest (or discount) rate. He defines a discount rate as

$$i' = \alpha_e r_e + (1-\tau)\alpha_d r_d, \tag{6.14}$$

where α_e, α_d are the fractions of investment from equity and debt, r_e, r_d are the corresponding annual rates of return, and τ is the corporate tax rate. Allowance may also be made for any investment tax credit rates τ_{CR}, property and other taxes τ_p, an insurance rate r_i and a capital replacement rate r_r (presumably related primarily to repairs). The capital charge factor is then

$$\beta' = \frac{\beta(i', N)}{(1-\tau)}\left[1 - \frac{\tau_{CR}}{(1+i)} - \frac{\tau}{N\beta}\right] + \tau_p + r_i + r_r. \tag{6.15}$$

Inflation can be allowed for in selecting a discount rate, or rates of return. This would involve setting the rate of return allowing for inflation correspondingly higher than that with zero inflation. For example Williams points out that the rates of return (with zero inflation) used in his studies ($r_e = 0.055$, $r_d = 0.028$) would correspond to $r_e = 0.12$ and $r_d = 0.09$ with 6% inflation. In making comparisons between plants it is more usual to assume that overall inflation does not take place and to set the discount rate at lower levels. However, allowance is then made for relative inflation (e.g. a rate of escalation in fuel costs) greater than the overall inflation rate.

In the U.K. the taxation system is different; writing down allowances are granted on the capital invested in the plant (usually in the next and subsequent years), the magnitude of the allowance varying from year to year. Tax calculated in this way (on capital written off) is deducted from the corporation tax charged on the project's profits. It is difficult to develop a value of β comparable to that used by Williams and a year by year analysis is usually made (see the next section).

6.4 Discounted Cash Flow

More detailed analyses take account of variations in income and costs from year to year (for example capital costs which gradually increase, as building of the plant proceeds). The discounted cash flow method is used, in which cash flows are discounted back to year zero.

6.4.1 Levelised Sums

The (positive) cash flow in a particular year (A_k in year k) is discounted back to a "present worth",

$$\frac{A_k}{(1+i)^k}.$$

The total present worth (or net present value) of such sums over a period of N years is

$$(\text{NPV}) = \sum_{k=1}^{N} \frac{A_k}{(1+i)^k} \tag{6.16}$$

If all the sums had been the same, year by year, $A_k = \bar{A}$; then the present worth would have been

$$\begin{aligned}\overline{(\text{NPV})} &= \bar{A}\left[\frac{1}{(1+i)^N} + \frac{1}{(1+i)^{N-1}} + \ldots + \frac{1}{(1+i)}\right] \\ &= \frac{\bar{A}}{{}_N f_{AP}},\end{aligned} \tag{6.17}$$

where

$${}_N f_{AP} = \frac{i(1+i)^N}{(1+i)^N - 1} = \beta.$$

From the identity of (NPV) and $\overline{(\text{NPV})}$, we obtain the levelised sum:

$$(\text{NPV}) = \sum_{k=1}^{N} \frac{A_k}{(1+i)^k} = \overline{(\text{NPV})} = \frac{\bar{A}}{{}_N f_{AP}}$$

so that

$$\begin{aligned}\bar{A} &= (\text{NPV})\,{}_N f_{AP} \\ &= \beta \sum_{k=1}^{N} \frac{A_k}{(1+i)^k}.\end{aligned} \tag{6.18}$$

This is how a levelised sum (cost, saving or cash flow) is defined, when actual values A_k vary from year to year.

Comparisons between plants, over a period of years, can be made by comparing levelised costs (or net present values). Examples of the use of both, for cogeneration plant, are given by Marshall *et al.* [4] and are quoted in a parallel volume to this on CHP plant (Horlock [3]).

6.4.2 Rates of Return and the Effects of Tax Systems

We indicated in Section 6.3 that discounted cash flow (DCF) methods, giving rates of return on capital, are critically affected by local tax systems. Various methods of calculating rates of return, allowing for taxation, are discussed in detail by Cowan [5]. We illustrate this aspect here with two examples—one given by Daudet and Trimble [6] involving the U.S. tax system, and the other involving the U.K. tax system, following an approach suggested by Lumby [7].

6.4.2.1 RATE OF RETURN UNDER THE U.S. TAX SYSTEM

Daudet and Trimble first illustrate their approach in a useful diagram, reproduced in Fig. 6.1. Revenue (SALES) arises from appropriate pricing of the electricity produced (P_E). The operating income (OI) is then obtained after meeting "operating costs" (OPCST)—including fuel costs (M), operation and maintenance costs (OM), *debt interest* (INT) and other costs such as insurance (OTHER)—plus an allowance for depreciation (DEP).

Thus

$$\begin{aligned} \mathrm{OI} &= \mathrm{SALES}-\mathrm{OPCST}-\mathrm{DEP} \\ &= P_E-(\mathrm{M}+\mathrm{OM}+\mathrm{INT}+\mathrm{OTHER})-\mathrm{DEP}. \end{aligned} \tag{6.19}$$

Tax is levied on this net operating income,

$$\begin{aligned} \mathrm{TX} &= \tau(\mathrm{OI}) \\ &= \tau(\mathrm{SALES}-\mathrm{OPCST}-\mathrm{DEP}), \end{aligned} \tag{6.20}$$

where τ is the corporate tax rate; note that DEP is thus included as a tax exemption in the U.S. tax system. Profit (PR) is then the net operating income less the tax levied,

$$\begin{aligned} \mathrm{PR} &= \mathrm{OI}-\mathrm{TX} \\ &= (1-\tau)(\mathrm{SALES}-\mathrm{OPCST}-\mathrm{DEP}). \end{aligned} \tag{6.21}$$

Daudet and Trimble define the cash flow as made up of the profit (PR) plus the depreciation (DEP), as shown in Fig. 6.1. This cash flow then goes to meet

(i) the internal rate of return (IR) on the equity capital invested, and
(ii) the provision of debt retirement (PDR),

$$PR + DEP = IR + PDR. \tag{6.22}$$

From equations (6.21) and (6.22) it follows that

$$SALES(1-\tau) = (OPCST)(1-\tau) + PDR + IR - \tau(DEP),$$

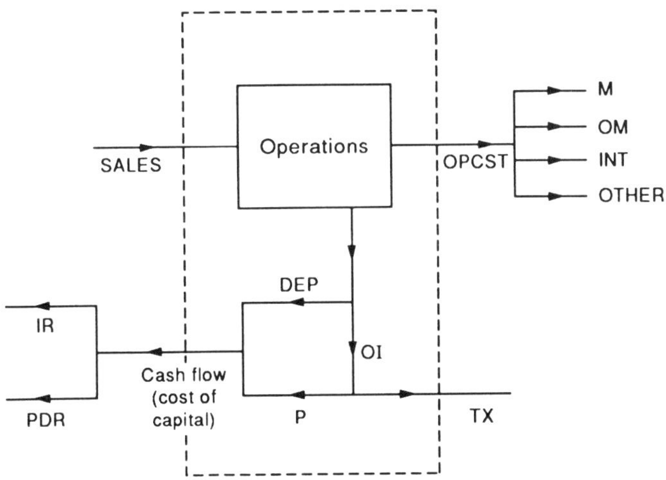

OPCST	Operating costs
M	Fuel cost
OM	Operational and maintenance costs
INT	Debt interest
OI	Operating income
TX	Tax
P	Profit after tax
DEP	Depreciation
PDR	Provision for debt retirement
IR	Internal return

FIGURE 6.1 Generalised cash flow diagram for the U.S. system (after Daudet and Trimble [6])

so that

$$\begin{aligned}\mathrm{IR} &= (1-\tau)(\mathrm{SALES}-\mathrm{OPCST})-\mathrm{PDR}-\tau(\mathrm{DEP})\\ &= (1-\tau)[P_{\mathrm{E}}-(\mathrm{M}+\mathrm{OM}+\mathrm{OTHER})-\mathrm{INT}]\\ &\quad -\mathrm{PDR}+\tau(\mathrm{DEP}).\end{aligned} \tag{6.23}$$

In then assessing the internal rate of return on equity capital, Daudet and Trimble take account of an "investment tax credit", for the U.S. internal revenue allowed a fraction (τ_{CR}) of the total capital (C_0) to be deducted before the internal rate of return on equity (r_{e}) was determined,

$$(1-\alpha_{\mathrm{d}}-\tau_{\mathrm{CR}})C_0=(\alpha_{\mathrm{e}}-\tau_{\mathrm{CR}})C_0=\sum_{n=1}^{N}\frac{\mathrm{IR}}{(1+r_{\mathrm{e}})^n} \tag{6.24}$$

where $\alpha_{\mathrm{e}}+\alpha_{\mathrm{d}}=1$.

Essentially this is an equation for the unknown rate of return (r_{e}) on the equity capital ($\alpha_{\mathrm{e}}C_0$). However, it should be emphasised that in comparison of two projects it is not sufficient simply to compare the two rates of return, as the relative advantage may change according to the project life N (see Brealey and Myers [8]). It is more usual to set a required company rate of return and require that it should be met or bettered.

The method outlined by Daudet and Trimble is similar (but not identical) to that summarised by Williams. Williams's approach is to specify a weighted discount rate [equation (6.15)] and to apply it to the after-tax cash flow.

6.4.2.2 RATE OF RETURN UNDER THE U.K. TAX SYSTEM

The U.K. tax system is different from the U.S. system in that depreciation is not specifically allowed in determining corporate tax as it was in equation (6.20). However, writing off allowances are allowed in the assessment of corporate tax in the years immediately after the capital investment is made. This is best shown in tabular form, year by year, and we illustrate this by modifying an example given by Lumby [7].

Let us suppose that a company's new project will require a capital expenditure $C_0=\alpha_{\mathrm{e}}C_0+\alpha_{\mathrm{d}}C_0$ and will have no salvage value after N years. By appropriate pricing (P_{E}) of the electricity produced, the project again generates a revenue (SALES) each year, and the operating costs are (OPCST) including interest. Corporate tax is charged at a rate τ on (SALES − OPCST) twelve months after the year end.

In the U.K. tax system writing down allowances ($\gamma\approx0.25$) on capital invested are allowed, year by year. Thus in the first year the tax relief is $\tau\gamma C_0$, and the capital invested is then written down to $(1-\gamma)C_0$ in the

TABLE 6.1

	Year 0	Year 1	Year 2	Year N
Debt capital	$-\alpha_d C_0$			
PDR		$-$PDR	$-$PDR	$-$PDR
Equity capital	$-\alpha_e C_0$			
Tax relief		$\gamma\tau C_0$	$\gamma\tau C_0(1-\gamma)$	$\gamma\tau C_0(1-\gamma)^{N-1}$
Revenue (SALES)		P_E	P_E	P_E
Operating cost (including interest)		$-$OPCST	$-$OPCST	$-$OPCST
Corporation tax			$\tau(P_E-\text{OPCST})$	$\tau(P_E-\text{OPCST})$
Cash flow (IR)	$-C_0$	$[(P_E-\text{OPCST})+\gamma\tau C_0-\text{PDR}]$	$[(P_E-\text{OPCST})(1-\tau)+\gamma\tau C_0(1-\gamma)-\text{PDR}]$	$[(P_E-\text{OPCST})(1-\tau)+\gamma\tau C_0(1-\gamma)^{N-1}-\text{PDR}]$

second year. The tax reliefs in the second year and subsequent years are then

$$\tau\gamma C_0(1-\gamma),\ \tau\gamma C_0(1-\gamma)^2,\ \text{etc.}$$

The discounted cash flows can be tabulated as follows (Table 6.1).

The net present value can be obtained by discounting the cash flows (IR) back to year zero, setting NPV to zero, determining the [unknown] rate of return and comparing it with a specified rate. Alternatively, a weighted discount rate can be specified and the resulting NPV assessed in comparison with that of other schemes.

6.5 Discussion

The last two sections have illustrated the complexity of calculating either the price of electricity or the internal rate of return on a new project, allowing for year-by-year variations in cash flows and local tax systems.

For practical purposes, economic assessment of a new project such as a combined power plant is usually based on a capital charge factor together with an assumed discount rate. Selection of that discount rate is a complex matter; it may be based on

(i) the type of approach implied in Williams's formula [equation (6.15)] or
(ii) the so-called capital asset pricing model (CAPM), which allows for the relative risks involved in financing the project (see, for example, Lumby and parallel references on financial management).

Here we assume that the selection of i has been made on one of these bases and we subsequently confine our attention to pricing of electricity assuming constant costs year by year (or equivalent levelised costs), and allowing for the time value of money through the capital charge factor $\beta(i, N)$. Further, in discussing the "exergoeconomic" analysis developed by Tsatsaronis and Winfold [9] (Section 6.7), we shall also use constant, or levelised, capital costs.

6.6 Comparative Pricing—Some Examples

6.6.1 Pricing of a Combined Plant Compared with a Basic Gas Turbine Plant

Wunsch [2] gives an example of pricing a combined power plant for comparison with that of a conventional plant as outlined in Section 6.2. The combined plant consists of three gas turbines rejecting heat to a steam turbine; the conventional plant consists of five gas turbines. The assumptions made are given in Table 6.2.

The electricity production costs for the two plants $(Y_E)_1$, $(Y_E)_2$(\$/kW h), obtained from equations (6.7) and (6.8), are shown in Fig. 6.2, for a utilisation period (H) of 800 hours per annum, plotted against an oil fuel cost, in \$/barrel (1 barrel = 0.135 tonnes on an average yield basis). The combined plant shows lower produced electricity costs only at high fuel prices. However, with higher utilisation ($H = 5000$ h/a), Fig. 6.3 shows that the picture changes completely and Fig. 6.4 shows a breakdown of $(Y_E)_1$, $(Y_E)_2$ for the latter case. Fuel costs dominate, the higher efficiency of the combined plant becoming the crucial factor, even though its capital cost is greater.

Wunsch develops his case for the combined power plant further by using another of the approaches outlined in Section 6.2. He considers repowering an original plant consisting of three gas turbines

TABLE 6.2

Assumptions	Plant 1	Plant 2
Configuration	5 gas turbines	3 gas turbines + 1 steam turbine
Net output	304 MW	277 MW
Overall efficiency (referred to lower calorific value of fuel)	0.293	0.437
Plant costs	250 \$/kW	350 \$/kW
Period of amortisation	15 a	15 a
Rate of interest	10% p.a.	10% p.a.
Design temperature	35° C	35° C
Condenser pressure	—	0.09 bar

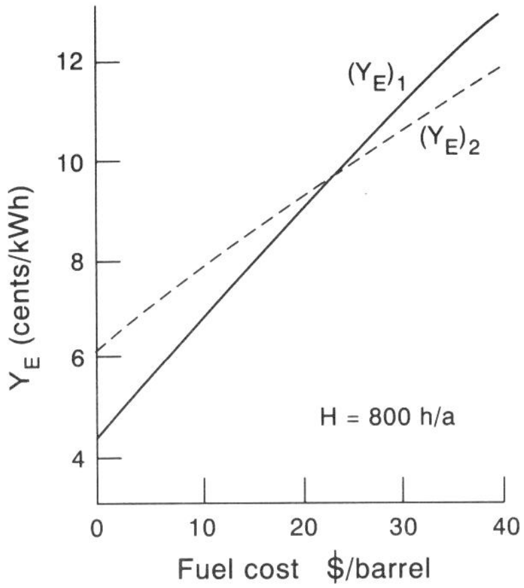

FIGURE 6.2 Electricity production costs for gas turbine plant (1) and combined gas turbine/steam turbine plant (2). Utilisation $H = 800$ h/a (after Wunsch [2])

(a) by adding two similar gas turbines, and
(b) by adding a steam turbine.

In case (b) the electricity production costs of the additional plant are not fuel related, an increase in power $\Delta \dot{W} = (\dot{W}_2 - \dot{W}_1)$ of some 50% of being possible by utilising the heat in the gas turbine exhaust to raise the steam, without burning additional fuel.

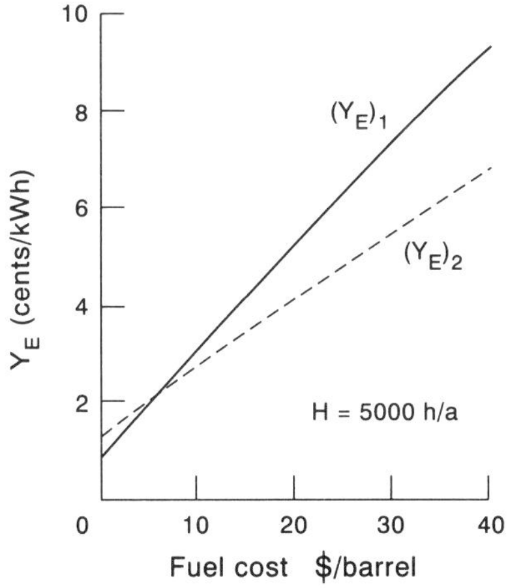

FIGURE 6.3 Electricity production costs for gas turbine plant (1) and combined gas turbine/stream turbine plant (2). Utilisation $H = 5000$ h/a (after Wunsch [2])

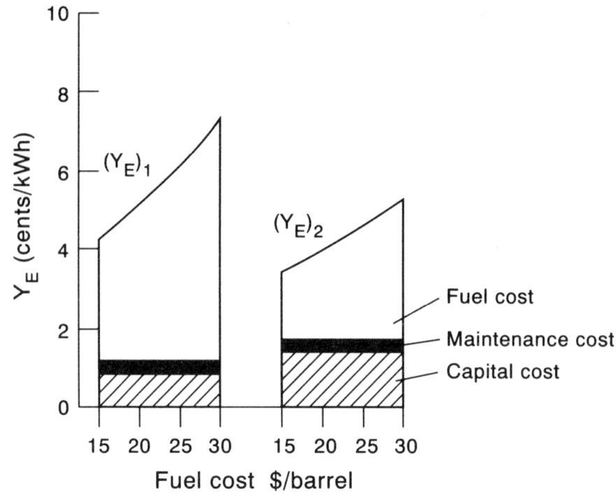

FIGURE 6.4 Breakdown of electricity costs shown in Fig. 6.2 (after Wunsch [2])

Case (b), that of the combined plant, then looks very strong; by adding two gas turbines, the unitised cost remains the same as that of the original gas turbines. By adding the steam turbine, the production costs are simply those associated with servicing the capital and meeting the operational and maintenance costs; there are no additional fuel costs.

This example illustrates both the advantages and the limitations of a combined cycle plant. Firstly the increased efficiency of a combined cycle plant can be critically advantageous, but only if the utilisation of the plant is high, i.e. if it is a base load plant. Secondly, the capital costs of a combined plant may be high, particularly in comparison with a basic gas turbine plant.

The economic comparison with the conventional plant is therefore dependent upon both the capital cost and the thermal efficiency of that conventional plant, both of which are low for the conventional gas turbine plant used by Wunsch in his comparison. Comparison of the combined power plant with a conventional steam plant would show less advantage on fuel costs (because of more comparable thermal efficiencies) but would also show some advantage on capital costs. The overall electricity prices would be more comparable, but the combined power plant generally shows to advantage if plant utilisation is high.

6.6.2 Pricing of a Range of Combined Plants

Dawkins *et al.* [10] have employed the basic electricity pricing method to compare a number of advanced power plants for the Electric Power Research Institute (EPRI). The levelised costs of electricity produced by eight different plants were assessed. With one exception these plants all burned clean fuel gas produced by a coal gasification system using the Texaco gasifier (the exception was a magnetohydrodynamic (MHD) steam cycle operating in a direct coal-fired mode).

These EPRI studies included initial assessment of (i) plant overall efficiency, for plant sizes in the range 500 MW to 700 MW, and (ii) sulphur emissions. Levelised costs were then calculated, essentially in the manner outlined in Section 6.4 above, but including detailed investigations of capital requirements (e.g. land costs, etc.) and operating and maintenance costs (labour, material, and disposal costs, etc.).

Three open circuit (gas turbine)/steam cycle combined plants were considered. The first involved a complex system for cooling the syngas before combustion (by radiation and convective cooling). There was no reheat employed in the gas turbine and no supplementary heating of the HRSG. The steam cycle involved one stage of reheating.

The second gas turbine involved a reheat stage together with a similar steam plant. The third also involved reheat but higher gas turbine top temperatures.

Two fuel cell plants were considered each in association with bottoming steam cycle plants.

Further, two closed cycle plants were analysed—a binary plant (sulphur vapour/steam) and a ternary plant (potassium/diphenyl/steam).

Finally, a direct coal-fired MHD/steam plant was studied (with a variant fired with clean gas from a Texaco gasifier).

The important result from the EPRI calculations (made in 1986, but based on mid-1984 dollars) was that although the fuel cell plants would achieve the highest efficiency (some 48%) and the two advanced gas turbines would also produce high efficiencies (of 41–41.5%), the first simplest, non-reheat gas turbine would produce electricity at the lowest levelised cost (4.7 cents/kWh) essentially because of lower capital costs. Their conclusion would appear to be that the steady development of present gas turbine/steam turbine plant offers the best prospect for reducing electricity costs and prices.

6.6.3 Pricing of IGCC Plants

Using levelised cost analysis, Marshall *et al.* [11] have also made cost and performance studies of Texaco-based Gasification Combined Power Plants for EPRI. Three basic IGCC plant configurations were considered, the major differences being in the level of heat recovery from the syngas downstream of the gasifier.

The first plant design incorporated both radiant and convective cooling of the gas (the configuration employed in the Cool Water plant described in Chapter 7). A second plant included a radiant cooler followed by water quenching. A third assumed no heat recovery from the raw syngas, i.e. direct water quenching of the hot gas.

The three configurations were thus in order of decreasing capital cost, as one (or both) of the costly high temperature syngas coolers was eliminated. However, against this reduction in capital cost had to be balanced an increased heat rate (decreased overall efficiency) which resulted from progressive reduction in the steam turbine power output as the heat recovery from the syngas was decreased. (This was offset to some extent by adding moisture to the fuel gas in the gas saturator and consequently increasing gas turbine mass flow and power output.)

Although there was some dependence on utilisation and coal cost, the broad conclusion from this work was that there were but small differences in levelised electricity cost between the three configurations. EPRI further claimed that the Texaco-based IGCC power plants have the potential to provide 15% reductions in levelised costs compared with conventional subcritical steam plants (which are now required to use flue gas desulphurisation to reduce SO_2 emissions).

6.6.4 A Graphical Presentation

An alternative presentation of an economic comparison, between an existing plant and a range of possible new plants, is given by Chester *et al.* [12]. The comparison is made on a plot of "specific capitalised cost" against "relative efficiency". The former is defined as the specific capital cost (or tender price) of the plant (£/kW for example), plus interest on funds used during construction, plus the capitalised cost of programmed replacement of components. The first of these corrections was allowed for simply by multiplying basic capital cost by 1.25, the second by obtaining the capitalised cost by discounted cash flow methods assuming a specified discount rate. The relative efficiency was obtained by determining the amount of fuel that could be purchased with the money spent on operating costs and correcting the plant overall efficiency accordingly.

These steps enabled the price of electricity to be written simply as a correction of equation (6.4),

$$Y_E^1 = \frac{\beta C_0^1}{\dot{W}H} + \frac{\zeta}{(\eta_O)^1} \tag{6.25}$$

where (C_0^1/W) is the "specific capitalised cost" and $(\eta_O)^1$ is the modified efficiency. The variation of $(C_0^1/\dot{W})$ with $(\eta_O)^1$ along a line of constant Y_E^1 is then given by

$$\frac{\partial(C_0^1/\dot{W})}{\partial(\eta_O)^1} = \frac{\zeta H}{[(\eta_O)^1]^2\beta} \tag{6.26}$$

Thus the change of the specific capitalised cost $\Delta(C_0^1/W)$ with a change of efficiency $\Delta(\eta_O)^1$ is

$$\frac{\Delta(C_0^1/\dot{W})}{(\Delta\eta_O)^1}=\left[\frac{\zeta(£/\mathrm{kW\,h})L\times 24.365(\mathrm{h})}{[(\eta_O)^1]^2\beta}\right]$$

$$=\frac{8760\zeta L}{[(\eta_O)^1]^2\beta}(£/\mathrm{kW}) \qquad (6.27\mathrm{a})$$

where $H=L\times 8760$, if L is a fractional load factor. [Chester *et al.* write this equation as a change of specific capitalised cost with a 1% point improvement in efficiency

$$\frac{\Delta(C_0^1/\dot{W})}{(\Delta\eta_O)^1}=\frac{31.5LX}{[(\eta_O)^1]^2\beta}\,£/\mathrm{kW} \qquad (6.27\mathrm{b})$$

where $(\eta_O)^1$ is a percentage efficiency (as opposed to fractional), and X is fuel cost, p/GJ.]

Figure 6.5 shows one of Chester's plots of the break even line, the slope of which is given by equation (6.26). The line passes through point A which corresponds to the price of electricity produced by a pulverised fuel station with three 660 MW units and flue gas desulphurisation. New plants located above the line give more expensive electricity, those below the line less expensive electricity. Points B and C show Chester's estimates for a cheaper IGCC scheme, with and without credit for the sale of sulphur produced.

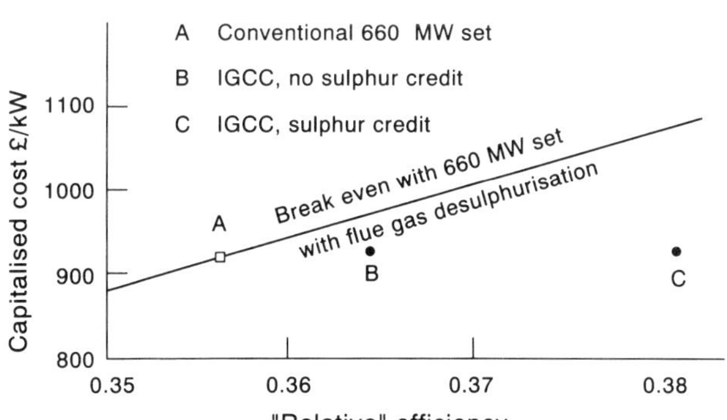

FIGURE 6.5 "Break even" plot for IGCC plant (5% discount rate) (after Chester *et al.* [8])

6.7 Exergoeconomic Analysis of a Combined Power Plant

A method of "exergoeconomic" analysis, based on a combination of exergy analysis with levelised fuel, capital and other costings, has been developed by Tsatsaronis, Winfold and their colleagues in a series of studies sponsored by EPRI [9, 13, 14, 15]. Here we describe the basis of the method and refer to some examples of its use.

In Chapters 1 and 5 we described the flux of exergy through a component i of a power plant (Fig. 6.6a). A rate equation can be derived of the form

$$\sum_{str} \dot{\mathrm{E}}_{\mathrm{X}} = \sum_{str} \dot{\mathrm{E}}_{\mathrm{Y}} + \dot{I}_i^{CR} \tag{6.28}$$

where $\dot{\mathrm{E}} = (\dot{E} + \dot{E}^{Q} + \dot{W})$, and the summation str is over the several streams entering and leaving the component. [For a single stream flowing through a component, producing work, not *receiving* heat but losing heat to the atmosphere,

$$\dot{E}_{\mathrm{X}} + \dot{\mathscr{W}} = \dot{E}_{\mathrm{Y}} + \dot{W} + \dot{I}_i^{CR} + \dot{I}_i^{Q} \tag{6.29}$$

where $\dot{I}_i^{Q}$ is numerically equal to $\dot{E}_{\mathrm{Y}}^{Q}$ and is the lost power due to external irreversibility.]

Tsatsaronis and Winfold parallel such an exergy rate equation with an economic equation, allocating unit costs (c) to the exergy and work flows

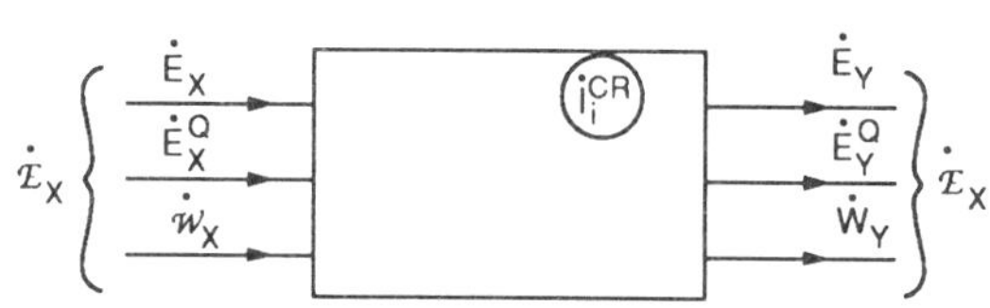

(a) Exergy flows

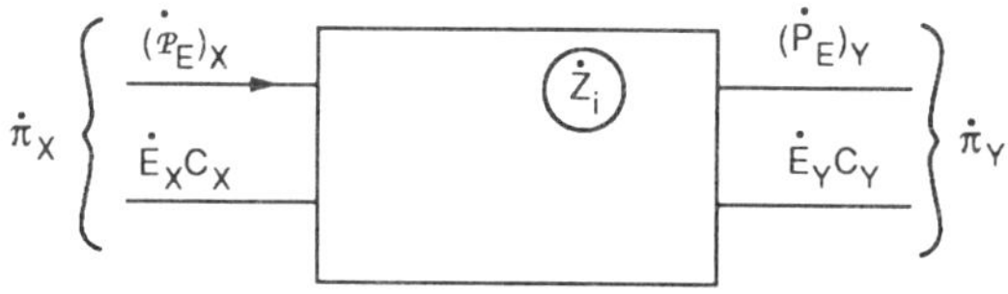

(b) Money flows

FIGURE 6.6 Exergy and money flows (after Tsatsaronis and Winfold [5])

where possible and inserting a capital cost rate $\dot{Z}_i$ for the component i (a selected fraction of the total capital cost rate (βC_0) for the entire plant). Thus

$$\dot{\mathscr{P}}_E + \sum_{str} (\dot{E}_X c_X) + \dot{Z}_i = \dot{P}_E + \sum_{str} (\dot{E}_Y c_Y) \tag{6.30a}$$

or

$$\dot{\mathscr{P}}_E + \sum_{str} \dot{\Pi}_X + \dot{Z}_i = \sum_{str} \dot{\Pi}_Y + \dot{P}_E \tag{6.30b}$$

where $\dot{\Pi} = \dot{E}c$, $\dot{\mathscr{P}}_E$ is the cost rate of electricity input, and $\dot{P}_E$ is the value of electricity output. Further $\dot{E}^Q$ is regarded as a form of exergy flux. Figure 6.6b shows the money flows corresponding to equations (6.30).

The steps in the exergoeconomic approach are as follows.

(i) Determine the exergy flows, work inputs and outputs and lost work terms $\dot{I}^{CR}$ and $\dot{I}^Q$ as outlined in Chapter 5.
(ii) Specify the exergy costs for inputs (mainly fuel) but also for electrical power input. Also specify exergy costs for disposal (e.g. ash, sulphur products).
(iii) Determine the output costs $\dot{\Pi}_Y$ and unit costs $c_Y = \dot{\Pi}_Y / \dot{E}_Y$. [The output cost thus reflects the cost of the lost work terms ($\dot{I}_i^{CR}$ and $\dot{I}_i^Q$) and capital costs $\dot{Z}_i$ in any component.]

For flow of a single stream these steps are straightforward, and an output exergy cost c_Y can be determined, which becomes c_X for the next component.

Two difficulties arise however: firstly, the allocation of the capital cost $\dot{Z}$; and secondly, the problem arising from more than one stream.

In their seminal paper (reference 9, reproduced in reference 13) Tsatsaronis and Winfold provide a detailed description of how the first problem is solved, i.e. how $\dot{Z}_i$ may be determined. Suppose the capital and maintenance costs for the whole plant $\dot{K} = \beta C_0 + (\dot{O}M)$ are known. The allocation $\dot{K}_i$ to the component is supposed to be partially dependent on the exergy flow rate

$$\dot{K}_i = B_i \dot{E}_i + \dot{D}_i = \dot{Z}_i + \dot{D}_i \tag{6.31}$$

where B_i and $\dot{D}_i$ are constants. Tsatsaronis and Winfold suggest that lower limit values of costs should be calculated for $\dot{Z}_i = 0$, and upper limit values for $\dot{Z}_i = \dot{K}_i$. With the first assumption, calculations include only the effect of the irreversibilities on exergy costs, not investment costs.

With the second assumption, investment costs are proportional to exergy flow rates.

The second problem arises if there are two or more streams leaving the component. The quantity $\Sigma\dot{\Pi}_Y = \dot{E}_Y c_Y$ may be determined from equation (6.30b), but before $c_1, c_2, c_3 \ldots$ for each of the exit streams can be found, an assumption has to be made about the allocation of $\dot{\Pi}_Y$ between the various streams. This problem may be solved by one of several methods; for example,

(a) setting the exergy costs of some products to market values;
(b) assuming unit costs of the leaving streams are the same, i.e. $c_1 = c_2 = c_3 \ldots$

After making various assumptions to meet these two major difficulties, Tsatsaronis and Winfold develop their analysis to determine the marginal cost of exergy losses in a component. The exit cost increase between input and output is

$$\Delta c_i = (c_i)_Y - (c_i)_X = \frac{\dot{Z}_i + (c_i)_X \dot{I}_i}{(\dot{E}_i)_Y} \tag{6.32a}$$

[Note that Δc_i is also given by

$$\Delta c_i = \frac{\dot{Z}_i + (c_i)_Y \dot{I}_i}{(\dot{E}_i)_X} \tag{6.32b}$$

where $(c_i)_Y$ is the marginal cost of exergy losses in the ith component.] That is to say, that if the exergy losses in components could be reduced by $\Delta\dot{I}_i$, without any other changes in plant costs, then $\Delta\dot{\Lambda}_i = (c_i)_Y \Delta\dot{I}_i$ [or $(c_i)_X \Delta\dot{I}_i$] could be saved (according to whether the total input or output exergy is held constant, but usually the former). Tsatsaronis and Winfold refer to $\dot{\Lambda}_i = c_i \dot{I}_i$ as the cost of the exergy losses in the plant component.

But the reduction of exergy losses will require additional capital expenditure. Tsatsaronis and Winfold wished to extend their approach to relate the marginal cost of exergy losses $(c_i)_Y$ to the marginal capital cost associated with the change $\Delta\dot{I}_i$ in the component, $Z_i = (\Delta\dot{K}_i/\Delta\dot{I}_i)$. This comparison would pick out those components which show a high potential to give improvement in plant performance. However, it is extremely difficult to determine Z_i, so Tsatsaronis and Winfold suggest that useful exergoeconomic factors are

$$f_i = \frac{\dot{Z}_i}{\dot{Z}_i + (\dot{\Lambda}_i)_Y} = \frac{\dot{Z}_i}{\dot{Z}_i + (c_i)_Y \dot{I}_i} \tag{6.33a}$$

or

$$f_i^1 = \frac{\dot{K}_i}{\dot{K}_i + (\dot{\Lambda}_i)_Y} = \frac{\dot{K}_i}{\dot{K}_i + (c_i)_Y \dot{I}_i} \tag{6.3b}$$

Tsatsaronis and Winfold first used their approach to study a conventional steam power plant [14]. Subsequently, Tsatsaronis, Winfold and Stojanoff [15] employed their exergoeconomic system of analysis in considerable detail to a gasification combined cycle plant. The plant analysed was the most complex IGCC configuration studied in the earlier EPRI work, in which levelised costs of three IGCC plants, of decreasing capital cost, were compared (see Section 6.6.3). A detailed account of this work is not given here, and the reader is referred to the original report for such details.

We summarise below the conclusion of Tsatsaronis and his colleagues.

1. The cost of exergy losses (irreversibilities) in gas turbine system and in the gasifier were very large, but only part of these losses were avoidable.
2. The large exergy losses in the gas turbine system could be reduced by adding a reheat stage. (No reheat was used in the basic plant studied.)
3. Capital costs of the syngas cooling units were high (these were reduced by elimination of up to two of the three units in the original EPRI economic studies described in Section 6.6.3).
4. There was potential for improvement in the overall heat exchanger network and in the steam conditions (i.e. reduction of irreversibility).

The power of this exergoeconomic analysis is well demonstrated in this study of the IGCC plant, but its complexity should not be underestimated. Several factors (e.g. the allocation of capital costs and maintenance costs to components) require a substantial degree of judgement. Nevertheless, analyses of this type for large-scale combined plant, involving major capital cost, would appear to be very worthwhile.

References

[1] Williams, R. H. Industrial Cogeneration. *Ann. Review Energy*, **3**, 313–356, 1982.

[2] Wunsch, A. Highest Efficiencies Possible by Converting Gas Turbine Plants into Combined Cycle Plants. *Brown Boveri Review*, **10**, 455–463, 1985.

[3] Horlock, J. H. *Cogeneration; Combined Heat and Power*. Pergamon Press, Oxford, 1987.

[4] Marshall, W. *et al. Combined Heat and Electrical Power Generation in the United Kingdom*. H.M.S.O. Dept. of Energy Report 35, 1979.

[5] Cowan, S. Taxation, The Cost of Capital and the Capital Asset Pricing Model. Private communication.

[6] Daudet, H. C. and Trimble, S. W. *Evaluation Method for Closed Cycle Gas Turbines in Cogeneration Applications*. ASME Paper 80-GT-176, 1980.

[7] Lumby, S. *Investment Appraisal and Financing Decisions*, 4th edition. Chapman and Hall, London, 1991.

[8] Brealey, R. and Myers, S. *Principles of Corporate Financing,* 3rd edition. McGraw Hill, 1988.

[9] Tsatsaronis, G. and Winfold, M. *Thermoeconomic Analysis of Power Plants.* EPRI AP3651, 1984.

[10] Dawkins, P. P., Rao, A. D. and Jacob, J. T. *Screening Evaluation of Advanced Power Plants.* EPRI 4826, 1986.

[11] Marshall, T. A., Rao, A. D., Ramanthan, V. and Sander, M. T. *Cost and Performance of Commercial Texaco-Based Gasification-Combined Cycle Plants.* EPRI 3486, 1984.

[12] Chester, P. F. *et al. Prospects for the Use of Advanced Coal Based Power Generation Plant in the United Kingdom.* H.M.S.O. Energy Paper 56, 1988.

[13] Tsatsaronis, G. and Winfold, M. Exergoeconomic Analysis and Evaluation of Energy Conversion Plants. Part 1. *Energy,* **10,** 1, 69–80, 1985.

[14] Tsatsaronis, G. and Winfold, M. Exergoeconomic Analysis and Evaluation of Energy Conversion Plants. Part 2. *Energy,* **10,** 1, 81–94, 1985.

[15] Tsatsaronis, G., Winfold, M. and Stojanoff, C. G. *Thermoeconomic Analysis of a Gasification-Combined Cycle Power Plant.* EPRI 4734, 1986.

CHAPTER 7

Some Practical Combined Power Plants

7.1 Introduction

In this chapter we describe some practical combined cycle power plants. Since the development of such plants, in various forms, spans some fifty years, we include some old plants of historical significance (which are now obsolete), some well developed and established plants and some more recent plants which are in the developmental stage rather than fully operational.

Examples are chosen of all the major combined cycle power plants. They are as follows.

(i) Binary Vapour Plant
The mercury/steam 40 MW plant at Schiller, New Hampshire, built in 1949 and now obsolete (General Electric)

(ii) CCGT (Gas/Steam Turbine) Plant (unfired)
The Korneuburg B 125 MW plant, with unfired heat recovery steam generator (Brown Boveri)

(iii) CCGT (Gas/Steam Turbine) Plant (supplementary fired)
The earlier Korneuburg A 60 MW plant, with a fired heat recovery steam generator (Brown Boveri)

(iv) CCGT (Gas/Steam Turbine) Plant (fully fired)
The 600 MW plant at Hemweg, which involved repowering a 500 MW steam plant. The original boiler is now supplied by the exhaust of a gas turbine which is maximally fired (Brown Boveri)

(v) IGCC (Integrated Gasifier Combined Cycle) Plant
The demonstration plant at Cool Water, U.S.A. using the Texaco gasifier process and a GE gas turbine

(vi) STIG (Steam Injection Gas Turbine) Plant
The STIG plant at Ripon Mill, California, using a GE/IHI gas turbine.

All these plants are combined power plants, providing electrical power only. The reader is referred to the parallel volume on cogeneration (Horlock [1]), and in particular to Chapter 5, Section 5.5, for a description of a combined *heat* and power plant, based on an open circuit gas turbine/closed steam turbine system, with useful heat supplied for district heating (the "STAG" plant at Saarbrücken).

7.2 The Mercury/Steam Binary Vapour Plant at Schiller, New Hampshire

A series of mercury/steam plants were developed in the 1940s by the General Electric Company of the U.S.A., culminating in the installation of the Schiller 40 MW plant at Portsmouth, New Hampshire, which went on load in 1950. Although the plant has now been dismantled, it is worthwhile reviewing its design and performance, as it was the first major combined cycle plant and represented a major engineering achievement. The development of the mercury boiler and turbines, and of the mercury condenser/steam evaporator, was a tremendous feat of engineering. But as Wood [2] has pointed out, developments in large conventional steam plant paralleled the thermal efficiency of this binary vapour plant, the capital cost of which was higher (in $ per kW).

Before describing the Schiller station, it is worthwhile briefly sketching the development of the mercury/steam plant. Although the search for alternative fluids "seems to be nearly as old as the steam engine itself" (Kearton [3]), the serious development of the mercury/steam binary plant can be traced back to proposals by Emmet before the First World War in a paper to the American Institute of Electrical Engineers in 1913. (Emmet also linked the name of Bradley to the idea of the basic principle of the cycle.) In 1925 Emmet [4] was able to describe substantial development work on mercury plant, boiler turbine and condenser/boiler (he did indeed refer to the system as "Emmet mercury-vapour process"). Figure 7.1 shows diagrams of the process, based on ideal vapour cycles, taken from Emmet's 1925 paper. The first (a) shows superheating of the lower steam cycle and ideal regenerative feed heating within that cycle; the second (b) shows similar pressure and temperature levels, but a lower "Rankine" steam cycle with no feed heating, and consequently a slightly lower overall efficiency.

The Schiller plant, installed some 26 years later, operated at higher mercury and steam pressures but with irreversibilities, did not achieve the ideal cycle efficiencies quoted by Emmet earlier. In 1951 a heat rate of 9000 Btu/kW h[$(\eta_O)_{CP}=0.370$] was reported (Hackett [5]), and a value of 9400 Btu/kW h[$(\eta_O)_{CP}=0.364$], if the power for auxiliaries was subtracted. The operating conditions for the Schiller station are given in Table 7.1, and a simple schematic diagram of the plant is shown in Fig. 7.2. However, the diagram gives only approximate values of (relative)

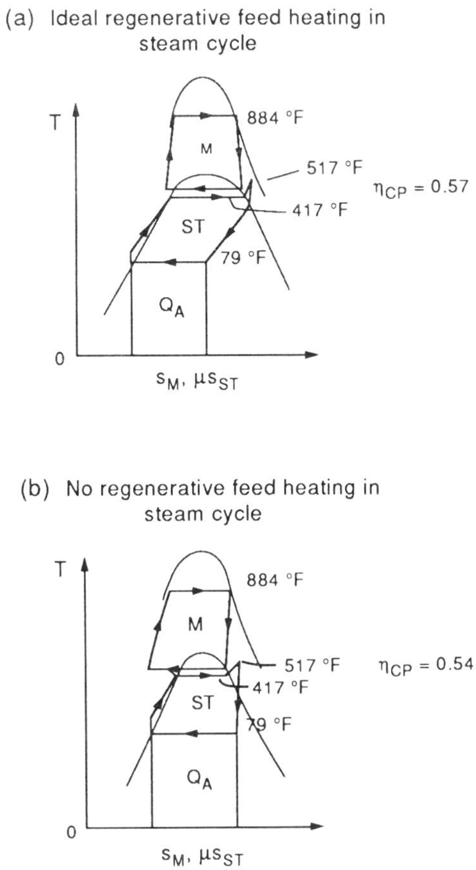

FIGURE 7.1 Theoretical possibilities of mercury/steam binary cycle (Emmet's original proposal [4])

TABLE 7.1 *Schiller Plant*

	Performance data (Hackett [5])
Plant rating	38.5 MW (net)
Mercury turbine (throttle conditions)	122 psia (0.84 MN/m^2) 937° F (502.8° C) (dry saturated)
Mercury turbine (exhaust conditions)	2.13 psia (14.2 kN/m^2) 508° F (264° C)
Steam conditions (superheater outlet)	615 psia (4.24 MN/m^2) 825° F (440° C)
Mercury flow rate (each of 2 turbines)	938,000 lb/h (118 kg/s)
Steam flow rate (approx)	250,000 lb/h (31.6 kg/s)
Mercury turbine output rating	2 × 7500 kW
Steam turbine output	25,000 kW
Condenser pressure (see Wood [2])	0.5 psia (3.5 kN/m^2)

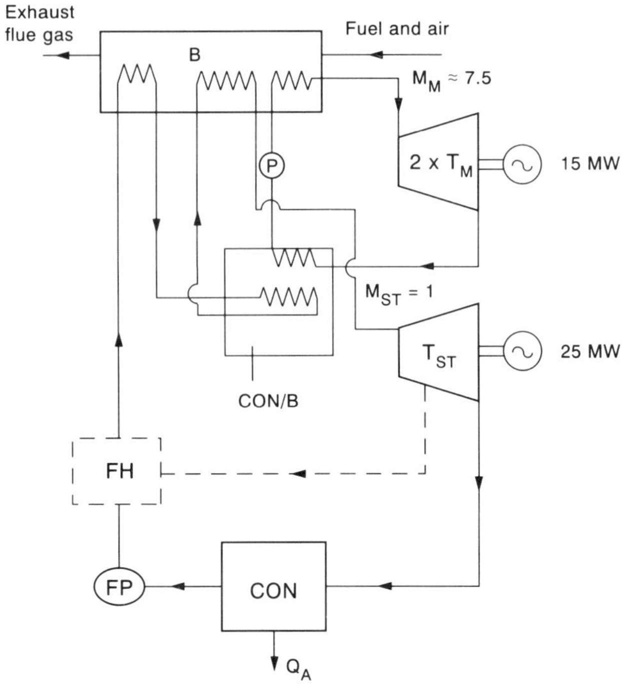

B	Main boiler
T_M	Mercury turbines
T_{ST}	Steam turbines
CON/B	Condenser/boiler
CON	Steam condenser
FP	Fuel pump
FH	Feed heaters

FIGURE 7.2 Schiller binary plant

mercury and steam flow rates; it shows regenerative feed heating by bled steam in the steam cycle, and no regenerative feed heating in the mercury cycle. Hackett and Douglass [6] describe the mercury boiler and mercury/steam condenser/boiler built for an earlier plant at South Meadow, Hartford (it appears this was a standard unit for the G.E. binary plants). They show feed water entering the economiser at 340° F (171° C), presumably after regenerative feed heating by bled steam; it is then heated by a further 70° F (39° C) in the economiser, located in the flue gas at exit from the mercury boiler, before entering the condenser/boiler. Oberly [7], in a general discussion of the G.E. mercury/steam combined plant, also refers to the lower steam cycle having "the usual feed heaters in the liquid region".

The example outlined in Chapter 5 was chosen to correspond closely to the Schiller plant (i.e. with bled steam feed heating to 170° C, further feed water heating in an economiser and superheating of the steam in the mercury boiler flue gases). The reader is referred to that example for a calculation of combined plant efficiency which is close to that achieved at Schiller. We do not assess its economics here; sufficient to say its capital cost was substantial, even if its efficiency was high.

7.3 The CCGT Plant (B) at Korneuburg, Austria (Unfired HRSG)

The Brown Boveri 125 MW (nominal) CCGT plant, Korneuburg B, is a good example of the open gas turbine/closed steam turbine system with no supplementary firing (Czermak and Wunsch [8]). The gas turbine supplies 81.1 MW and the steam turbine 48.7 MW and the overall efficiency of the plant, based on the lower calorific value of the fuel (natural gas), is 0.470 at the generator terminals (0.466 at the transformer terminals, after deduction of power for station auxiliaries), both for an air entry temperature of 10° C.

Figure 7.3 is a simplified diagram of the plant, showing flow rates, pressures and temperatures. The gas turbine (Brown Boveri type 13) is open circuit; the mass flow rate is 357 kg/s, the pressure ratio is 10.0 and the turbine inlet temperature is of the order of 1000° C (cooling of the turbine rotor limits local maximum temperatures to 540° C). The overall efficiency of the gas turbine, based on the lower calorific value of the fuel (36.7 MJ/m^3N), is 0.294.

The temperature of the gas turbine exhaust is 491°C, and this gas is fed to a heat recovery steam generator, which raises steam at two pressure levels in a dual pressure system. The gas leaves the HRSG at a low temperature of 95° C, which is permissible for burning of natural gas with low sulphur content. This means that only one stage of feed heating is needed in the steam cycle, a direct contact feed heater/aerator operating at below atmospheric pressure (0.16 bar) and raising the water temperature from 28° C to 54° C. After this regenerative feed heating the water is first heated in an economiser in the HRSG. Part of the feed water is then evaporated at low pressure (4.4 bar) and superheated to 180° C and part pumped to higher pressure (33.2 bar), further heated as water in an HP economiser, evaporated and then superheated to 433° C. Further performance data for the plant are shown in Table 7.2.

An approximate thermodynamic analysis of the plant is as follows.

$$\text{Heating value of fuel} = 276.3\,\text{MW}$$

$$\text{Gas turbine work output} = 81.1\,\text{MW}$$

$$\text{Gas turbine efficiency } (\eta_O)_G = \frac{81.1}{276.3} = 0.294$$

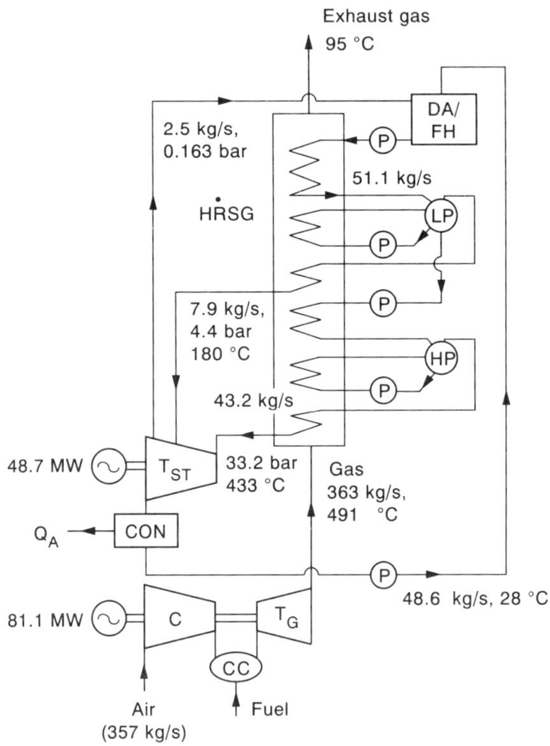

C	Compressor
T_G	Gas turbine
T_{ST}	Steam turbines
P	Pumps
HRSG	Heat recovery steam generator
DA/FH	Deaerator/feed heater

FIGURE 7.3 Korneuburg B open circuit (gas turbine)/closed cycle (steam turbine) combined plant (after Czermak and Wunsch [8])

TABLE 7.2

	Reading	Units
Gas turboset		
Gas turbine	Type 13	—
Fuel Natural Gas	—	—
Calorific value LHV	36.7	MJ/m³ N
Power supplied in the fuel	276.3	MW
Efficiency of the gas turbine (overall)	0.294	(—)
Output at generator terminals	81.1	MW
Exhaust gas mass flow	363	kg/s
Exhaust gas temperature (T_4)	491	°C
Heat-recovery steam generator		
Exhaust gas temperature at HRSG outlet (T_S)	95	°C
Exhaust gas pressure drop in HRSG	20	mbar
Feedwater mass flow	50.9	kg/s
Feedwater temperature	54	°C
LP steam mass flow	7.9	kg/s
LP steam temperature	180	°C
LP steam pressure	4.4	bar
HP steam mass flow	43.2	kg/s
HP steam temperature	433	°C
HP steam pressure	33.2	bar
Steam turboset		
Vacuum in condenser	0.036	bar
Cooling water temperature	8.0	°C
Cooling water volume flow	7280	m³/h
Output at the generator terminals	48.87	MW
Combined-cycle unit total		
Total power output (gas and steam turbines)	129.79	MW
Consumption of station auxiliaries	−0.96	MW
Net output of the plant	128.83	MW
Net heat rate	7721.8	kJ/kW h
Net overall efficiency of plant	0.466	(—)

Heat received by the steam flow in the HRSG

$$
\begin{aligned}
&= (\dot{M}_{HP}h_{HP} + \dot{M}_{LP}h_{LP}) - (\dot{M}_{HP} + \dot{M}_{LP})h_w \\
&= (43.2 \times 3302 + 7.9 \times 2873) - 51.1 \times 226 \\
&= 153.4\,\text{MW}
\end{aligned}
$$

A check using an approximate gas specific heat of $(\bar{c}_p)_G = 1.1\,\text{kJ/kg K}$ gives the heat transferred from the exhaust gas as

$$
\begin{aligned}
\dot{M}_G(\bar{c}_p)_G(T_4 - T_S) &= 363 \times 1.1(491 - 95) \\
&= 158.1\,\text{MW}
\end{aligned}
$$

The work output at the turbine = 48.7 MW.

The heat rejected by steam to cooling water is approximately

$$\dot{M}_{\mathrm{CON}}(\Delta h) = \left(\frac{7280 \times 10^3}{3600 \times 10^3}\right)(83.9 - 33.6)$$
$$= 101.7\,\mathrm{MW}$$

(A check is given by (heat supplied)—(work done) = 153.4 − 48.7 = 104.7 MW.) Steam plant thermal efficiency

$$\eta_{\mathrm{ST}} = \frac{48.7}{153.4}$$
$$= 0.317$$

The energy lost in the exhaust stack

$$\dot{M}_{\mathrm{G}}(c_p)_{\mathrm{G}}(T_{\mathrm{S}} - T_0) = 363 \times 1.1\,(95 - 10)$$
$$= 33.9\,\mathrm{MW},$$

$$\nu_{\mathrm{UN}} = \frac{\text{“heat” unused}}{\text{“heat” supplied}}$$
$$= \frac{33.9}{276.3} = 0.123.$$

Thus an estimate of overall efficiency is given by

$$(\eta_{\mathrm{O}})_{\mathrm{CP}} = (\eta_{\mathrm{O}})_{\mathrm{G}} + \eta_{\mathrm{ST}} - (\eta_{\mathrm{O}})_{\mathrm{GT}}\eta_{\mathrm{ST}} - \nu_{\mathrm{UN}}\eta_{\mathrm{ST}} \tag{7.1}$$
$$= 0.294 + 0.317 - 0.294 \times 0.317 - 0.123 \times 0.317$$
$$= 0.294 + 0.317\,(0.583)$$
$$= 0.479,$$

cf. the measured total efficiency $(\eta_{\mathrm{O}})_{\mathrm{CP}} = 129.8/276.3 = 0.470$.

The plant attains high thermal efficiency from a simple cycle. It was originally designed to operate as a base load plant, but has more recently been in service as a medium load plant. It has achieved a high level of availability.

In Chapter 6 we referred to calculations of electricity production costs by Wunsch [9] for combined cycle plants. If we take his typical combined cycle (1978) capital cost of 350$/kW for the Korneuburg B plant, together with a fuel price of say 1 c/kW h and maintenance costs of 1.0 c/kW h (Williams [10]) we would calculate a 1978 unit price for electricity of

$$
\begin{aligned}
Y_{\mathrm{E}} &= \frac{\beta C_0}{\dot{W}H} + \frac{\zeta}{(\eta_{\mathrm{O}})_{\mathrm{CP}}} + \frac{\mathrm{OM}}{\dot{W}H} \\
&= \frac{0.1315 \times 350 \times 10}{5000} + \frac{1}{0.466} + 1.0 \\
&= 0.92 + 2.15 + 1.0 \\
&= 4.07\ \mathrm{c/kW\,h}
\end{aligned}
$$

using Wunsch's annuity factor of 0.1315 (10% discount over 15 years). (This compares with his 1978 estimate of about 4.2 c/kW h for a rather larger combined plant burning oil at 20$ per barrel.) Both figures are very low, in comparison with prices in the 1990s.

7.4 The CCGT Plant (A) at Korneuburg, Austria (Supplementary Fired HRSG)

The original Brown Boveri CCGT plant at Korneuburg (Korneuburg A) was a 75 MW plant, consisting of two 25 MW gas turbosets, a heat recovery steam generator with supplementary firing and a 25 MW steam turbine (all figures nominal). It was installed in 1960 and for fourteen years operated as a base-load plant at an average utilisation of 6000 hours per year.

Korneuburg A (Auer [11]) is included here as an example of a gas/steam turbine plant with supplementary firing of the gas turbine exhaust (as opposed to maximum firing). It was a remarkable plant in that it involved two complex gas turbines (with intercooling and reheating, but no heat exchanger), each of maximum overall efficiency of about 0.26 (net) at a turbine inlet temperature (and reheat turbine temperature) of 625° C (998 K). With the steam turbine added an overall combined plant efficiency of some 0.32 (net) was achieved. The gain in efficiency over the basic gas turbine was therefore marginal, but the increase of power by adding the steam turbine was some 43%. A point of historical interest is that Brown Boveri made the design choice of supplementary "heating" for Korneuburg A, with a relatively low turbine entry temperature (of 625° C); they subsequently dispensed with supplementary heating for Korneuburg B, with a much higher turbine inlet temperature of 1000° C, with no intercooling and no reheat. The two plants together provide a classic illustration of combined plant development over the period between the early sixties and the late seventies, in that both complexity in the gas turbine plant and supplementary firing of the exhaust were replaced by a simple (CBT) plant with a much higher turbine inlet temperature. It is perhaps significant that the author has had to turn back

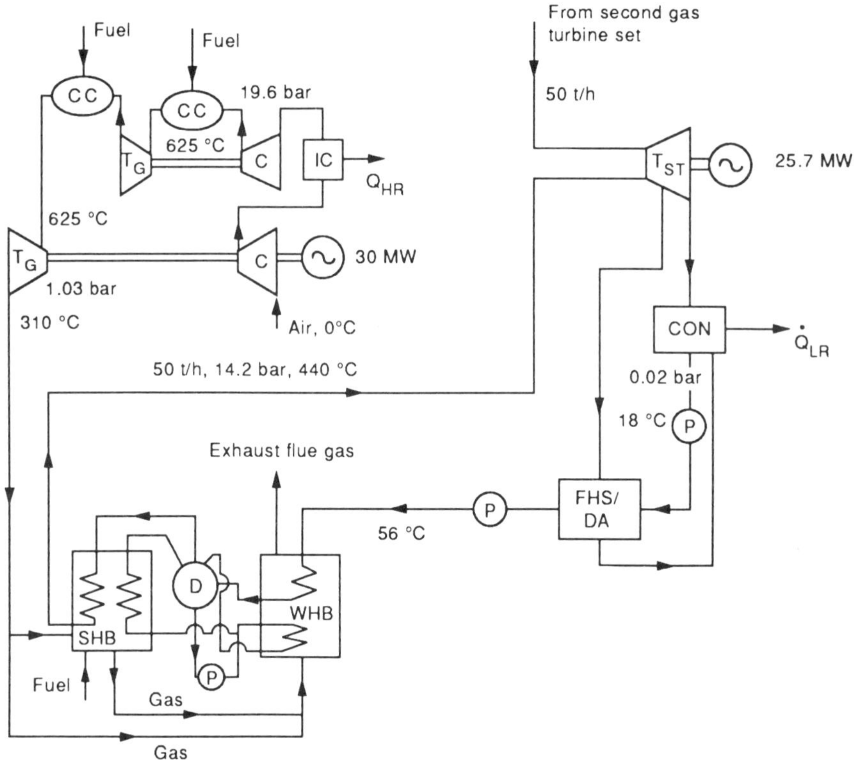

C	Compressor	CON	Condenser
T_G	Gas turbine	FHS/DA	Feed heaters/deaerator
CC	Combustion chambers	P	Pumps
IC	Intercooler	D	Steam separation drum
SHB	Superheater/ boiler (supplementary fired)		
WHB	Wasteheat boiler (economiser, evaporator)		
T_{ST}	Steam turbine		

FIGURE 7.4 Korneuburg A open circuit (gas turbine)/closed cycle (steam turbine) combined plant with supplementary firing (after Auer [11])

the clock thirty years to find an example of supplementary firing for this volume.

A "thermal" diagram of the Korneuburg A plant was given by Auer and is shown in Fig. 7.4, with approximate design values of properties round the "cycle" and of mass flow rates. The overall pressure ratio in the gas turbine, achieved in two stage compression, was a little below 20. The temperature after the main combustion chamber and after reheat was 625° C, and the temperature after full expansion in the second turbine was 310° C. The recuperative boiler was in two parts—a waste heat boiler with economiser and evaporator sections and a superheater/boiler, which also contained some 18% of evaporator surface; steam was delivered to the steam turbine at 14.2 bar and 440° C. A single stage of bled steam feed heating followed by deaeration raised the condensate from a temperature of 18° C (at 0.0206 bar) to 56° C.

Czermak and Lahr [12] present performance information for the plant. They give the gas turbine thermal efficiency as 0.257 under combined plant operation, producing 55.3 MW (net) at the generator terminals. The steam turbine delivers an additional 25.9 MW, for which an estimated thermal efficiency is 0.312 (cf. 0.298 proposed by Auer) and the mechanical and electrical losses are 10.1 MW and 5.7 MW respectively (15.8 MW total).

Approximate estimates of energy flows are shown in Fig. 7.5. The fuel energy input to the gas turbine is 215.3 MW and the fuel energy to the total plant is 253.9 MW. The supplementary fuel supply is 38.6 MW (253.9 − 215.3) and the parameter v_S is therefore 38.6/253.9 = 0.152; the heat rejected in the gas turbine intercooler is 53.5 MW.

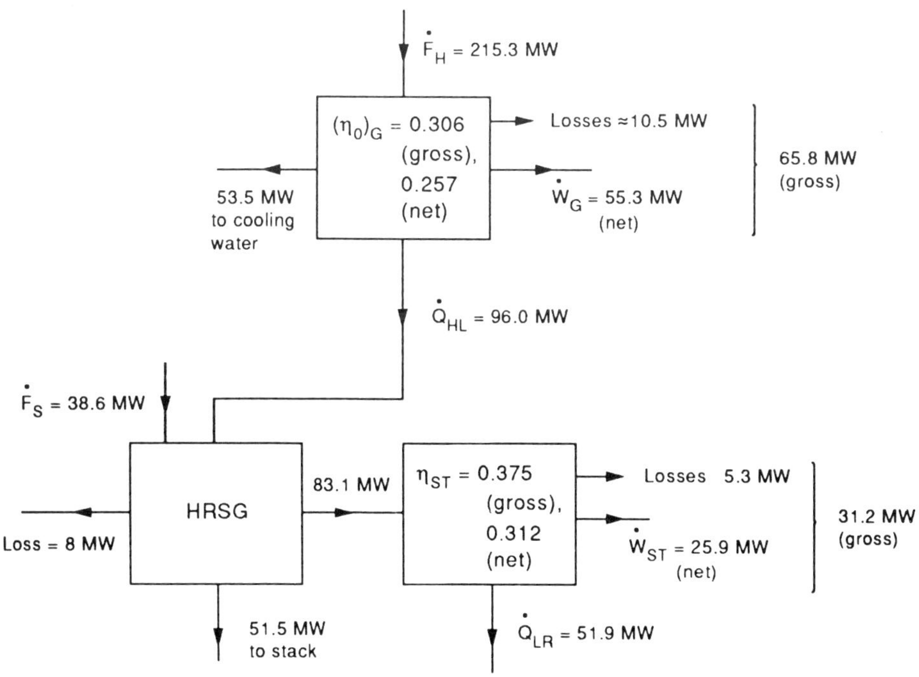

FIGURE 7.5 Energy flows for Korneuburg A (estimated from Czermak and Lahr [12])

The heat transferred to the HRSG from the exhaust is estimated as 96.0 MW, where we have assumed two thirds of the mechanical and electrical losses are debited to the gas turbine ($215.3-55.3-53.5-10.5=96$ MW). Additional fuel energy of 38.6 MW is added in supplementary firing, but the stack loss is 51.5 MW. The net heat supplied to the steam cycle is therefore 83.1 MW ($96.0+38.6-51.5$) and the lower cycle net efficiency is $25.9/83.1=0.312$. The total unused "heat" rejected by the gas turbine is 105 MW ($53.5+51.5$) and the parameter ν_{UN} is therefore $105/253.9=0.414$.

We can now check on the validity of the equation

$$(\eta_O)_{CP}=[(\eta_O)_G+\eta_{ST}-(\eta_O)_G\eta_{ST}]-\nu_{UN}\eta_{ST}$$
$$-\nu_S(1-\eta_{ST})(\eta_O)_G \tag{7.3}$$

applied to the Korneuburg A plant using gross efficiencies. With $(\eta_O)_G=0.306$, $\eta_{ST}=0.375$, $\nu_S=0.152$ and $\nu_{UN}=0.414$, it follows that

$$\begin{aligned}(\eta_O)_{CP}&=(0.306+0.375-0.306\times0.375)\\&\quad-0.414\times0.375-0.152\times0.625\times0.306\\&=0.566-0.155-0.029\\&=0.392\end{aligned}$$

The measured gross efficiency is $(55.3+25.9+15.8)/253.9=0.382$.

It is important to note the large drop in $(\eta_O)_{CP}$ compared with the ideal series plant efficiency $(\eta_O)_{CPS}=0.566$, arising mainly from the term $\nu_{UN}\eta_{ST}$. The parameter ν_{UN} was large for Korneuburg A because a large fraction of the heat rejected by the gas turbines was lost in the intercoolers and was not available to raise steam in the HRSG. The mechanical and electrical losses appear to be substantial.

An efficiency-load diagram given by Czermak and Lahr is illustrated in Fig. 7.6, although the figures shown do not quite correspond with those given above.

Czermak and Wunsch [8] reported in 1982 that the Korneuburg A plant had become uneconomic; hence the decision to replace it by Korneuburg B generating an increased power of 125 MW.

7.5 The CCGT Plant at Hemweg, Netherlands (Maximally Fired)

An example of the "maximally-fired" CCGT plant (open circuit/closed cycle) is the Brown Boveri installation at Hemweg in Amsterdam, Pjipker and Keppel [13]. The original 500 MW steam plant has been repowered by the addition of a Type 13E gas turbine, the exhaust from the gas

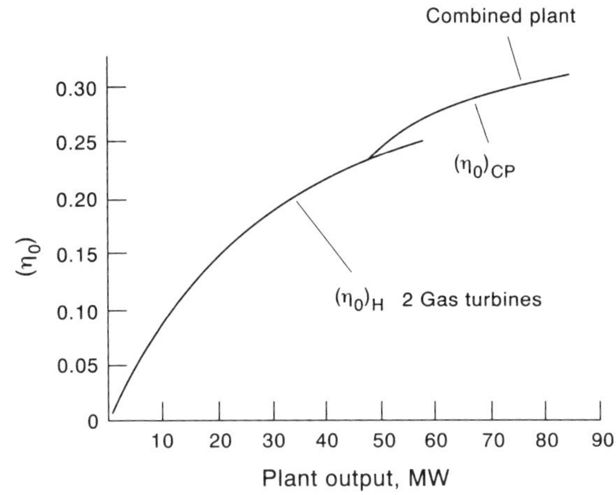

FIGURE 7.6 Efficiency-power output for Korneuburg A plant (Czermak and Lahr [12])

turbine feeding the steam plant boiler and providing the oxygen to support further combustion. The boiler fuel (natural gas or light distillate oil) is burnt with the 15% excess oxygen content in the exhaust gas; the stack gas leaving the boiler still contains 3% of oxygen. The full plant layout is shown in Fig. 7.7, and performance data is presented in Table 7.3.

With the high gas turbine exhaust temperature (534° C), the air preheaters installed in the original steam plant (to preheat air for combustion) are not required in combined cycle operation. However, the heating capacity of the boiler flue gas is utilised in high pressure and low pressure economisers. But not all the boiler feed water is passed through the economisers; a parallel arrangement, with part of the feed water bypassing four of the seven heaters to flow through the economisers, is employed, as recommended by Seippel and Bereuter [14]. The remaining feed water flows directly through all seven feed heaters. A direct contact feed heater is located between the L.P. and H.P. surface heaters, and the L.P. economiser bypass follows the first two heaters in the feed heating train. The steam produced in the boiler expands through the H.P. turbine and is reheated before entry to the I.P. turbine. The top point for bleeding steam is after the H.P. expansion, as is usual for high performance steam plant.

The overall efficiency of the steam turbine plant before repowering was 0.413 so that an improvement of 4.6% is achieved, with a 28% increase in power output.

It is difficult to assess the performance of the combined plant in terms of its two components through the equation

$$(\eta_O)_{CP} = (\eta_O)_G - \eta_{ST} - (\eta_O)_G \eta_{ST} - \nu_{UN} \eta_{ST}$$
$$- \nu_S(1 - \eta_{ST})(\eta_O)_G. \quad (7.3)$$

This is because the thermal efficiency of the steam turbine plant has been somewhat reduced in the repowering operation (some of the feed water is no longer heated regeneratively, but by the exhaust gases from the boiler). Similarly the boiler efficiency is not unchanged, since air preheaters have been replaced by the two economisers. However, we can interpret equation (7.3) approximately.

Let us assume that the steam turbine plant efficiency *is* unchanged at 0.413; the gas turbine plant efficiency is 0.314. With distillate fuel the boiler exhaust temperature is set at 145° C (and the water temperature at L.P. economiser inlet at 120° C), so that an approximate estimate of the factor ν_{UN} is

$$\nu_{UN} = \frac{\text{"heat" loss}}{\text{"heat" supplied}}$$
$$= \frac{\dot{M}_G(c_p)_G(145 - 15)}{1289.3}$$
$$= 0.101\ \dot{M}_G(\bar{c}_p)_G,$$

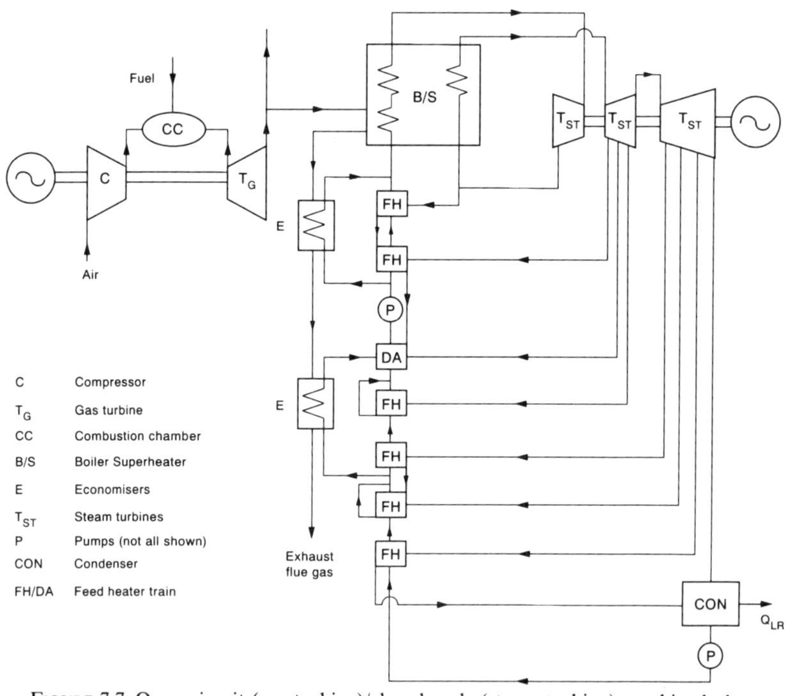

FIGURE 7.7 Open circuit (gas turbine)/closed cycle (steam turbine) combined plant at Hemweg (maximally fired)—simplified flow chart (after Pjipker and Keppel [13])

TABLE 7.3 *Performance Data at Design Operation*

Gas turbine	
Efficiency (overall)	0.314
Exhaust gas temperature	534° C
Pressure ratio	14.6
Maximum temperature	1070° C
Fuel	Natural gas
Lower calorific value	44,100 kJ/kg
Output	134.9 MW
Inlet pressure loss	5 mbar
Exhaust system pressure loss	90 mbar
Steam turbine	
Steam flow	1240 t/h
Live steam conditions	186 bar, 540° C
Reheat steam conditions	46 bar, 535° C
Fuel supply to HRSG	Natural gas or light distillate oil
Output	465.1 MW
Overall plant	
Gross output	600 MW
In-house load	8.7 MW
Net output	591 MW
Heat capacity of fuel	1289.3 MW
Efficiency (overall)	0.459

where $\dot{M}_G(\bar{c}_p)_G$ is the heat capacity of the boiler exhaust gases. If this heat capacity is unchanged from that of the gases leaving the gas turbine under normal operation (it will in fact be less because of the lower temperature) then since

$$(\eta_O)_G = 1 - \frac{\text{heat rejected}}{\text{heat supplied}}$$

$$= 1 - \frac{\dot{M}_G(\bar{c}_p)_G(534-15)}{(134.9)/(\eta_O)_G},$$

$$\dot{M}_G(\bar{c}_p)_G = \left[\frac{1-(\eta_O)_G}{(\eta_O)_G}\right]\frac{134.9}{519}$$

$$= \left(\frac{0.586}{0.314}\right)\frac{134.9}{519}$$

$$= 0.485 \text{ MW/K},$$

and

$$\nu_{UN} = 0.101 \times 0.485$$

$$= 0.0490.$$

The factor ν_S is approximately given by

$$\nu_S = 1 - \frac{\text{"heat" supplied to gas turbine}}{\text{total "heat" supplied}}$$

$$= 1 - \frac{(134.9/0.314)}{1289.3}$$

$$= 0.669.$$

Substituting these values for ν_{UN} and ν_S into equation (7.3) we obtain an approximate estimate of the overall efficiency of the combined plant

$$(\eta_O)_{CP} = 0.314 + 0.413 - 0.314 \times 0.413$$

$$- 0.049 \times 0.413 - 0.669 \times 0.587 \times 0.314$$

$$= 0.597 - 0.020 - 0.123$$

$$= 0.453,$$

compared with the stated plant efficiency of 0.459.

It is of interest to note that the "ideal" series efficiency (0.597) is but little reduced by the loss of "heat" in the flue gases, which has been reduced substantially by the use of the economisers. (With natural gas firing the exhaust temperature is dropped to about 100° C. ν_{UN} is reduced from 0.049 to about 0.032 which suggests that $(\eta_O)_{CP}$ would increase by between $\frac{1}{2}$% and 1%.) However, the supplementary "heating", which is a large fraction (about a third) of the overall "heating", causes a substantial reduction from that ideal efficiency.

Wunsch [9] has presented strong economic arguments for repowering a plant in this way (see Chapter 6, Section 6.6.1).

7.6 The IGCC Plant at Cool Water, California

The most advanced practical demonstration of the IGCC system yet undertaken is the plant built by the General Electric Company, U.S.A., in conjunction with the Electric Power Research Institute, for Southern California Edison at Cool Water, Daggett, California. It uses the Texaco coal gasification process together with a General Electric gas turbine, heat recovery and steam turbine components.

Figure 7.8 is a simplified diagram of the plant (see Plumley [15]). It is essentially an open cycle gas turbine/closed steam turbine combined system using an unfired heat recovery steam generator. However, the difference from a plant of similar size such as the Brown Boveri plant Korneuburg B (described in Section 7.3) lies in the gasification plant which produces syngas (instead of the natural gas supplied to Korneu-

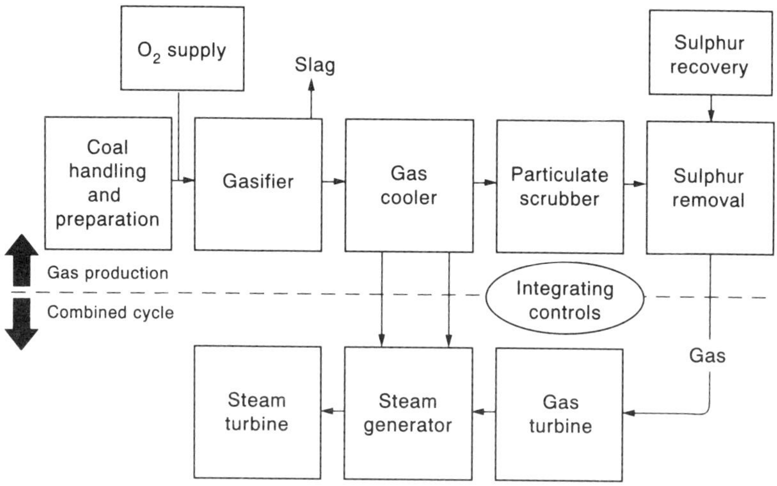

FIGURE 7.8 Simplified diagram of Cool Water IGCC plant (Plumley [15])

burg) and the close integration between the gasifier and the HRSG in the generation of steam. The gas turbine, HRSG and steam turbine components are standard, so it is the performance of the gasifier which is the critical new development. Cleaning of the gas is particularly important.

Figure 7.9 shows more detail of what is a very complex plant (see Spencer *et al.* [16]). Coal is ground and mixed with water to form a slurry and this is fed into the gasifier through a burner in which partial combustion takes place (with oxygen). During gasification the coal ash is melted into slag, quenched with water and removed as a solid.

Following the high temperature reactions of the coal and water with the oxygen (which is supplied from a separate plant), the raw synthetic gas (syngas), consisting primarily of hydrogen and carbon monoxide, is cooled in radiant and convection coolers in which saturated steam is generated. The gas is then passed through a particulate scrubber and further cooled to near ambient temperature, prior to sulphur removal. Table 7.4 gives an analysis of the clean gas (Plumley [15]).

The gas is then saturated before it enters the combined plant; this reduces the subsequent combustion temperature and NO_x production. The fuel gas is burned in the gas turbine combustion chamber with air from the compressor. The combustion gases supply the gas turbine (which drives the air compressor and a generator) and then exhaust into the heat recovery steam generator (HRSG—unfired), raising superheated steam. The by-product steam from the gasifier (40% of the total steam supply) is also superheated in the HRSG. Together with superheated

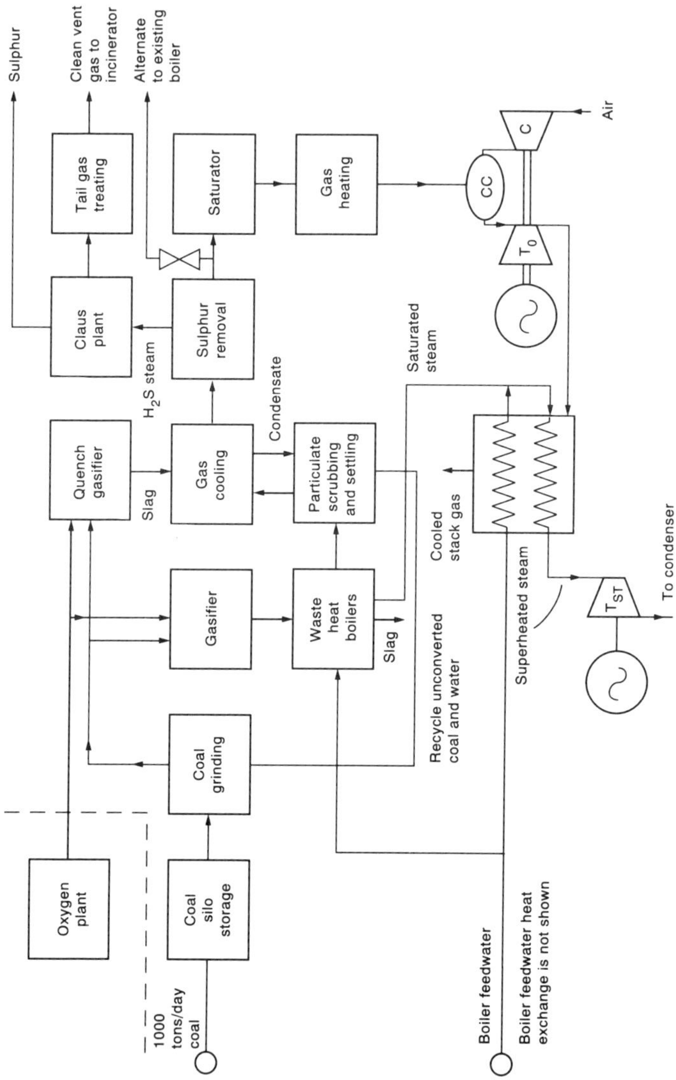

FIGURE 7.9 Cool Water coal gasification plant (Plumley [15])

TABLE 7.4 *Clean Syngas Analysis* (mol. %)

CO	43.53
H_2	38.33
CO_2	17.39
CH_4	0.11
Ar	0.17
N_2	0.47
H_2S	15 ppm
COS	55 ppm

steam raised directly in the HRSG, it is delivered to the steam turbine, which drives its own generator.

Table 7.5 shows the design operating conditions of the plant (Plumley [15]) and the performance achieved (EPRI report [17]).

In spite of the saturation before combustion, the flame temperature is higher than that for natural gas, ensuring complete combustion of carbon monoxide. The exhaust gases have a low environmental impact, SO_2 and NO_x emissions being of the order of 0.2 and 0.06 kg/10^6 kJ of coal input. The cold gas efficiency (higher calorific value of the syngas divided by the higher calorific value of the coal) is of the order of 72%. The heat rates and efficiencies shown in Table 7.5 are based on the higher calorific value of the coal supplied. The net heat rates allow for the power supply to the oxygen plant and the auxiliaries.

TABLE 7.5 *Cool Water IGCC Plant*

	Design operating conditions (Plumley [15])	Performance achieved with SUFCO coal (EPRI [17])
Gasifier pressure	4.137 MN/m^2	
temperature	1200° C–1500° C	
Higher calorific value of syngas	260–280 Btu/SCF 9700–10,400 kJ/m^3	263.6 Btu/SCF 9833 kJ/m^3
Gas turbine power output	65 MW	69.8 MW
Steam turbine entry pressure	1350 psia (9.3 MN/m^2)	
entry temperature	950° F (510° C)	
power output	55 MW	49.2 MW
Coal supply	1000 t/d	1000 t/d
Syngas flow		84 × 10^4 SCM/h
Gross output	117 MW	119 MW
Gross heat rate	9507 kJ/kW h	9242 kJ/kW h
Gross efficiency	0.379	0.389
Net output (allowing for power to auxiliaries and oxygen plant)	91.9 MW	95.2 MW
Net heat rate	12,112 kJ/kW h	11,552 kJ/kW h
Net efficiency	0.297	0.312

TABLE 7.6 *Performance Projections*

	Cool Water data	Mature IGCC plant
Net capacity (MW)	93	360
Net heat rate (Btu/kW h)	11,300	9000
Capital cost ($/kW)	2828	1530
Operating and maintenance costs (c/kW h)	10.6	2.3

(costs in 1983 dollars)

The Cool Water project has thus demonstrated the practical feasibility of IGCC plants achieving overall efficiencies comparable to those of coal-fired steam plants, and it was argued by Spencer *et al.* [16] that unit capital cost ($/kW) and the levelised cost of the electricity produced may be somewhat less. In 1981, on the basis of mid-1978 dollars, they estimated the latter at about 5.1 cents per kW h, assuming an overall efficiency of 35.8% (heat rate of 9530 Btu/kW h). Table 7.6 shows the corresponding 1986 costs estimated in EPRI's 1988 progress report [17] for the actual Cool Water plant, and a commercial plant projected from the Cool Water performance. The detailed justification for the apparently optimistic projections are given in the report. If accepted, then for a 30 year life with interest at 10% ($\beta = 0.106$), from equation (7.2), the 1986 unit price of electricity for the projected IGCC plant would be

$$Y_E = \frac{\beta C_0}{\dot{W}H} + \frac{\zeta}{(\eta_O)_{CP}} + \frac{(OM)}{\dot{W}H} \tag{7.2}$$

$$= \frac{0.106 \times 1530 \times 10^2}{0.65 \times 24 \times 365} + \frac{1.59}{0.379} + 2.3$$

$$= 2.8 + 4.2 + 2.3$$

$$= 9.3 \text{ c/kW h}.$$

Here, we have used EPRI's price for coal (1.59 c/kW h in 1986 dollars) and assumed a 65% capacity factor.

Finally we should note the magnitude of CO_2 production. In terms of kg/kW h this will not be much different from a conventional power station of comparable high efficiency. The original design estimate of coal requirement was 1000 t/d for a plant of about 100 MW. This suggests a "unit" production of CO_2 in excess of 1 kg/kW h, which we shall refer to again in Chapter 8.

7.7 The STIG Plant at Ripon, California

The Simpson Paper Company has installed a cogeneration plant for its paper producing mill at Ripon, California, employing steam injection in a GE LM 5000 gas generator. An IHI ITA 1203 power turbine produces 4000 kW and process steam is delivered at 20,400 kg/h when operating at full steady-state capacity.

The plant has been adapted so it can operate as a power producer alone, with full steam injection, a "full STIG" plant which we describe here, following Kolp and Moeller [18].

The basic circuit is as described earlier in Chapter 4—the exhaust from the complete gas turbine (referred to as an IHI IM 5000) passes to a heat recovery steam generator, before discharge to a stack. Superheated steam produced in the HRSG is injected into the gas generator.

However, the actual plant is quite complex as Fig. 7.10 shows. There are three stages of steam generation in the HRSG. For the full STIG operation the L.P. steam is sufficient to heat the feed water from 21° C to 121° C in a deaerator/direct contact feed heater (the L.P. economiser of the cogeneration plant is not used). The stack temperature is 135° C in this case of full STIG.

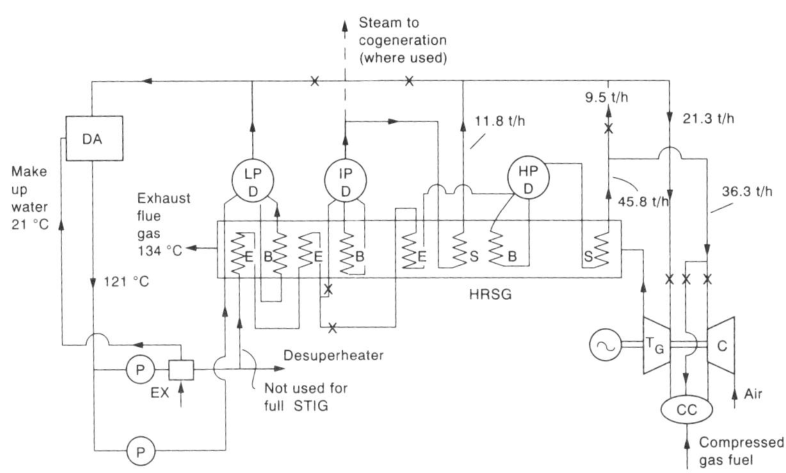

C	Compressor
CC	Combustion chamber
T_G	Gas turbine (with steam injection)
E	Economisers
B	Boilers
S	Superheaters
D	Separation drums
HRSG	Heat recovery steam generator
EX	Feed water heat exchanger

FIGURE 7.10 STIG Plant at Ripon, California (simplified diagram after Kolp and Moeller [18])

The I.P. boiler operates at 2070 kN/m^2 producing 11,800 kg/h of saturated steam. The H.P. boiler produces 45,800 kg/h of superheated steam (at 5170 kN/m^2, 315° C). The maximum allowable (high pressure) steam injection rate into the compressor discharge posts and the fuel nozzles is 36,280 kg/h. The balance of the steam available from the H.P. boiler is throttled to the I.P. pressure, and with the output from I.P. boiler supplies the low pressure injection between the H.P. and L.P. turbine of the gas generator.

Other features of the plant include a make-up/feed water heat exchanger which raises the temperature of the make-up water before it enters the deaerator; and a selective catalytic reactor which reduces the NO_x level from 25 ppm to 6–13 ppm volumetric.

The improvement in the performance of the plant with steam injection is impressive. Figure 7.11, taken from Kolp and Moeller, shows how the heat rate varies with power output—for no injection and with full STIG. The efficiency improves from the basic gas turbine efficiency of 36% to 43% (STIG) and the power output from 32.5 MW to 49.5 MW. The effect of different levels of steam injection on power output is shown in Fig. 7.12 plotted against inlet temperature. It is the higher pressure steam injection (into the compressor discharge and fuel nozzles)] which produces the major effect.

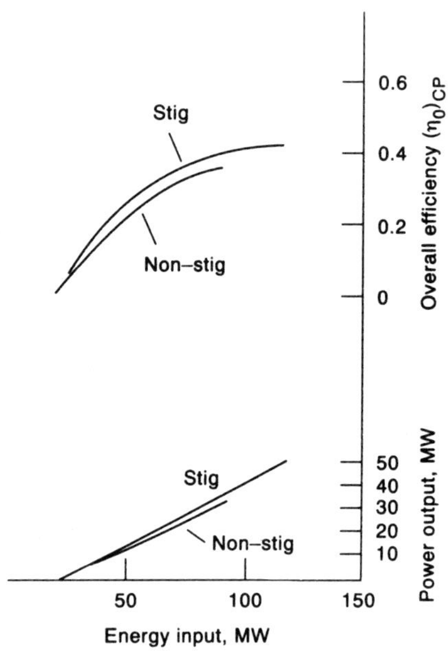

FIGURE 7.11 Improvement in power output and efficiency with steam injection (Kolp and Moeller [18])

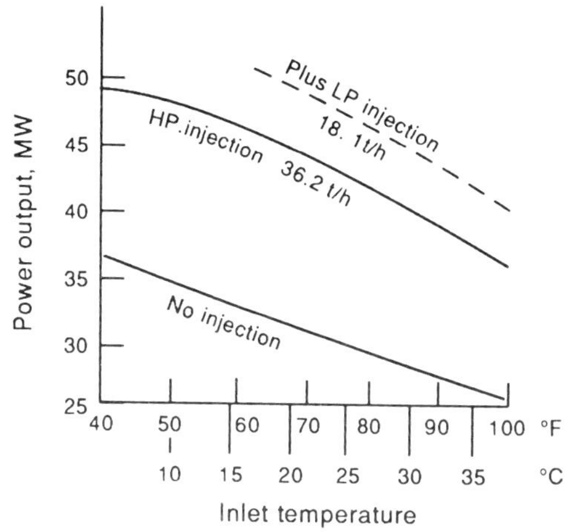

FIGURE 7.12 Variation in power output of STIG plant with L.P. and H.P. steam injection (after Kolp and Moeller [18])

Kolp and Moeller argue that the capital and operating costs of the full STIG plant are only 30% of the equivalent costs of a comparable LM5000 combined cycle plant. They make a comparison on the basis of net present value, but here we can use their figures (given in 1989) to calculate the cost of electricity production:

Capital cost	$\$30 \times 10^6$ (625 \$/kW)
Capital cost factor (30 years, 10% discount)	0.106
Efficiency (LHV)	0.43
Fuel cost	\$250/(10^6 Btu) (0.85 c/kW h)
Net power output	48.1 MW
Utilisation	8000 hours per annum

If operating costs were 2.0 cents per kW h then the 1989 cost of electricity would be

$$Y_E = \frac{\beta C_0}{\dot{W}H} + \frac{\zeta}{(\eta_O)_H} + \frac{(OM)}{\dot{W}H} \tag{7.2}$$

$$= \frac{0.106 \times 30 \times 10^8}{48.1 \times 10^3 \times 8000} + \frac{0.85}{0.43} + 2.0$$

$$= 0.8 + 2.0 + 2.0$$

$$= 4.8\ \text{c/kW h}.$$

This appears to be very low, because of the low fuel cost and high utilisation assumed.

References

[1] Horlock, J. H. *Cogeneration: Combined Heat and Power*. Pergamon Press, Oxford, 1987.

[2] Wood, B. Alternative Fluids for Power Generation. *Proc. I.Mech.E.*, **184,** 1, 40, 713, 1969–1970.

[3] Kearton, W. J. Discussion of reference 2.

[4] Emmet, W. L. R. The Emmet Mercury-Vapour Process. *Trans. ASME*, **46,** 253–285, 1925.

[5] Hackett, H. N. Mercury-Steam Power Plants. *Mechanical Engineering*, July, 559–564, 1951.

[6] Hackett, H. N. and Douglass, D. Modern Mercury Unit Power Plant Design. *Trans. ASME*, **72,** 3, 89, 1950.

[7] Oberly, W. N. The Mercury-Vapour Cycle. *Power Generation*, March, 56–57, 1950.

[8] Czermak, H. and Wunsch, A. *The 125 MW Combined Cycle Plant Korneuburg B; Design Features, Plant Performance and Operating Experience*. ASME Paper 82-GT-323, 1982.

[9] Wunsch, A. Highest Efficiencies Possible by Converting Gas Turbine Plants into Combined Cycle Plants. *Brown Boveri Review*, **10,** 455–463, 1985.

[10] Williams, R. H. Industrial Cogeneration. *Ann. Review Energy*, **3,** 313–356, 1982.

[11] Auer, W. P. Practical Examples of Utilizing the Waste Heat of Gas Turbines in Combined Installations. *Brown Boveri Review*, **47,** 12, 800–825, 1960.

[12] Czermak, H. and Lahr, G. Möglichkeiten des Einsatzes von Gasturbinen zur öffentlichen Energieversorgung. *Electrotechnik und Maschinenbau* (edited by Sequenz, H. und Smola, F.), 305–316. Springer-Verlag, Wien, 1967.

[13] Pjipker, B. B. and Keppel, W. E. Amsterdam to Have First 140 MWe Gas Turbine. *Modern Power Systems*, May, 28–33, 1986.

[14] Seippel, C. and Bereuter, R. The Theory of Combined Steam and Gas Turbine Installations. *Brown Boveri Review*, **47,** 783–789, 1960.

[15] Plumley, D. R. Cool Water Coal Gasification 1—a Progress Report. *Trans. ASME Journal of Engineering for Gas Turbines and Power*, **107,** 4, 856–860, 1985.

[16] Spencer, D. F., Alpert, S. B., O'Shea, T. P., Ahner, D. J. and Plumley, D. R. The Cool Water Coal Gasification Project. *Proc. American Power Conference*, 300–308, 1981.

[17] EPRI. Cool Water Coal Gasification Project. 5th Annual Progress Report, AP-5931, 1988.

[18] Kolp, D. A. and Moeller, D. J. World's First Full STIG LM 5000 Installed at Simpson Paper Company. *Trans. ASME, Journal of Engineering for Gas Turbines and Power*, **111,** 2, 200–210, 1989.

CHAPTER 8

Summary—Historical and Current Perspectives of Combined Power Plants

8.1 Introduction

In this summary chapter we briefly review the history of combined power plants, referring to four major papers (by Seippel and Bereuter [1], Wood [2], Wunsch [3], Davidson and Keeley [4]) published at approximately ten-year intervals over the period 1960–1990. In particular, we summarise the development of the so-called CCGT combined gas/steam turbine plant, which has become dominant, relating this to the basic thermodynamics and economics that we have outlined in previous chapters.

8.2 History up to the Early Seventies

The feasibility of the combined power plant was demonstrated in the binary (double cycle) mercury/steam plants built in the period 1928–1950 by General Electric in the U.S.A. These plants achieved high thermal efficiency, but they were expensive because of the substantial construction costs, associated particularly with the "higher" mercury cycle plant.

The early development of the gas/steam turbine combined plant was described in two seminal papers by Seippel and Bereuter (in 1960, [1]) and Wood (1971, [2]). The remarkable paper by Seippel and Bereuter reviewed many possible combinations of gas turbine and steam turbine plant (see Section 3.6 and Fig. 3.13), including the original concept of the pressurised boiler (the Velox), in which the gas turbine compressor charged the steam boiler, and which had been the subject of development by Brown Boveri in the 1930s. Seippel and Bereuter outlined a comparison between the efficiencies that could be achieved in a combined plant (with and without supplementary heating) and those of a basic steam plant. If that comparison has now been superseded by more complex analyses and computer studies, it was nevertheless a remarkably forward looking and original piece of thermodynamic analysis for that period (the late fifties and early sixties).

Wood's paper gave a broad review of the position as it was ten years later in 1971 (he had earlier seen the improbability of further development of the binary plant in his review of alternative fluids for power generation [5]). In addition to recognising the demise of the "superposition" (mercury/steam) binary cycle, he also virtually rejected the substitution of other fluids for water in a "subposition" binary plant, although he did also consider, without great enthusiasm, the possible use of ammonia, or the refrigerant R113, in a low temperature vapour cycle below a gas turbine.

In his 1971 paper, however, Wood compared the performance of the combined gas/steam turbine plants which had been installed to that date (mainly in the sixties) against that of current steam plant. He gave separate attention to the "maximally fired" and recuperative plants, recognising that the gap between them could be narrowed by supplementary firing of the HRSG in the latter. Previously, in the early fifties, Wood had been doubtful about the advantages of the maximally fired plant because he had considered that the prospective gain in efficiency would be severely restricted by the necessity of dispensing with the low temperature bled steam feed heaters and air pre-heating for the boiler. But in 1971 he recognised the importance of the proposal by Seippel and Bereuter to pass about one-third of the feed water through the economiser, retaining most of the regenerative feed heating. Thus in 1971 Wood could refer to several projected large maximally fired plants (e.g. the Robert Frank plant to be commissioned in Germany in 1973).

Also by 1971 the recuperative gas/steam turbine plant had become established, primarily by General Electric in the United States and by Brown Boveri in Europe. Wood listed many plants (mostly small plants of 15–20 MW) installed in the U.S.A., mainly unfired and with single pressure steam cycles. One of the biggest was a combined heat and power plant, the Dow Chemical plant in Texas, producing 63 MW [43 MW from the gas turbines and 20 MW from the steam turbines, with supplementary firing providing steam at 1200 psia (8.3 MN/m^2) and 950° F (511° C), exhausting to process at 185 psia (1.3 MN/m^2)]. In Europe the original Korneuburg A plant was the biggest plant in service at that time (75 MW at 0.326 overall efficiency, based on lower calorific value), and was an excellent example of the recuperative plant with supplementary heating of the HRSG.

Wood gave a simple argument for the level of efficiency of the regenerative plant and its change with supplementary heating. If the gas turbine efficiency were 0.26, and 50% extra power could come from the steam turbine, then the overall efficiency would be $(1.5 \times 0.26) = 0.39$. He argued that output from additional fuel would be obtained at low steam plant efficiency [say $(\eta_O)_L = 0.176$ overall, the product of a thermal efficiency of $\eta_L = 0.26$ and a boiler efficiency of $(\eta_B) = 0.677$]. Thus the overall combined efficiency of the plant would decrease with supplementary heating as the proportion of steam turbine output increased. Here he was neglecting any possible increase in the steam cycle thermal efficiency (η_L) with increased temperature and pressure, which we shall refer to again later.

To amplify Wood's example in terms of the analysis given earlier in this book in Chapter 3, the boiler efficiency of an HRSG (unfired) would be

$$(\eta_B) = \frac{Q_B[1-(\eta_O)_H] - Q_{UN}}{Q_B[1-(\eta_O)_H]} = 1 - \frac{\nu_{UN}}{[1-(\eta_O)_H]}, \tag{8.1}$$

where

$$\nu_{UN} = [\text{“stack loss”}]/[\text{total heat supplied}]$$
$$= Q_{UN}/Q_B,$$

so the work output from the steam turbine would be

$$W_L = (\eta_O)_L Q_B[1-(\eta_O)_H], \tag{8.2}$$

where

$$(\eta_O)_L = \eta_L(\eta_B).$$

The combined plant overall efficiency would then be

$$\begin{aligned}(\eta_O)_{CP} &= (\eta_O)_H + (\eta_O)_L - (\eta_O)_H(\eta_O)_L \\ &= (\eta_O)_H + (\eta_B)\eta_L[1-(\eta_O)_H] \\ &= (\eta_O)_H + \eta_L - \eta_L(\eta_O)_H - \nu_{UN}\eta_L.\end{aligned} \tag{8.3}$$

Wood's estimates were $(\eta_O)_H = 0.26$, $\eta_L = 0.26$, $(\eta_B) = 0.677$,

$$(\eta_O)_L = 0.677 \times 0.26 = 0.176,$$

so that

$$(\eta_O)_{CP} = 0.26 + 0.176(1-0.26) = 0.39,$$

which is Wood's figure for overall efficiency of the combined plant $(\eta_O)_{CP}$. He considered that the thermal efficiency for the steam plant $(\eta_L = 0.26)$ was attainable at 200 psia (1.38 MN/m^2) and 600° F (316° C).

These were simple calculations, but enlightening nevertheless. They illustrated a potential for increase in overall efficiency of the simple recuperative plant, if η_L could be increased sufficiently as supplementary heating was introduced. However, it has been the increase in the gas turbine maximum temperature, leading to increased specific work in the higher level plant, that has enabled the simple recuperative combined plant to become increasingly competitive with conventional high pressure steam plant (with reheating between the steam turbines).

It is of interest that it was the simple open circuit gas turbine that became the favoured early choice for the higher level plant. The closed cycle gas turbine, in spite of its capacity for increased power obtainable from pressuring the cycle, was not developed for the combined plant, because of the increasing ability to design and build reliable cooled turbines for open circuit plant (burning natural gas or light fuel oil). Neither did the heat exchanger find favour in the higher gas turbine plant; the exit temperature from the so-called CBTX (open circuit) plant is lower than in the corresponding CBT plant, and any increase in efficiency of the higher level plant is counter-balanced by a drop in efficiency of the lower level steam cycle, because of the lower steam turbine entry temperature.

8.3 Subsequent Development of the Combined Gas/Steam Turbine (CCGT) Plant—1970–1990

We next consider the subsequent development of the open circuit (gas turbine)/closed cycle (steam turbine) combined (CCGT) plant. Wunsch, of Brown Boveri, reviewed the position in 1978; more recently, Davidson and Keeley (in 1990) have provided a useful summary of the current position.

Like others, Wunsch classified these combined gas/steam turbine plants into three types:

(i) the recuperative plant with an unfired HRSG;
(ii) the recuperative plant with a supplementary fired HRSG (up to 760° C exit temperature);
(iii) the plant with a maximally fired boiler (using most of the available oxygen in the gas turbine exhaust), usually for base load.

Figure 8.1 shows Wunsch's summary of the first two of these combined plants and Table 8.1 shows the historical development of some of Brown Boveri's gas/steam turbine combined plants, including examples of all three types—unfired HRSG, supplementary fired HRSG, maximally fired boiler.

8.3.1 The Unfired Recuperative Plant

As indicated above, the first and simplest of these plants—the unfired recuperative combined plant—had received most attention in the 1950s and 1960s. We have described in detail many of the performance analyses that have been made of this plant (see Chapter 4). Here we summarise the main features that follow from those analyses, most of which have been adopted in the development of practical plant (e.g. the new Korneuburg B plant referred to by Czermak and Wunsch in 1982 [6] and described in Chapter 7).

(i) Reference has already been made to the importance of high maximum temperature (and high specific work) in the gas turbine plant. This leads to selection of fairly high pressure ratio, although it is a "flat" optimisation—see Fig. 4.43 which showed the comprehensive calculations by Cerri [7]. Davidson and Keeley [4] give a current comparison between a gas turbine plant and a combined plant (Fig. 8.2); for the latter, modern practice would suggest a design pressure ratio of about 14 for the combined plant operating with a gas turbine entry temperature of

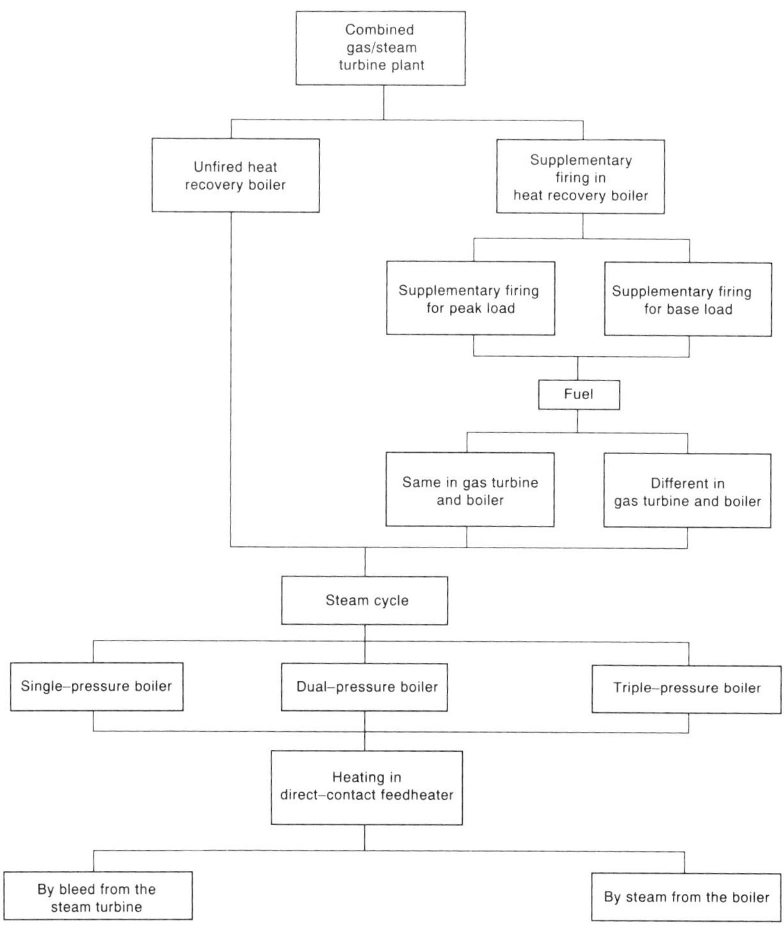

FIGURE 8.1 Classification of open circuit (gas turbine)/closed cycle (steam turbine) combined plant (Wunsch [3], 1978)

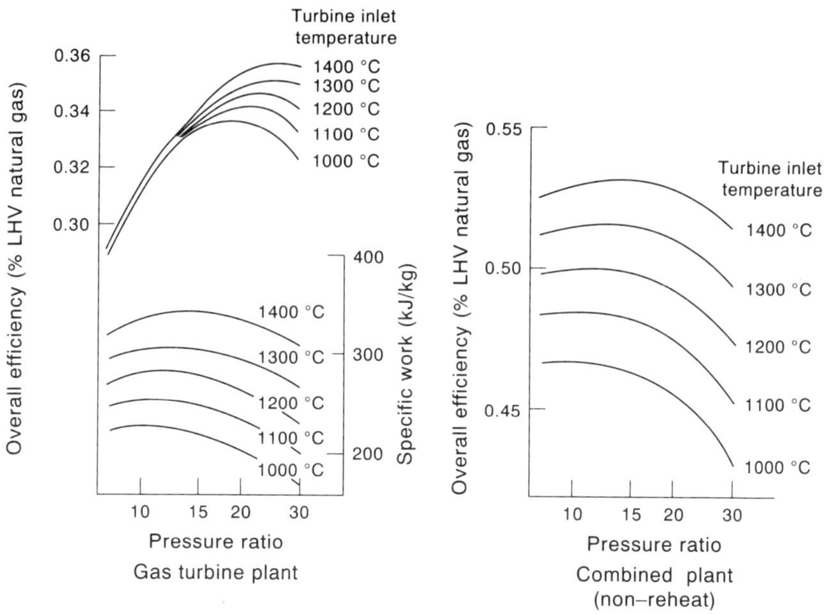

FIGURE 8.2 Efficiency of open circuit (gas turbine)/closed cycle (steam turbine) combined plant, compared with efficiency and specific power of basic gas turbine plant (Davidson and Keeley [4])

1350° C, and about 12 for a turbine entry temperature of 1100° C. Figure 8.2 shows that these pressure ratios are close to those for maximum specific work in the gas turbine operating alone. Boulter [8], also in 1990, agrees that there is little point in going to higher pressure ratios than these, although his calculations do not show the clear optimum indicated by Davidson and Keeley.

(ii) In order to minimise the irreversibility in the HRSG, the steam is usually raised at two pressure levels. Davidson and Keeley point out in 1990 that the non-reheat steam cycle is common for large modern plant. However, they argue that as gas turbine temperatures increase, together with turbine exhaust temperatures (approaching 590° C), so the reheat steam cycle becomes more attractive. This is seen in the ABB plant for UNA in the Netherlands (Pegus 12) which has a three-pressure level HRSG, including a reheater at exit from the H.P. turbine. Other three-pressure level plants with reheat are proposed by GEC/Alsthom (see Paren and Parietti [9]).

(iii) The choice of the feed water heating arrangements (and the allowable stack temperature, T_S) is closely associated with the fuel to be used; the feed water temperature (T_b) at entry to the HRSG must not be below the dew point of the exhaust gases (Kehlhofer [10]). Use of natural gas with a low sulphur content leads to an allowable T_s of about 95° C in the Korneuburg B plant, whereas for light oil fuels the allowable T_s is about 150° C. The lower T_S the better, for there will be a smaller "enthalpy loss" in the exhaust stack. Referring to single pressure steam cycles, if there were no minimum limit on the feed water entry temperature T_b and regenerative feed heating were not required, then no turbine work would be lost by bleeding steam. However, a limit on the feed water temperature T_b at entry to the HRSG (leading to a T_S of about 150° C for fuels with sulphur) means that the feed water must be raised from condenser temperature to T_b either by regenerative heating using bled steam or in a preheater loop (which gives the lower T_S and the higher overall efficiency $(\eta_O)_{CP}$ since steam is not bled from the turbine). In a dual pressure steam cycle in a CCGT plant using fuel containing sulphur, substantial feed heating may be used (up to three bled steam heaters), but for sulphur-free fuels a single-stage feed heater/deaerator is used, because a L.P. economiser can subsequently be introduced, reducing the flue gases to a lower T_S (90–100° C). In the recuperative plant, the choice of feed heating arrangements thus depends on which fuel is to be used and the number of pressure levels selected (see the discussion of Section 4.3.2.2.2 for further details).

8.3.2 The Supplementary Fired Plant

Supplementary firing of the exhaust gases in the HRSG was an option recommended by several authors in the past, but is now less popular as gas turbine entry temperatures have increased (see also later). Both Pfenninger [11], in 1973, and Wunsch [3], in 1978, both of Brown Boveri, argued that the overall efficiency of the simple non-heated recuperative plant could be improved by supplementary firing up to about 760° C, because of the increased thermal efficiency of the lower steam cycle and the increased steam generation that could be achieved.

However, relatively few if any of the more recent plants use supplementary fired HRSGs. Increased efficiency has been obtained by introducing dual pressure steam raising, rather than increasing the temperature to entry to the HRSG by supplementary firing.

The thermodynamic argument for gas turbine reheat advanced so strongly by Rice in several papers—that it is better to supply the extra "heat" at higher temperature through reheat in the gas turbine rather than in supplementary "heating" in the HRSG—appears to be sound.

8.3.3 The Maximally Fired Plant

The maximally fired plant is clearly well proven—most recently in the repowering schemes undertaken by Brown Boveri, e.g. Hemweg (see Table 8.1 and Chapter 7). Here a 28% increase in power output was obtained by modifying the boiler of the original steam plant to receive and fully fire the exhaust from a gas turbine. An increase of 4.6 percentage points overall efficiency (LHV) was obtained, from 41.3% to 45.9%. At Hemweg a train of feed heaters is used, although part of the feed water is bypassed round some heaters, fed to an economiser and heated by the exhaust gases.

8.3.4 The Present Position on After Firing of the Exhaust

Davidson and Keeley [4] suggest that advances in gas turbine design (to higher turbine entry temperatures) now enable the simple recuperative plant (with no steam cycle reheat but two pressure levels of evaporation) to match the efficiencies of plant with after firing (usually with steam cycle reheat). This is an important conclusion and is illustrated in

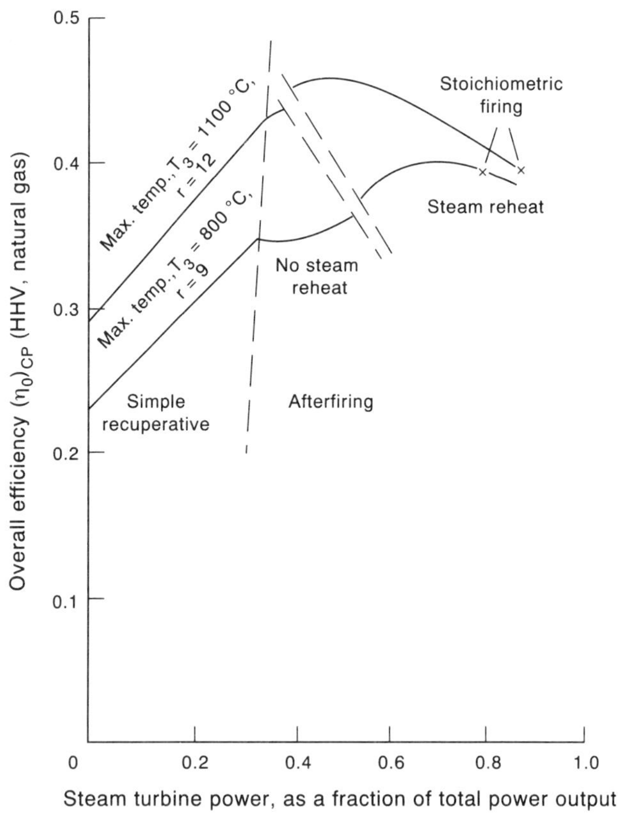

FIGURE 8.3 Combined plant efficiency—effect of maximum temperature and after firing (Davidson and Keeley [4])

TABLE 8.1 *Historical development of some Brown Boveri combined gas/steam turbine power plants*

Plant/(reference)	Date of commissioning	Fuel	Gas turbine		
			Turbine inlet temp. (°C)	Pressure ratio	Turbine exit temp. (°C)
Korneuburg A (Pfenninger, 1973 [11], Auer, 1960 [15])	1961	Natural gas	625 after combustion chamber, 625 after reheat	20	310
Geertruidenberg (Pfenninger, 1973 [11], Wunsch, 1978 [3])	1975	Natural gas	950	9.7	480
Korneuburg B (Czermak and Wunsch, 1982 [6])	1980	Natural gas	990	12.5	491
Rooscote (Stonard, 1990 [16])	1991	Natural gas	1070	14.3	514
Hemweg (Pjipker and Keppel, 1986 [17])	1988	Natural gas/natural gas or light distillate oil in boiler	1070	14.3	534
Pegus 12 (Jeffs, 1990 [18])	1986	Natural gas	1070	14.3	525

TABLE 8.1 *continued*

Plant/(reference)	Date of commissioning	Fuel	HRSG: Fired or unfired HRSG (°C)	HRSG: HRSG inlet temp. (°C)	HRSG: HRSG exit temp. to stack (°C)	HRSG: Single or dual pressure	HRSG: Feed water inlet temp. (°C)
Korneuburg A (Pfenninger, 1973 [11], Auer, 1960 [15])	1961	Natural gas	Supplementary fired	—	145 (estimate)	Single	56
Geertruidenberg (Pfenninger, 1973 [11], Wunsch, 1978 [3])	1975	Natural gas	Unfired	480	108	Dual	—
Korneuburg B (Czermak and Wunsch, 1982 [6])	1980	Natural gas	Unfired	491	95	Dual	54
Rooscote (Stonard, 1990 [16])	1991	Natural gas	Unfired	514	121	Dual	75
Hemweg (Pjipker and Keppel, 1986 [17])	1988	Natural gas/ natural gas or light distillate oil in boiler	Maximally fired	—	100/145	Single plus reheat	–/120
Pegus 12 (Jeffs, 1990 [18])	1986	Natural gas	Unfired	525	—	Tri pressure plus reheat	—

TABLE 8.1 *continued*

Plant/(reference)	Date of commissioning	Fuel	Steam turbine: Entry temp. (°C)	Steam turbine: Entry pressure (bar)	Type of feed heating	Quoted power output (MW)	Power ratio. Gas turbine/steam turbine (MW)/(MW)	Overall efficiency
Korneuburg A (Pfenninger, 1973 [11], Auer, 1960 [15])	1961	Natural gas	440	14.2	Single bled steam surface heater. Separate deaerator	75	$\frac{60}{25.7}=2.33$	0.326
Geertruidenberg (Pfenninger, 1973 [11], Wunsch, 1978 [3])	1975	Natural gas	395	30	Single bled steam direct contact feed heater/deaerator	120	$\frac{76.7}{47.3}=1.62$	0.461
Korneuburg B (Czermak and Wunsch, 1982 [6])	1980	Natural gas	433	33.2	Single bled steam direct contact feed heater/deaerator	128	$\frac{81.1}{48.7}=1.67$	0.466
Rooscote (Stonard, 1990 [16])	1991	Natural gas	490	61.7	Single bled steam direct contact feed heater/deaerator	224	$\frac{165.6}{63.1}=2.62$	0.491 (with steam injection)
Hemweg (Pjipker and Keppel, 1986 [17])	1988	Natural gas/natural gas or light distillate oil in boiler	535/535	161 (H.P.) 46 (after reheat)	Six bled steam surface heaters and one direct contact feed heater/deaerator (two heaters bypassed)	600	$\frac{134.9}{465.1}=0.290$	0.459
Pegus 12 (Jeffs, 1990 [18])	1986	Natural gas	494 448 268	68.3 (H.P.) 18 (I.P.) 768 (L.P.)	L.P. feed heater (in HRSG) plus feed heater/deaerator	225	$\frac{144.5}{81.6}=1.77$	0.518

Fig. 8.3, which shows that at low turbine entry temperature (800° C), efficiency continues to increase with after firing, first with no reheat in the steam cycle and then with reheat. However, with a higher turbine entry temperature (of 1100° C) there is but little gain in efficiency with after firing, reheat in the steam cycle being required even to maintain peak efficiency.

8.3.5 Advances in CCGT Plant

Table 8.1 shows the development of some Brown Boveri (ABB) CCGT plants over the past thirty years, together with the associated changes in the main thermodynamic properties (e.g. gas turbine entry temperature and pressure ratio, steam turbine entry temperature and pressure, feed water temperature, etc.) and overall performance. The reader is referred to the cited references for complete details of the various plants.

8.4 The IGCC Plant and Other Combined Plants

The use of natural gas in new large combined power plants leads to reduced levels of CO_2 produced in the exhaust, in comparison with conventional coal-fired stations. Figure 8.4 shows the amount of CO_2

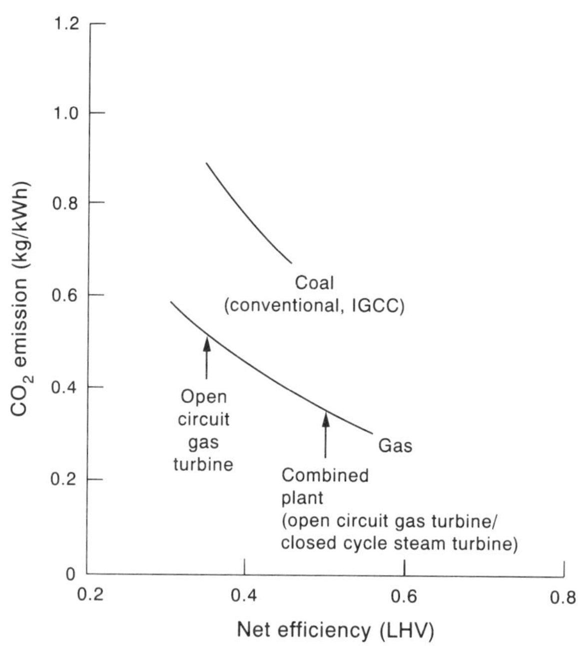

FIGURE 8.4 CO_2 emission for various plants (Davidson and Keeley [4])

produced (kg/kW h) for various plants (after Davidson and Keeley). As a result of increased concern about the amount of CO_2 being discharged into the atmosphere, and the global warming that results, there may prove to be less emphasis in future on the IGCC plant (the integrated coal gasification combined "cycle") although, like the natural gas-fired plant, it produces low levels of SO_2 and NO_x.

However, a prototype of the IGCC plant has been successfully demonstrated at Cool Water in the United States (see Chapter 7). The thermodynamics of the IGCC plant has been the subject of intensive study by EPRI and its sub-contractors, and the potentialities for improvement in plant economics assessed by Tsatsaronis, Winfold and their colleagues [12], by exergoeconomic methods. Further development may involve linking fluidised bed combustion of coal with combined cycle gas/steam turbine plant (see Section 2.4.2.2).

That coal-fired gasification/combined power plants can achieve good rational efficiency (but with some penalty compared with recuperative CCGT plants fired by natural gas) must be accepted, but two major questions remain about their widespread acceptance:

(i) the complexity (and associated cost), compared with that of the open circuit (gas turbine)/closed cycle (steam turbine) CCGT plant;
(ii) whether the level of CO_2 production in the exhaust, together with the by-products of gasification, will prove acceptable environmentally.

At the time of writing (the early 1990s) other forms of combined power plant (e.g. those involving fuel cells and magneto-gasdynamics) seem to be as far away as ever, in spite of their high "paper" overall efficiencies. However, the potential for substantial increases of both power output and efficiency into gas turbines by steam injection (the doubly open circuit plant) has been amply demonstrated (e.g. by Kolp and Moeller [13] at the Ripon STIG plant described in Chapter 7).

8.5 A Note on Cogeneration (Combined Heat and Power) Plants

It has not been the object of this book to describe the thermodynamics and economics of combined heat and power plant, as that was the subject of an earlier, parallel volume (Horlock [14]). But this summary would not be complete without reference to the successful development of cogeneration plant, providing both electrical power and useful heat, in which the power production is based on a combination of higher and lower level power plants. For example, the STAG (steam and gas turbine) plant at Saarbrücken (described in Chapter 5 of the 1987 cogeneration book) involves a gas turbine with exhaust gases supplying an HRSG, the steam produced generating further power in a steam turbine and usefully rejecting heat in a condenser, to a district heating plant. The Brown Boveri Pegus 12 plant at Utrecht is another example of a cogeneration plant involving a higher level gas turbine and a lower level steam turbine cycle, the latter rejecting heat to district heating. The new natural gas fired plant at Teesside, being built by Enron, is a combined power plant which also supplies steam for process heat. Such CHP plants achieve very high utilisation of the fuel energy supplied. The energy utilisation factor (EUF), defined as work output plus useful heat output divided by fuel energy supplied (see Horlock [14]), may reach over 80%.

8.6 Conclusions

It is concluded that the open circuit (gas turbine)/closed cycle (steam turbine) combined power plant (CCGT)

(i) is well proven, even as a base load plant;
(ii) is more than competitive in overall (or rational) efficiency with highly developed coal-fired steam power stations;
(iii) has low levels of pollutant production;
(iv) can be built quickly and relatively cheaply, which substantially adds to its overall economics.

The combined gas/steam turbine plant is therefore here to stay as a major producer of electrical power. It may well become the dominant one by the twenty-first century.

References

[1] Seippel, C. and Bereuter, R. The Theory of Combined Steam and Gas Turbine Installations. *Brown Boveri Review*, **47**, 783–789, 1960.
[2] Wood, B. Combined Cycles—A General Review of Achievements. Modern Steam Practice. *Proc.I.Mech.E.*, 75–86, April 1971.
[3] Wunsch, A. Combined Gas/Steam Turbine Power Plants—The Present State of Progress and Future Developments. *Brown Boveri Review*, **65**, 10, 646–655, 1978.
[4] Davidson, B. J. and Keeley, K. R. The Thermodynamics of Practical Combined Cycles. Combined Cycle Gas Turbines. *Proc. I.Mech.E. Seminar*, 23–50, 1990.
[5] Wood, B. Alternative Fluids for Power Generation. *Proc. I.Mech.E.*, **184**, 1, 40, 713, 1969–1970.
[6] Czermak, H. and Wunsch, A. *The 125 MW Combined Cycle Plant Korneuburg B; Design Features, Plant Performance and Operating Experience*. ASME Paper 82-GT-323, 1982.
[7] Cerri, G. Parametric Analysis of Combined Gas-Steam Cycles. *Trans. ASME Journal of Engineering for Gas Turbines and Power*, **109**, 1, 46–55, 1987.
[8] Boulter, D. G. The Future, Combined or Open Cycle Gas Turbines. Combined Cycle Gas Turbines. *Proc. I.Mech.E. Seminar*, 89, 1990.
[9] Paren, J. and Parietti, C. *Combined Cycle Plants—Three Pressure Reheat VEG 109F*. Technical Review GEC/Alsthom 4, 15–26.
[10] Kehlhofer, R. *Combined Cycle Gas and Steam Turbine Power Plants*. Fairmont Press, Lilburn GA, 1991.

[11] Pfenninger, H. Combined Steam and Gas Turbine Power Stations. *Brown Boveri Review*, **60,** 9, 389–397, 1973.
[12] Tsatsaronis, G., Winfold, M. and Stojanoff, C. G. *Thermoeconomic Analysis of a Gasification-Combined Cycle Power Plant*. EPRI 4734, 1986.
[13] Kolp, D. A. and Moeller, D. J. World's First Full STIG LM 5000 Installed at Simpson Paper Company. *Trans. ASME, Journal of Engineering for Gas Turbines and Power*, **111,** 2, 200–210, 1989.
[14] Horlock, J. H. *Cogeneration; Combined Heat and Power*. Pergamon Press, Oxford, 1987.
[15] Auer, W. Practical Examples of Utilizing the Waste Heat of Gas Turbines in Combined Installations. *Brown Boveri Review*, **47,** 12, 800–825, 1960.
[16] Stonard, M. Recent Combustion Development Control Concept and Cycle Example of ABB Gas Turbines in Combined Cycle. Combined Cycle Gas Turbines, *Proc. I.Mech.E. Seminar*, 69–80, 1990.
[17] Pjipker, B. B. and Keppel, W. E. Amsterdam to Have First 140 MWe Gas Turbine. *Modern Power Systems*, May, 28–33, 1986.
[18] Jeffs, E. Utrecht 225 MW Combined Cycle Certified at 52% Efficiency. *Gas Turbine World*, Jan./Feb. 1990, 1–7.

Index